Sarcocystosis of Animals and Man

Authors

J. P. Dubey
Zoonotic Diseases Laboratory
Livestock and Poultry Sciences Institute
Agricultural Research Service
U.S. Department of Agriculture
BARC-East
Beltsville, Maryland

C. A. Speer
Veterinary Research Laboratory
Montana State University
Bozeman, Montana

R. Fayer
Zoonotic Diseases Laboratory
Livestock and Poultry Sciences Institute
Agricultural Research Service
U.S. Department of Agriculture
BARC-East
Beltsville, Maryland

CRC Press, Inc.
Boca Raton, Florida

Library of Congress Cataloging-in-Publication Data

Dubey, J. P.
   Sarcocystosis of animals and man/authors, J. P. Dubey, C. A.
Speer, R. Fayer
      p.   cm.
   Bibliography: p.
   Includes index.
   ISBN 0-8493-6364-0
   1. Sarcocystosis—Diagnosis. 2. Sarcocystosis in animals—
Diagnosis. 3. Sarcocystis. 4. Sarcocystosis in animals.
I. Speer, C. A. II. Fayer, R.
QR201.S27D83 1989
636.089'6936—dc19   88-15684
                                                       CIP

This book represents information obtained from authentic and highly regarded sources. Reprinted material is quoted with permission, and sources are indicated. A wide variety of references are listed. Every reasonable effort has been made to give reliable data and information, but the author and the publisher cannot assume responsibility for the validity of all materials or for the consequences of their use.

All rights reserved. This book, or any parts thereof, may not be reproduced in any form without written consent from the publisher.

Direct all inquiries to CRC Press, Inc., 2000 Corporate Blvd., N.W., Boca Raton, Florida, 33431.

©1989 by CRC Press, Inc.

International Standard Book Number 0-8493-6364-0

Library of Congress Card Number 88-15684
Printed in the United States

PREFACE

Sarcocystis is one of the most prevalent parasites of livestock. In some hosts, such as domestic cattle and sheep, virtually all adult animals are infected. It also infects many wild mammals, birds, cold-blooded animals, and man.

Sarcocystis is not a newly discovered organism. It was first reported in the muscles of the house mouse in 1843, and until the early 1970s it was regarded as innocuous. In the past 15 years, over 600 scientific publications have elucidated basic information on the life cycle, pathogenicity, clinical signs, diagnosis, immunity, and treatment.

Completion of the life cycle requires two host species: an intermediate (or prey) host and a definitive (or predator) host. Hosts can harbor more than one species of *Sarcocystis*. Some species of *Sarcocystis* can cause reduced weight gain, poor feed efficiency, anorexia, fever, anemia, muscle weakness, reduced milk yield, abortion, and death in intermediate hosts such as cattle, sheep, goats, and pigs. Some species of *Sarcocystis* can cause digestive disturbances including nausea, vomiting, and diarrhea in man.

Diagnosis can now be made antemortem, based on clinical signs and serologic findings. The techniques and interpretations for serodiagnosis are primitive and require refinement.

We provide a current and comprehensive review of *Sarcocystis* and sarcocystosis in animals and man. The book is divided into 16 chapters. The first chapter is a general review of history, structure, life cycle, pathogenesis, lesions, clinical signs, diagnosis, immunity, epidemiology, treatment, prevention, and control. The second chapter emphasizes techniques and includes the maintenance of the parasite in the laboratory, collection and purification of sporozoites, merozoites, and bradyzoites, preparation of antigens, serologic techniques, clinicopathologic techniques, and *in vitro* and *in vivo* cultivation. Successive chapters concern sarcocystosis in cattle, sheep, goats, pigs, equids, humans and primates, camels, buffalo, birds, wild ruminants and zoo animals, and poikilothermic animals. For each major host species, information of the structure, life cycle, clinical signs, diagnosis, immunity, and economics is critically reviewed. The last chapter summarizes information on the related coccidian genus *Frenkelia* and other cyst-forming coccidians. There is a comprehensive bibliography for each host.

It is hoped that this book will be useful to biologists, veterinarians, physicians, and researchers.

J. P. Dubey
C. A. Speer
R. Fayer

THE AUTHORS

J. P. Dubey, B.V.Sc. and A.H., M.V.Sc., Ph.D., received his veterinary degree in 1960 and Masters in Veterinary Parasitology in 1963 from India. He obtained a Ph.D. in Medical Microbiology in 1966 from the University of Sheffield, England. He obtained postdoctoral training with Dr. J. K. Frenkel, Department of Pathology and Oncology, University of Kansas Medical Center, Kansas City, from 1968 to 1973. From 1973 to 1978, he was Associate Professor of Veterinary Parasitology, Department of Pathobiology, Ohio State University, Columbus. He was then Professor of Veterinary Parasitology, Department of Veterinary Science, Montana State University, Bozeman, from 1978 to 1982. He is presently Project Leader, Zoonotic Diseases Laboratory, Livestock and Poultry Sciences Institute, Agricultural Research Service, U.S. Department of Agriculture, Beltsville, Maryland.

Dr. Dubey has spent over 25 years researching *Sarcocystis* and related cyst-forming coccidian parasites of man and animals. He has published over 250 refereed research papers in international journals. In 1985 he was chosen to be the first recipient of the Distinguished Veterinary Parasitologist Award by the American Association of Veterinary Parasitologists. He is an established researcher and distinguished teacher.

C. A. Speer, Ph.D., received a B.S. degree in zoology from Colorado State University in 1967, and M.S. and Ph.D. degrees in zoology from Utah State University in 1969 and 1972. During 1972 to 1973, Dr. Speer was Assistant Professor of Histology at the University of Texas Medical Center, Houston. From 1973 to 1975, he was a research associate of the U.S.-A.I.D. Malaria Immunity and Vaccination Project at the University of New Mexico, Albuquerque. From 1975 to 1983, Dr. Speer was Assistant, and then Associate Professor of Microbiology at the University of Montana, Missoula. During 1983 to 1986, he was Associate Professor and Director of the electron microscope facility in the Department of Veterinary Science/Veterinary Research Laboratory at Montana State University, Bozeman. In 1986, Speer was appointed Professor and Head of the Department of Veterinary Science and Director of the Veterinary Research Laboratory at Montana State University.

Dr. Speer has published more than 100 refereed research papers on the *in vitro* cultivation, immunity, ultrastructure, pathogenesis, and experimental infections of coccidian, trypanosomal, and malarial parasites. His current research is aimed at identifying, characterizing, and cloning coccidian antigens that induce protective immune responses. Dr. Speer received the Mershon Award for Excellence in Research in 1986 from the Montana Academy of Sciences, and Honorary Alumnus Award in 1987 from the College of Natural Sciences, Colorado State University, Fort Collins.

R. Fayer, Ph.D., received his B.S. degree in 1962 from the University of Alaska, and his M.S. and Ph.D. degrees in zoology from Utah State University in 1964 and 1968. From 1968 to 1972 he was Zoologist with the Coccidiosis Project, Beltsville Parasitology Laboratory, Agricultural Research Service (ARS), U.S. Department of Agriculture (USDA), Beltsville, Maryland. He was Project Leader for the *Sarcocystis* Project from 1972 to 1978 at the Animal Parasitology Institute (API), ARS, USDA, and then Laboratory Chief for Ruminant Parasites Laboratory, API, ARS, USDA, from 1978 to 1984. He was Director of the API, ARS, USDA, from 1984 to 1988 and now serves as Research Leader, Zoonotic Diseases Laboratory, Livestock and Poultry Sciences Institute, ARS, USDA, Beltsville, Maryland.

Dr. Fayer has spent over 25 years researching *Sarcocystis* and related coccidia of animals and of man. He has published over 150 papers in scientific journals. He was instrumental in developing cell culture systems for coccidia and was first to grow *Sarcocystis in vitro*. With colleagues, he was the first to discover the severe pathological features of *Sarcocystis* in the intermediate host. He received the USDA Superior Service Award and Medal as well as the American Society of Parasitologists H. B. Ward Medal in 1978. He is an internationally recognized scientist specializing in research on protozoan diseases of livestock.

ACKNOWLEDGMENTS

We wish to thank Joan Haynes and Anne Angermeyr for secretarial assistance, and Andy Blixt, Dianne Friesen, and Robert B. Ewing for technical assistance with illustrations. We also thank Drs. Richard J. Cawthorn, David S. Lindsay, Norman D. Levine, Edwin C. Powell, and Michel Rommel for reviewing the manuscript. We are grateful to many other scientists and publishers for contributing illustrations. This book was supported by funds from the Montana State University Agricultural Experiment Station (MSUAES), Bozeman, MT. MSUAES Journal Series No. J2123.

TABLE OF CONTENTS

Chapter 1
General Biology 1
- I. Introduction and History 1
- II. Structure and Life Cycle 2
  - A. Structure 2
  - B. Life Cycle 5
- III. Ultrastructure 13
  - A. Sarcocysts 13
  - B. Metrocytes 15
  - C. Bradyzoites 16
    1. Pellicle 16
    2. Apical Rings 17
    3. Polar Rings 17
    4. Microtubules 20
    5. Micropore 21
    6. Rhoptries and Micronemes 22
    7. Conoid 23
    8. Golgi Complex 24
    9. Endoplasmic Reticulum 24
    10. Ribosomes 24
    11. Mitochondria 24
    12. Multivesicular Bodies 26
    13. Inclusion Bodies 27
  - D. Endodyogeny 27
  - E. Types of Sarcocyst Walls 30
  - F. Gametogenesis 37
  - G. Fertilization 44
  - H. Oocyst 46
  - I. Sporocyst 46
  - J. Sporozoites 47
  - K. Schizogony 48
- IV. Taxonomic Criteria 53
  - A. Sarcocysts 54
  - B. Schizonts 59
  - C. Oocysts and Sporocysts 59
  - D. Host Specificity 60
  - E. Isoenzymes 60
- V. Pathogenicity 62
  - A. Intermediate Hosts 62
    1. Clinicopathological Findings 62
    2. Gross Lesions 64
    3. Microscopic Lesions 65
  - B. Definitive Hosts 67
- VI. Pathogenesis 68
  - A. Tissue Necrosis 68
  - B. Inflammation 69
  - C. Immune Regulation of Pathogenesis 69
  - D. Edema 70
  - E. Fever 70
  - F. Anemia 70

	G.	Abortion 71
		1. Placental Infection and Lesions 73
		2. Possible Mechanisms Causing Abortion or Fetal Death 74
	H.	Eosinophilic Myositis and Sarcocystosis 77
	I.	Chronic Sarcocystosis and Toxins 78
VII.	Immunity 79
	A.	Antigenic Structure 79
	B.	Humoral Responses 82
	C.	Cellular Responses and Immunosuppression 84
	D.	Protective Immunity and Vaccination 84
VIII.	Diagnosis 86
IX.	Economic Losses 90
X.	Transmission 90
XI.	Epidemiology 90
XII.	Control 91
XIII.	Chemoprophylaxis and Chemotherapy 91

Chapter 2
Techniques 93
I. Experimental Infection of Intermediate and Definitive Hosts 93
II. Isolation, Purification, and Preservation 93
    A. Bradyzoites 93
    B. Sporocysts 94
III. Diagnostic Techniques 94
    A. Examination of Feces for Sporocysts 94
    B. Examination of Muscles for Sarcocysts 95
    C. Serological Techniques 97
        1. Preparation of Soluble Antigen from Bradyzoites 97
        2. Method for ELISA Test 97
IV. *In Vitro* Cultivation 97
    A. Excystation and Cultivation of Schizonts 98
    B. Cultivation of Gamonts 101

Chapter 3
Sarcocystosis in Cattle (*Bos taurus*) 105
I. *Sarcocystis cruzi* (Hasselmann, 1926) Wenyon, 1926 105
    A. Structure and Life Cycle 105
    B. Pathogenicity 105
    C. Natural Outbreaks 105
    D. Protective Immunity 108
II. *Sarcocystis hirsuta* Moulé, 1888 109
    A. Structure and Life Cycle 109
    B. Pathogenicity 109
III. *Sarcocystis hominis* (Railliet and Lucet, 1891) Dubey, 1976 109
    A. Structure and Life Cycle 109
IV. Prevalence of Sarcocysts in Cattle 110
V. Sarcocystosis-Like Encephalitis in Cattle 112
Bibliography 112

Chapter 4
Sarcocystosis in Sheep (*Ovis aries*) 113
I. *Sarcocystis tenella* (Railliet, 1886) Moulé, 1886 113

	A.	Structure and Life Cycle 113
	B.	Pathogenicity 113
		1. Effect on Clinical Disease. 113
		2. Effect on Parturition and Reproductivity 114
	C.	Immunity and Protection 116
	D.	Clinical Sarcocystosis in Naturally Infected Sheep 116
II.	*Sarcocystis arieticanis* Heydorn, 1985 118
	A.	Structure and Life Cycle 118
	B.	Pathogenicity 118
III.	*Sarcocystis gigantea* (Railliet, 1886) Ashford, 1977 118
	A.	Structure and Life Cycle 118
	B.	Pathogenicity 118
IV.	*Sarcocystis medusiformis* Collins, Atkinson, and Charleston, 1979 119
	A.	Structure and Life Cycle 119
V.	*Sarcocystis* Infection and Carcass Condemnation 119
VI.	*Sarcocystis*-Like Encephalomyelitis in Sheep 119
Bibliography 120

Chapter 5
Sarcocystosis in Goats (*Capra hircus*) 121
I.	*Sarcocystis capracanis* Fischer, 1979 121
	A. Structure and Life Cycle 121
	B. Pathogenicity 121
	C. Protective Immunity 121
II.	*Sarcocystis hircicanis* Heydorn and Unterholzner, 1983 121
	A. Structure and Life Cycle 121
III.	*Sarcocystis moulei* Neveu-Lemaire,1912, and Other Macroscopic Sarcocysts in Domestic and Mountain Goats 121
Bibliography 125

Chapter 6
Sarcocystosis in Pigs (*Sus scrofa*) 127
I.	*Sarcocystis miescheriana* (Kühn, 1865) Labbé, 1899 127
	A. Structure and Life Cycle 127
	B. Pathogenicity 127
	C. Protective Immunity 127
II.	*Sarcocystis suihominis* (Tadros and Laarman, 1976) Heydorn, 1977 ... 128
	A. Structure and Life Cycle 128
	B. Pathogenicity 129
III.	*Sarcocystis porcifelis* Dubey, 1976 129
IV.	Prevalence of *Sarcocystis* and Economic Impact 129
Bibliography 130

Chapter 7
Sarcocystosis in Equids (*Equus* sp.) 131
I.	Introduction 131
II.	*Sarcocystis bertrami* Doflein, 1901 131
III.	*Sarcocystis equicanis* Rommel and Geisel, 1975 131
IV.	*Sarcocystis fayeri* Dubey, Streitel, Stromberg, and Toussant, 1977 131
V.	Pathogenicity 132
VI.	Prevalence of Natural Infections and Clinical Disease 132

VII.	Remarks on Validity of *Sarcocystis* Species in Equids	132
VIII.	Equine Protozoal Myelitis (EPM)	133
Bibliography		134

Chapter 8
Sarcocystosis in Water Buffalo (*Bubalus bubalis*) 137
- I. *Sarcocystis levinei* Dissanaike and Kan, 1978 137
  - A. Structure and Life Cycle 137
- II. *Sarcocystis fusiformis* (Raillet, 1897) Bernard and Bauche, 1912 137
  - A. Structure and Life Cycle 137
- Bibliography 137

Chapter 9
Sarcocystosis in Camels (*Camelus dromedarius* and *C. bactrianus*) 141
- I. Introduction 141
- II. *Sarcocystis cameli* Mason, 1910 141
  - A. Structure and Life Cycle 141
- Bibliography 141

Chapter 10
Sarcocystosis in Humans and Other Primates 143
- I. Humans (*Homo sapiens*) 143
  - A. Intestinal Sarcocystosis 143
    1. *Sarcocystis hominis* (Raillet and Lucet, 1891) Dubey, 1976 143
    2. *Sarcocystis suihominis* (Tadros and Laarman, 1976) Heydorn, 1977 143
    3. Natural Prevalence 143
  - B. Muscular Sarcocystosis 143
- II. Other Primates 144
  - A. *Sarcocystis kortei* Castellani and Chalmers, 1909 144
  - B. *Sarcocystis nesbiti* Mandour, 1969 144
- Bibliography 144

Chapter 11
Sarcocystosis in Dogs, Cats, and Other Carnivores 145
- I. Muscular Sarcocystosis 145
- II. Intestinal Sarcocystosis 145

Chapter 12
Sarcocystosis in Wild Ruminants and Other Large Animals 149
- I. Introduction 149
- II. Roe Deer (*Capreolus capreolus*) 149
  - A. *Sarcocystis gracilis* Ratz, 1908 149
  - B. *Sarcocystis capreoli* Levchenko, 1963 (Syn. *S. capreolicanis* Erber, Boch, and Barth, 1978) 149
  - C. *Sarcocystis sibrica* Machuslkij, 1947 149
- III. Fallow Deer (*Cervus dama*) 149
- IV. Reindeer (*Rangifer tarandus tarandus*) 150
- V. White-Tailed Deer (*Odocoileus virginianus*) 150
  - A. *Sarcocystis odocoileocanis* Crum, Fayer, and Prestwood, 1981 151

	B.	*Sarcocystis odoi* Dubey and Lozier, 1983 151
	C.	*Sarcocystis* sp. Dubey and Lozier, 1983 151
VI.	Red Deer or Wapiti (*Cervus elaphus*) 152	
	A.	*Sarcocystis wapiti* Speer and Dubey, 1982 152
	B.	*Sarcocystis sybillensis* Dubey, Jolley, and Thorne, 1983 152
	C.	*Sarcocystis cervicanis* Hernandez-Rodriquez, Navarrete, and Martinez-Gomez, 1981 153
VII.	Mule Deer (*Odocoileus hemionus*) 153	
	A.	*Sarcocystis hemionilatrantis* Hudkins and Kistner, 1977 153
VIII.	Moose (*Alces alces*) 154	
	A.	*Sarcocystis alceslatrans* Dubey, 1980 154
	B.	*Sarcocystis* sp. Dubey, 1980 154
	C.	*Sarcocystis* sp. Colwell and Mahrt, 1981 154
IX.	Pronghorn (*Antilocapra americana*) 154	
X.	Unnamed Deer 154	
XI.	Bighorn Sheep (*Ovis canadensis*) 154	
XII.	North American Mountain Goat (*Oreamnos americanus*) 155	
XIII.	Bison (*Bison bison*) 155	
XIV.	African and Other Antelope 155	
XV.	Llamas (*Lama glama*) 155	
XVI.	Yak (*Poephagus grunniens*) 156	
XVII.	Sea Mammals 156	

Chapter 13
Sarcocystosis in Birds 157
I. Introduction 157
II. Chickens (*Gallus gallus*) 157
III. Ducks 157
IV. Other Birds 158
V. Raptorial Birds 159

Chapter 14
Sarcocystosis in Miscellaneous Homoiothermic Animals 161

Chapter 15
Sarcocystosis in Poikilothermic Animals 167
I. Reptiles 167
II. Fishes 167

Chapter 16
***Frenkelia* and Related Genera** 171
I. Genus *Frenkelia* Biocca, 1965 171
    A. *Frenkelia microti* (Findlay and Middleton, 1934) Biocca, 1965 171
    B. *Frenkelia glareoli* (Erhardova, 1955) Biocca, 1965 171
    C. Pathogenicity of *Frenkelia* 171
II. Related Genera. 173
Bibliography 173

References 177

Index 207

Chapter 1

GENERAL BIOLOGY

I. INTRODUCTION AND HISTORY

Sarcocystis was first reported by Miescher in 1843 as "milky white threads" in the skeletal muscle of a house mouse, *Mus musculus,* in Switzerland.[500] During the succeeding 20 years, in which there was no generic name, the parasite was known simply as Miescher's tubules. Kühn[417] found a similar parasite in the pig and named it *Synchytrium miescherianum.* However, this genus was occupied and Lankester[428] introduced the genus name *Sarcocystis* (sarco = muscle). Labbé[422] then changed the name *Synchytrium miescherianum* to *Sarcocystis miescheriana.* Thus, the parasite *Sarcocystis miescheriana* (Kühn, 1865) Labbé 1899 became the type species of the genus.[443] The parasite originally found by Miescher in the mouse was described by Blanchard in 1885[32a] and was named *Sarcocystis muris* by Railliet in 1886.[11a] Thus, controversy arose as to whether *S. muris* or *S. miescheriana* was the correct type species of the genus. *S. miescheriana* is the correct type species of the genus because it was named before *S. muris.*

Between 1885 and 1972, numerous species of *Sarcocystis* were named based upon the finding of sarcocysts in the host. Because the life cycle of *Sarcocystis* was unknown until 1972 (Table 1), there was no way to validate the identity of the different species. Heydorn et al.[353] first provided conclusive evidence that there were three species of *Sarcocystis* with sarcocysts in cattle and sexual stages in dogs, cats, and man, respectively; there were two species with sarcocysts in sheep, and sexual stages in dogs and cats. The sarcocysts of these species were structurally different from one another. Previously, any sarcocyst in cattle was considered to be one species, *S. fusiformis.* Furthermore, *S. fusiformis* was also thought to parasitize the water buffalo. Equipped with knowledge of the life cycle of *Sarcocystis*, various authorities searched the literature for clues that earlier investigators recognized structural differences among sarcocysts and named them accordingly. Moulé[504a] had named *S. hirsuta* for a sarcocyst with hairy projections he observed in tissue from cattle. Hasselmann[342b] had named *S. cruzi* based on the observation of a sarcocyst without hairy projections from the heart of cattle.[440] Another parasite, *Isospora hominis* (Railliet and Lucet, 1891) which had been identified in feces of man, was found to initiate the development of yet another type of sarcocyst now named *Sarcocystis hominis.*[440] Because the original descriptions were inadequate, and no type specimens were available, it was impossible to confirm that the originally named species were identical to those species wherein the life cycles were documented.

Because of this inability to make a positive identification, Heydorn et al.[353] proposed new names for the three species in cattle: *S. bovicanis, S. bovifelis,* and *S. bovihominis*; and two species in sheep: *S. ovifelis* and *S. ovicanis*, combining the names of the intermediate and definitive hosts. Levine[440] extensively searched the literature and said that the old names, *S. cruzi, S. hirsuta,* and *S. hominis* for sarcocysts in cattle, and *S. tenella* and *S. gigantea* for sarcocysts in sheep, were valid and should not be replaced by new names, following the International Code of Zoological Nomenclature (ICZN). Levine[443] states, "A name is or remains available even though it is found that the original description relates to more than one taxonomic unit. The species must be simply redescribed."

These two divergent views, one supporting the new names and one opposing the new names, caused considerable confusion. Heydorn and associates collaborated with Frenkel[288-290,495] to review the life cycles and sarcocyst structures of several species. They deposited neohepantotypes of sarcocysts with 25 institutions and investigators, endeavoring to reduce confusion concerning the old and new names. They argued that there was no scientific validity in arbitrarily assigning the old names to their new *Sarcocystis* species because the original descriptions are inadequate and because no type specimens exist. They applied to the Secretary of ICZN to have

Table 1
HISTORICAL LANDMARKS CONCERNING *SARCOCYSTIS*

Year	Findings	Ref.
1843	Sarcocysts found in muscles of a house mouse	Miescher[500]
1882	Genus *Sarcocystis* named	Lankester[428]
1972	Sexual phase cultured *in vitro*	Fayer[238]
1972	Two-host life cycle found	Rommel and Heydorn;[583] Rommel et al.[582]
1973	Vascular phase recognized and pathogenicity demonstrated	Fayer and Johnson[239]
1975	Multiple *Sarcocystis* species within a given host recognized	Heydorn et al.[353]
1975	Chemotherapy demonstrated	Fayer and Johnson[245]
1976	Abortion due to sarcocystosis recognized	Fayer et al.[248]
1981	Protective immunity demonstrated	Dubey[155a]
1986	Vascular phase cultured *in vitro*	Speer and Dubey[640]

the new names replace the old ones. However, ICZN has rejected their application.[497a,498] Frenkel et al.[288] also proposed changing the type species of *Sarcocystis* from *S. miescheriana* to *S. muris* because in their view the original description of *S. miescheriana* was inadequate.

The views proposed by Heydorn and associates are rational and easy to use, but we believe that the question of validity of scientific names must be the prerogative of the ICZN. Consequently, this debate[291,445] concerning the old and new names for *Sarcocystis* should end and caution should be used in designating new species without an adequate description.

In this book we have used the names identified by Levine[443] and Levine and Tadros[442] and supported by the ICZN. The valuable contributions made by Heydorn et al.[353] in elucidating the structure and life cycle of *Sarcocystis* is in no way diminished by the use of the ICZN-sanctioned names. As far as proposing future names for species of *Sarcocystis*, adherence to ICZN rules is the only constraint.

Additional historical accounts of *Sarcocystis* are given by Levine.[440,443,446]

Sarcocystis species are coccidian parasites and belong to

Phylum — Apicomplexa; Levine, 1979
Class — Sporozoasida; Leuckart, 1879
Subclass — Coccidiasina; Leuckart, 1879
Order — Eucoccidiorida; Leger and Duboseq, 1910
Suborder — Eimeriorina; Leger, 1911
Family — Sarcocystidae; Poche, 1913
Subfamily — Sarcocystinae; Poche, 1913
Genus — *Sarcocystis*; Lankester, 1882

Generic diagnosis — Sarcocysts in muscles and central nervous systems (CNS) of homiothermic and poikilothermic animals. Obligatorily two hosts, asexual multiplication in intermediate host, "sarcocysts" in muscles and CNS, gamonts in the intestine of definitive hosts, endogenous sporulation of oocysts.

Type species — *S. miescheriana* (Kühn, 1865) Labbé, 1899.

II. STRUCTURE AND LIFE CYCLE

A. Structure

Sarcocysts (in Greek, *Sarkos* = flesh, k*ystis* = bladder) are the terminal asexual stage found

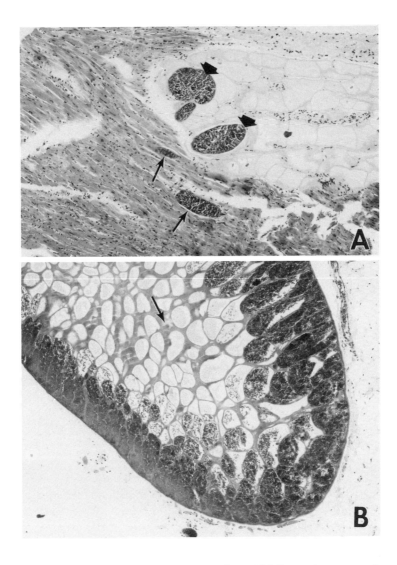

FIGURE 1. Sarcocysts in tissues of llama (*Lama glama*). (A) Microscopic sarcocysts of an unnamed species in myocytes (arrow) and in Purkinje fibers (arrowheads) in the myocardium. (B) A macroscopic *S. aucheniae* sarcocyst in the esophagus. Note prominent septa (arrow), zoites at the periphery, and empty areas in the center without zoites. (H. & E.; magnification × 75.)

encysted, primarily in striated muscles of mammals, birds, and poikilothermic animals (intermediate hosts). The number and distribution of sarcocysts throughout the body vary greatly from host to host. Factors affecting number and distribution of sarcocysts include the number of organisms (sporocysts) ingested, the species of *Sarcocystis*, the species of host, and the immunological state of the host. Although most sarcocysts develop in striated muscles of the heart, tongue, esophagus, diaphragm, and skeletal muscles, some sarcocysts have been found in smooth muscles. Sarcocysts of *S. mucosa* have been found in the gut of marsupials[32a,533a] in what appears to be smooth muscle cells. Infrequently, immature sarcocysts of other species (*S. cruzi* and *S. tenella*) also have been found in the gut.[160,164] Sarcocysts also have been found in the CNS and in Purkinje fibers of the heart and muscle bundles, but always in relatively low numbers (Figure 1). They are found in different types of myofibers.[565]

Sarcocysts vary in size and shape, depending on the species of the parasite (Figures 2 to 4). Some always remain microscopic (e.g., *S. cruzi*), whereas others become macroscopic (e.g., *S.*

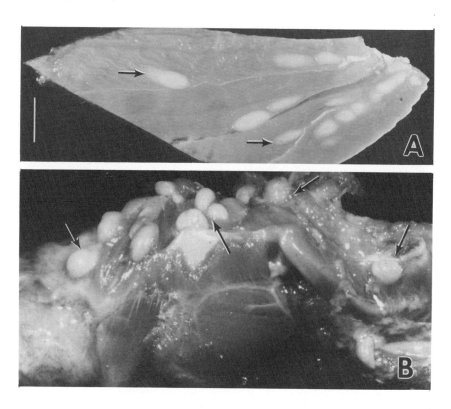

FIGURE 2. Macroscopic sarcocysts. (A) Rice grain-like *S. rileyi* sarcocysts in breast muscles of a duck. (Bar = 1 cm.) (From Dubey, J. P., *J. Am. Vet. Med. Assoc.*, 169,1061, 1976. With permission.) (B) *S. gigantea* (arrows) in laryngeal area of sheep. (From Dubey, J. P., Leek, R. G., and Fayer, R., *J. Am. Vet. Med. Assoc.*, 188, 151, 1986. With permission.)

gigantea, S. muris). Microscopic sarcocysts vary from very long and narrow to short and wide. Macroscopic sarcocysts, which are nearly always in skeletal muscles or esophageal muscles, appear filamentous (e.g., *S. muris*), like rice grains (e.g., *S. rileyi*), or globular (e.g., *S. gigantea*).

Sarcocysts are always located within a parasitophorous vacuole (PV) in the host cell cytoplasm (Figure 5). More than one sarcocyst may be found in one host cell. The sarcocyst consists of a cyst wall that surrounds the parasitic metrocyte or zoite stages. The structure and thickness of the cyst wall varies among species of *Sarcocystis* and within each species as the sarcocyst matures. A connective tissue wall (secondary sarcocyst wall) surrounds the *S. gigantea*, *S. hardangeri*, and *S. rangiferi* sarcocysts[320,322,474a,485a] (Figure 6). Histologically, the sarcocyst wall may be smooth, striated or hirsute or may possess complex branched protrusions (Figure 7). Internally, groups of zoites may be divided into compartments by septa that originate from the sarcocyst wall or may not be compartmentalized. Septa are usually less than 2 µm thick and are eosinophilic. They are present in all but a few[191,397] species of *Sarcocystis*. The structure of the parasites within the sarcocysts varies with the maturation of the sarcocyst. Immature sarcocysts contain globular parasites called metrocytes (mother cells). Each metrocyte produces two progeny by an internal form of multiplication called endodyogeny (described in Section III.D). After what appears to be several such generations, some of the metrocytes, through the process of endodyogeny, produce banana-shaped zoites called bradyzoites (also called cystozoites). Within the sarcocysts, metrocytes are generally located in the cortex, whereas bradyzoites are located in the medulla. In old, large sarcocysts, bradyzoites near the center of the sarcocyst are sometimes degenerate and are replaced by granules or globules (Figure 1B). Live sarcocysts are probably fluid filled with metrocytes and bradyzoites moving within the fluid. In histologic

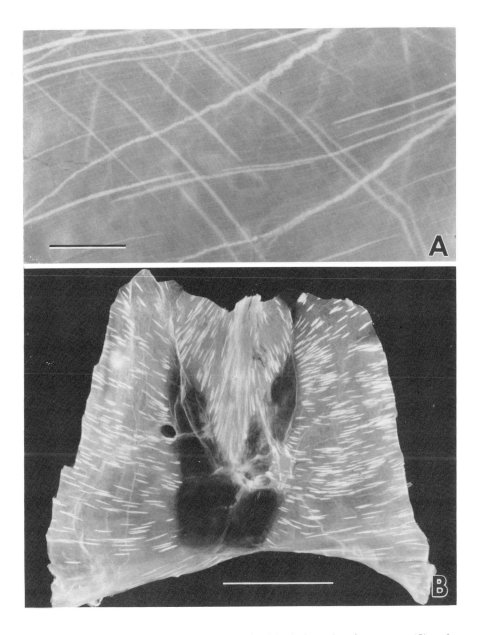

FIGURE 3. Macroscopic sarcocysts. (A) Sarcocysts in abdominal muscles of a cotton rat (*Sigmodon hispidus*). (Bar = 1 cm.) (From Dubey, J. P. and Sheffield, H. G., *J. Parasitol.*, in press. With permission.) (B) Sarcocysts in diaphragm of a Richardson's ground squirrel (*Spermophilus richardsonii*). (Bar = 1 cm.) (From Cawthorn, R. J., Wobeser, G. A., and Gajadhar, A. A., *Can. J. Zool.*, 61, 370, 1983. With permission.)

sections stained with hematoxylin and eosin (H. & E.), metrocytes are paler than bradyzoites. The bradyzoites contain prominent amylopectin granules that stain bright red with periodic acid Schiff (PAS) reaction.

B. Life Cycle

Sarcocystis has an obligatory prey-predator two-host life cycle (Figure 8). Asexual stages develop only in the intermediate host which in nature is often a prey animal. Sexual stages develop only in the definitive host which is carnivorous. Intermediate and definitive hosts vary

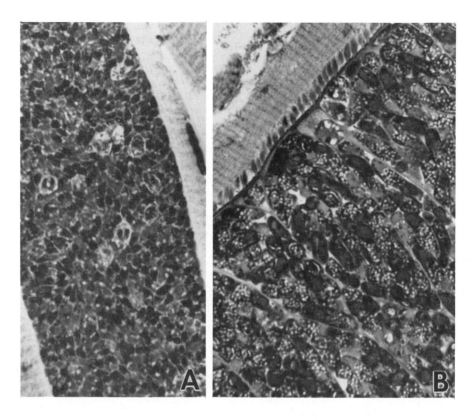

FIGURE 4. (A) *S. bozemanensis* and (B) *S. campestris* sarcocysts in skeletal muscles of *S. richardsonii*. The former has a thin sarcocyst wall and tiny bradyzoites, whereas the latter has a thick sarcocyst wall and large bradyzoites. (Giemsa stain; magnification × 1000.) (From Dubey, J. P., *Can. J. Zool.*, 61, 942, 1983. With permission.)

for each species of *Sarcocystis*. For example, there are three named species of *Sarcocystis* in cattle: *S. cruzi*, *S. hirsuta*, and *S. hominis*. The definitive host for these species are canids, felids, and primates, respectively. For the following description of the life cycle and structure of *Sarcocystis*, *S. cruzi* will serve as the example. Examples from other species are discussed when information on *S. cruzi* is deficient or at variance.

Dogs, coyotes, red foxes, and possibly wolves, jackals, and raccoons are the definitive hosts, whereas bison (*Bison bison*) and cattle (*Bos taurus*) are the intermediate hosts for *S. cruzi*.[163,258] The definitive host becomes infected by ingesting muscular or neural tissue containing mature sarcocysts. Bradyzoites are liberated from the sarcocyst by digestion in the stomach and intestine. Bradyzoites move actively, penetrate the mucosa of the small intestine, and transform into male (micro) and female (macro) gamonts. Within 6 h of ingesting infected tissue, gamonts were found within a PV in goblet cells near the tips of villi (Figure 9A). In certain species of *Sarcocystis* (e.g., *S. idahoensis*), gametogony is delayed for several days after ingestion of sarcocysts.[34,35] The ratio of macrogamonts to microgamonts is approximately 95:5. Macrogamonts are ovoid to round, 10 to 20 μm in diameter, and contain a single nucleus with compact chromatin. Microgamonts are ovoid to elongated and contain one to several nuclei. The microgamont vesicular nucleus divides into several nuclei (usually up to 15), and as the microgamont matures, the nuclei move towards the periphery of the gamont[246] (Figure 9B to E). In *S. cruzi*, mature microgamonts which are about 7 × 5 μm, contain 3 to 11 slender gametes (Figure 9E). The microgametes, which are about 4 × 0.5 μm in size, consist of a compact nucleus and two flagella. Microgametes liberated (Figure 9F) from the microgamont actively move to

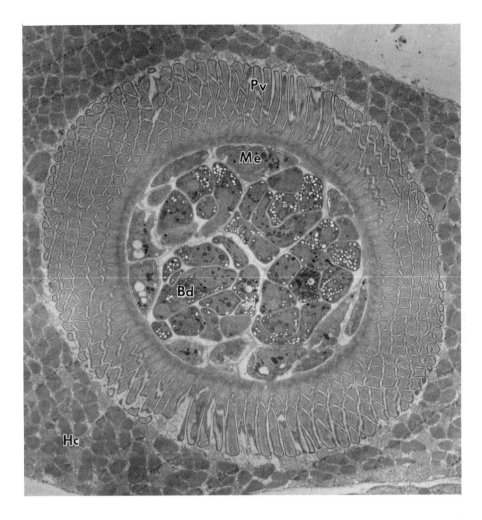

FIGURE 5. Cross-section of *S. hemioni* in skeletal muscles of mule deer. The host cell (Hc) cytoplasm is separated from the sarcocyst by a parasitophorous vacuole (Pv). The sarcocyst contains metrocytes (Me) and bradyzoites (Bz). (Magnification × 3055.)

the periphery of the macrogamont. After fertilization,[624] a wall develops around the zygote and the oocyst is formed. The entire process of gametogony and fertilization can be completed within 24 h, but it is asynchronous. Thus, gamonts and oocysts may be found at the same time. The location of gametogony and the type of cell parasitized varies with species of *Sarcocystis* and stage of gametogenesis. For example, gamonts of *S. muris*, *S. hirsuta*, and *S. cruzi* initially develop in goblet cells at or near the surface of the intestine, whereas those of *S. idahoensis* develop in enterocytes next to the basement membrane adjoining the lamina propria. Occasionally oocysts have been found in the mesenteric lymph nodes.[46a,625]

Oocysts sporulate in the lamina propria (Figure 10). Initially, the sporont was granular, eosinophilic, and filled the young oocysts. During the initial stages of sporont condensation, a clear cap-like structure was found in *S. cruzi*[160] and *S. idahoensis*.[34] When the sporont condensed further, a similar clear area appeared at the opposite end. The oocyst contained one large nucleus with one or two prominent nucleoli and several PAS-positive granules. As sporulation progressed, the nucleus elongated and became parallel to the longitudinal axis of the sporont.[34,67,633] The elongate nucleus divided into two nuclei, one at each pole of the sporont. These nuclei began a second transverse division, and the sporont cytoplasm divided transversely into

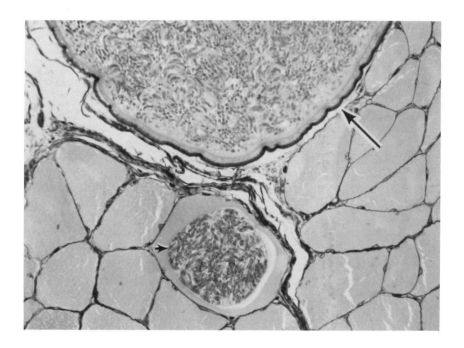

FIGURE 6. Sarcocysts of *S. gigantea* (top) and *S. arieticanis* (bottom) in skeletal muscles of sheep. *S. gigantea* (arrow) is surrounded by a connective tissue capsule (secondary sarcocyst wall), while *S. arieticanis* (arrowhead) has no secondary sarcocyst wall. (Gomori's methanine silver stain; magnification × 300.)

two sporoblasts (Figure 10). Each sporoblast contained two nuclei. The nuclei moved towards the opposite poles of each sporoblast, and the sporoblasts became surrounded by an eosinophilic wall; sporoblasts became sporocysts. Four sporozoites formed in each sporocyst, apparently by a third nuclear division. Because sporulation is asynchronous, unsporulated and sporulated oocysts are found simultaneously (Figure 10).

Sporulated oocysts are generally colorless, thin-walled (<1 μm), and contain two elongate sporocysts. An oocyst residuum and a micropyle are absent. Each sporocyst contains four elongated sporozoites and a granular sporocyst residuum which may be compact or dispersed, but a Stieda body is absent. Because sporozoites are flexed in the sporocyst, all four are often not seen in a single plane of focus. Each sporozoite has a central to terminal nucleus and several cytoplasmic granules, but there is no refractile body.

The oocyst wall is thin and often ruptures (Figure 11). Free sporocysts, released into the intestinal lumen, are passed in feces (Figure 12). Occasionally unsporulated and partially sporulated oocysts are shed in feces. The prepatent and patent periods vary, but for most *Sarcocystis* species oocysts are first shed in the feces between 7 and 14 d after ingesting sarcocysts.

The intermediate host becomes infected by ingesting sporocysts in food or water. Sporozoites excyst from sporocysts in the small intestine. The fate of the sporozoite from the time of ingestion of the sporocyst until initial development in mesenteric lymph node arteries is not known. Sporozoites of *S. cruzi* were first found in the lumen and in the endothelium of arteries 4 to 7 days postinoculation (DPI). At such time, free zoites have been seen in arteries in mesenteric lymph nodes. First-generation schizogony or merogony begins in endothelial cells as early as 7 DPI and may be completed as early as 15 DPI (Figure 13). Second-generation schizonts or meronts have been seen in endothelium from 19 to 46 DPI, predominantly in capillaries, but also in small arteries, virtually throughout the body. These schizonts were most

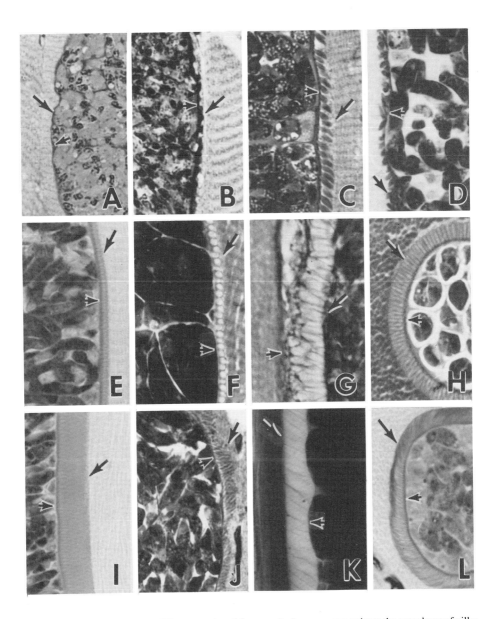

FIGURE 7. Sarcocyst walls in different species of *Sarcocystis*. Large arrows point to the outer layer of villar protrusions, and the small arrows point to inner layer of the sarcocyst wall. (A and B) Thin-walled *S. idahoensis*, deer mouse. The wall is smooth in A and irregular in B. (C) *S. campestris* with villar protrusions, Richardson's ground squirrel; (D) *S. microti* with small villi of uneven length, meadow vole; (E) *S. odocoileocanis* with small stubby villi, white-tailed deer. (F to I) Mule deer: (F) *S. hemionilatrantis* with inverted T-like villi; (G) *S. youngi* with villi of uneven width; (H and I) *S. hemioni* with thick wall with long villi; (J) *S. peromysci* with fine hair protrusions, deer mouse; (K) *S. arieticanis* with long, hair-like protrusions, sheep; (L) *S. sybillensis* with thick tufts of hairy protrusions, elk. (Magnification × 1000.)

numerous in the glomeruli of the kidney (Figure 14). Immature schizonts stained with H. & E. are basophilic and stain lighter than the host tissue. The nucleoplasm of the schizont is diffuse and difficult to recognize in 6-μm sections, but appears clear in 1- to 3-μm sections (Figure 14A). The nucleus becomes lobulated (Figure 14B) and divides into several nuclei (up to 37). Merozoites form at the periphery (Figure 14C). The shape and size of schizonts vary considerably. Schizonts in skeletal muscle are longer than those in other tissues. Both first- and second-generation schizonts are located within the host cytoplasm and are not surrounded by a PV.

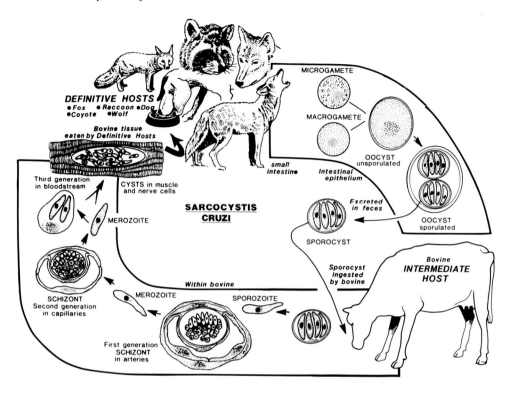

FIGURE 8. Life cycle of *S. cruzi*.

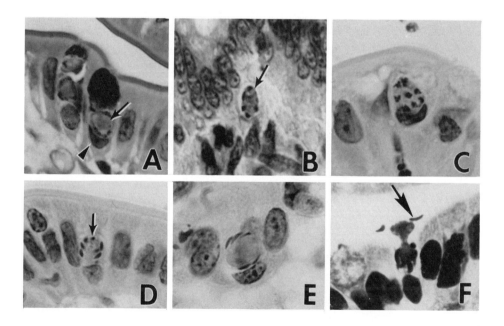

FIGURE 9. Gamonts of *S. hirsuta* (A, C, D, and F) and *S. cruzi* (B and E) in the small intestine of cat (*S. hirsuta*) and coyote (*S. cruzi*), respectively. (A) Four oval early gamonts in two adjacent goblet cells. One gamont (arrow) is located between the indented host cell nucleus (arrowhead) and deeply stained mucin; (B) microgamont (arrow) with five nuclei; (C) microgamont with eight nuclei; (D) microgamont (arrow) with seven peripheral nuclei; (E) mature microgamont with gametes; (F) ruptured microgamont liberating microgametes (arrow). (Magnification × 1000.) (From Dubey, J. P., *J. Protozool.*, 29, 591, 1982; and Dubey, J. P., *Parasitology*, 86, 7, 1983. With permissions.)

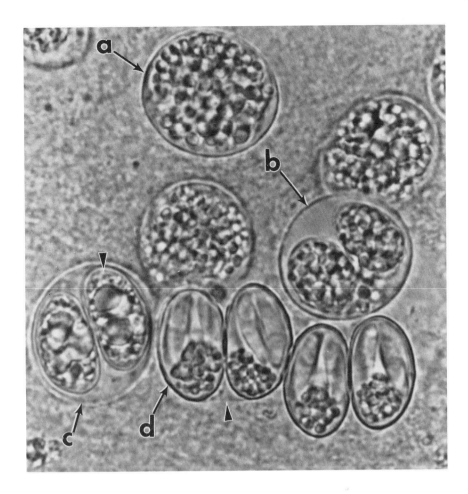

FIGURE 10. Sporulating *S. cruzi* oocysts in the intestine of an experimentally infected coyote. (a) Unsporulated oocyst; (b) oocyst with two sporoblasts; (c) oocysts with two sporocysts with polar "nucleated" areas; (d) sporulated oocyst with two sporocysts and a thin oocyst wall (arrowhead). (Magnification × 1200.)

Merozoites are found in peripheral blood 24 to 46 DPI, coincident with the maturation of second-generation schizonts (Figure 14). Merozoites in blood are extracellular or located within unidentified mononuclear cells (Figure 14E). Intracellular merozoites contain one or two nuclei, and some divide into two, apparently by endodyogeny[254] (Figures 14F and G). Extracellular merozoites are often degenerate. Division in blood cells has not been seen in some species of *Sarcocystis* (e.g., *S. hirsuta*). Individual merozoites were seen in macrophage-like cells in tissues of animals infected with some species.[20,159,177,542]

The number of generations of schizogony and the type of host cell may vary with each species of *Sarcocystis*, but trends are apparent. For example, all species of *Sarcocystis* of large domestic animals (sheep, goat, cattle, pigs, horses) form first- and second-generation schizonts in vascular endothelium, whereas only a single precystic generation of schizogony has been found in *Sarcocystis* species of small mammals (mice, deer mice), and this is generally in hepatocytes. *S. hemionilatrantis* represents a variation with the first two generations of schizogony in vascular endothelium and an additional third or fourth generation in macrophages in muscles (Figure 15). Still other variations are seen in cattle where the second-generation schizont of *S. cruzi* develops in virtually every organ, whereas those of *S. hirsuta* are restricted to muscles.

Usually, schizonts are time-limited stages that disappear before sarcocysts form. However,

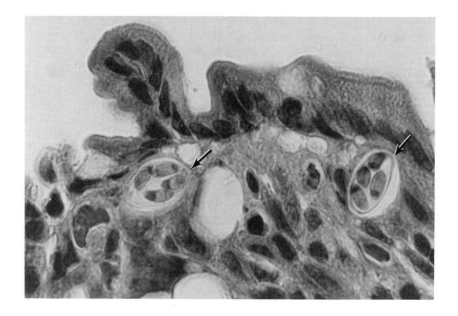

FIGURE 11. Two *S. cruzi* sporocysts (arrows), each with four sporozoites, in the lamina propria of the small intestine of a dog. (H. & E.; magnification × 1200.) (From Dubey, J. P., *J. Am. Vet. Med. Assoc.*, 169,1061, 1976. With permission.)

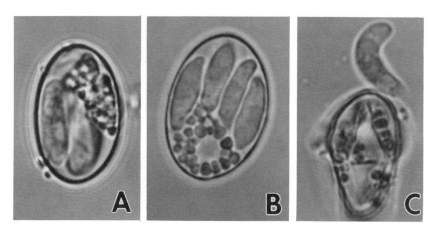

FIGURE 12. *In vitro* excystation of sporozoites of *S. cruzi* in trypsin-bile excystation fluid. (A) Untreated sporocyst; (B) treated sporocyst with four sporozoites and sporocyst residuum; (C) one excysted sporozoite and collapsed sporocyst wall. (Magnification × 1000.) (From Fayer, R. and Leek, R. G., *Proc. Helminthol. Soc. Wash.*, 40, 294, 1973. With permission.)

schizonts of *S. falcatula* develop in blood vessels of birds (budgerigar) from day 2 to 5.5 months postinoculation,[631a] with two distinct peaks of schizogony at 7 to 8 DPI and 28 DPI. All schizonts are structurally similar. The site of schizogony shifts progressively from capillaries to venules to veins.[631a]

Merozoites liberated from the terminal generation of schizogony initiate the sarcocyst formation. The intracellular merozoite surrounded by a PV becomes round to ovoid (metrocyte) (Figure 16). After repeated divisions, the sarcocyst is filled with bradyzoites, and the presence of which is the infective stage for the predator. Sarcocysts generally become infectious at about 75 DPI, but there is considerable variation among species of *Sarcocystis*. Immature sarcocysts and schizonts are not infectious for the definitive host.

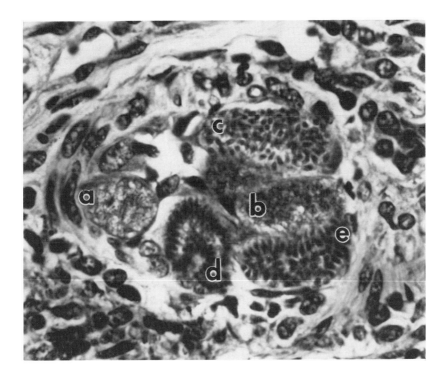

FIGURE 13. Development of first-generation schizonts of *S. cruzi* in a mesenteric lymph node artery of an experimentally infected calf. Progressive developing stages are (a) lobulated nucleus; (b) undifferentiated nuclei; (c) differentiated nuclei; (d) developing merozoites; (e) mature with merozoites. (Iron hematoxylin plus Giemsa stain; magnification × 800.) (From Dubey, J. P., *J. Protozool.*, 29, 591, 1982. With permission.)

Sarcocysts may persist for the life of the host, but many begin to disappear after 3 months postinoculation (Figures 17 and 18).

III. ULTRASTRUCTURE

Sarcocystis spp. are single-cell, eukaryotic organisms that contain a nucleus, nucleolus, endoplasmic reticulum, ribosomes, Golgi complex, and mitochondria. They also have organelles that are characteristic of the phylum Apicomplexa, such as apical rings (also called conoidal or preconoidal rings), polar rings, conoid (may be absent in some apicomplexans, i.e., *Plasmodium*), pellicle, subpellicular microtubules, micropores, rhoptries, and micronemes.

There are numerous reports on the ultrastructure of *Sarcocystis* spp. sarcocysts (see References 9, 28a, 29, 47, 48, 56, 61, 64, 65, 97, 105a, 119, 133, 160, 164, 172, 173, 175, 176, 178a, 181, 186, 194, 195, 201, 206, 209, 212, 213, 216 to 218, 220, 303a, 305, 306, 318 to 326, 347, 354, 354a, 368, 373, 376, 394, 395, 398 to 400, 409 to 410a, 461, 468, 469, 472, 473, 485a, 485c to 493, 496, 527, 532 to 533a, 535, 542, 545, 558, 562 to 563a, 599, 607, 609a to c, 611, 612, 614, 617 to 619, 623, 627a, 627b, 637, 641, 655, 667, 675, 676, 690 to 692, 695, and 696), schizonts (see References 9, 20, 72, 155, 158, 165, 211, 360, 472, 496, 544, 619, 631a, 631b, and 634 to 636), and gamonts (see References 25, 208, 214, 215, 218 to 220, 472, 485b, 486, 494, 496, 610, 624, 633, 673a, and 692), but many details are still missing. Electron micrographs and data reported herein will fill some of the missing gaps concerning the ultrastructure of *Sarcocystis*.

A. Sarcocysts

The ultrastructure section begins with sarcocysts because of the importance of this stage as

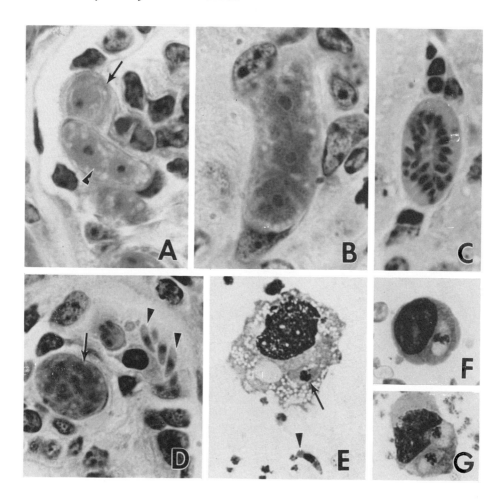

FIGURE 14. Second-generation schizonts (A to D) and intravascular zoites (E to G) of *S. cruzi* in experimentally infected calves. (A) Immature schizont with one (arrow) or two (arrowheads) nucleoli; (B) lobed nucleus with three prominent nucleoli; (C) peripherally arranged merozoites; (D) individual merozoites (arrowheads) and a multinucleated schizont (arrow). (E to G) Merozoites in peripheral blood: (E) intracellular merozoite (arrow) in a monocyte and an extracellular merozoite (arrowhead); (F) binucleated merozoite in a mononuclear cell; (G) mononuclear cell with two merozoites. (Magnification × 1000.) (From Dubey, J. P., *J. Protozool.*, 29, 591, 1982. With permission.)

a taxonomic aid. Each *Sarcocystis* sp. produces a sarcocyst wall that usually has unique ultrastructural characteristics which can be used to distinguish it from other species within the same intermediate host.

There are only a few reports on the ultrastructural development of sarcocysts and sarcocyst walls,[164,176,487–490,492,493,545,623] and none described the complete development.

Sarcocyst development begins when a merozoite enters a myocyte (Figure 19) or neural cell. The merozoite resides in a parasitophorous vacuole and is surrounded by a parasitophorous vacuolar membrane (Pm) which appears to develop immediately into a primary sarcocyst wall (Pw). The Pw consists of a Pm plus an underlying electron-dense layer. As a merozoite transforms into a metrocyte, many of the organelles of the apical complex, such as micronemes, conoid, polar, and apical rings, disappear (Figure 19) while making ribosomes, endoplasmic reticulum, and mitochondria become more abundant and the nucleus larger. A granular layer located immediately beneath the Pw develops early, with the acquisition of minute granules and vesicles which pinch off from invaginations in the Pm (Figure 20). (Figures 20 to 23 are a series of transmission electron micrographs [TEMs] showing development of the primary sarcocyst wall of *S. singaporensis*.)

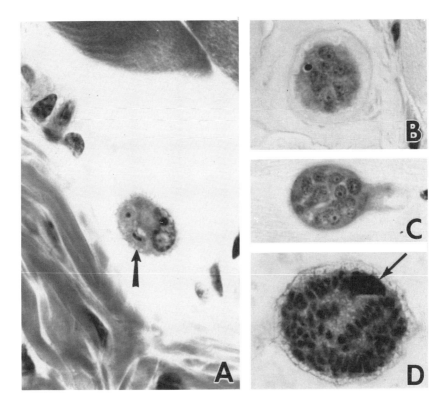

FIGURE 15. Third-generation schizonts of *S. hemionilatrantis* in macrophages within skeletal muscles of mule deer. (A) Binucleate schizont with one dividing nucleus (arrow). Myocytes are on top right and connective tissue cells at the bottom left and the infected macrophage is in the edematous perimysial space. (B) Multinucleate schizont; (C) multinucleate schizont; (D) nearly mature schizont. The host cell nucleus (arrow) is indented. (Magnification × 1000.) (From Dubey, J. P., Kistner, T. P., and Callis, G., *Can. J. Zool.*, 61, 2904, 1983. With permission.)

Mature sarcocysts of all *Sarcocystis* spp. have minute undulations in the Pw. In some species, the Pw may remain relatively simple (i.e., *S. rauschorum*), whereas other species may have highly complex walls in which the primary sarcocyst wall is folded or branched to form protrusions that project outwardly from the sarcocyst. The protrusions may contain minute granules, electron-dense bodies, microfilaments, microtubules, and small vesicles. *S. singaporensis* has a complex wall which requires nearly 200 d to become fully formed (Figures 20 to 23). At 25 to 30 DPI, the Pw initially consists of minute undulations only (Figure 20A), which soon becomes folded to form short villar protrusions (Figure 20B), elongate protrusions with parallel sides (Figure 20C), and then much larger club-shaped protrusions that are approximately 2.3 µm long (Figure 20D). At 176 DPI, the protrusions are sausage-shaped with a narrow stalk (Figure 21) and are 10 µm long. At 184 DPI, most sarcocysts have protrusions that are shorter and usually irregular, measuring approximately 3 µm long.

As the sarcocyst develops, the contents of the villar protrusions also change, which probably reflects a change in the metabolic rate within the sarcocysts. The villar protrusions of early and intermediate sarcocysts are filled with small vesicles and membranous whorls which pinch off from the Pm at indentations in the Pw (Figures 20B to D and 21A and B). In mature sarcocysts, the protrusions contain granules and clusters of vesicles (Figures 21B to 23).

B. Metrocytes

Metrocytes are ovoid, rapidly multiplying forms which have an electron-lucent cytoplasmic matrix, numerous ribosomes, several mitochondria, one or more micropores, a pellicle consist-

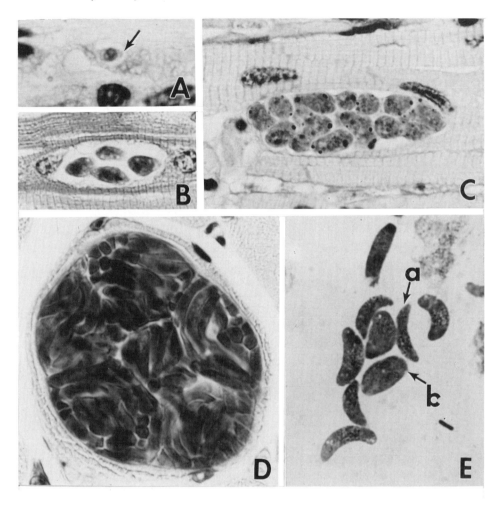

FIGURE 16. *S. cruzi* sarcocysts in myocytes of experimentally infected calves. (A) Merozoite in a parasitophorous vacuole (arrow), probably a unizoite sarcocyst (H. & E.); (B) sarcocyst with four metrocytes (H. & E.); (C) sarcocyst with numerous metrocytes (PAS); the darkly stained granules are amylopectin; (D) mature sarcocyst with banana-shaped bradyzoites (H. & E.); (E) bradyzoites (a) and metrocyte (b) with two organisms (Giemsa stain). (Magnification × 1000.) (From Dubey, J. P., *J. Protozool.*, 29, 591, 1982. With permission.)

ing of two or three membranes, subpellicular microtubules, endoplasmic reticulum, one or two Golgi complexes, a few amylopectin granules, centrioles, electron-dense bodies, an occasional lipid body, and a diffuse nucleus (Figures 19, 24, and 25). Metrocytes may also contain anlagen of two progeny formed by a process called endodyogeny (Section III.D). Metrocytes may arise from metrocytes or from dedifferentiating merozoites. Metrocytes in early stages of endodyogeny may also contain remnants of the conoid, micronemes, and rhoptries. Early sarcocysts contain only metrocytes (Figure 25), whereas relatively few metrocytes are contained in intermediate and mature sarcocysts (Figure 26).

C. Bradyzoites
1. Pellicle

The pellicle consists of three membranes: an outer plasmalemma and an inner double membrane complex (Figures 27 and 28). The plasmalemma is a continuous unit that completely encloses the whole parasite, whereas the inner pellicular membranes are interrupted at the anterior and posterior ends and at micropores. The plasmalemma is separated from the inner

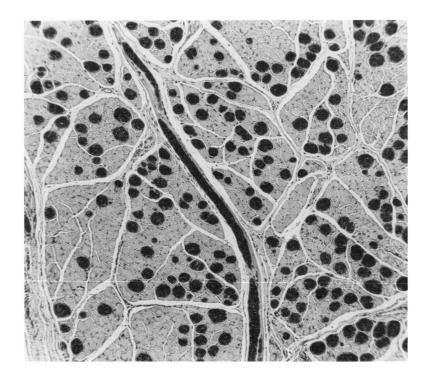

FIGURE 17. Numerous *S. cruzi* sarcocysts in tongue of an experimentally infected calf. Note absence of inflammation. (H. & E.; magnification × 160.) (From Dubey, J. P., Speer, C. A., and Epling, G. P., *Am. J. Vet. Res.*, 43, 2147, 1982. With permission.)

membranes by an electron-lucent space, 15 to 20 nm thick. In *S. tenella*, the intramembranous particles are randomly arranged in the plasmalemma, but have an orderly arrangement in the inner pellicular membranes.[563a] The inner pellicular membrane complex is similar to tight intercellular junctions of higher organisms and is composed of 11 rows of rectangular contiguous plaques (0.5 × 2 μm in size) that are arranged in a loosely coiled helical pattern. The plaques converge at the posterior end to form the posterior pore (=posterior ring) which represents an interruption in the inner membrane complex. At 1.5 to 2 μm from the anterior tip, the rectangular plaques abutt tightly to a conical inner membrane complex called the anterior cape. The anterior cape has an interruption at its anterior tip which appears electron dense in TEM and forms the polar ring. At the anterior tip, the plasmalemma contains a rosette of intramembranous particles (one central and eight peripheral particles). The rosette of intramembranous particles may be involved in rhoptry secretion during host cell penetration or serve as a receptor-processor system which enables the parasite to recognize and penetrate host cells.[563a]

2. Apical Rings

Two apical rings (also called preconoidal or conoid rings) are situated at the anterior tip immediately beneath the plasmalemma and above the conoid (Figures 27 to 29). Bradyzoites of *S. tenella* may have three apical rings.[563a] The function of these apical rings is not known.

3. Polar Rings

The polar rings of *Sarcocystis* spp. have been studied less extensively than those of other coccidia. A single polar ring appears to be present in *S. tenella* bradyzoites,[563] whereas other species have two polar rings. Polar ring 1 appears as an electron-dense thickening at the anterior termination of the anterior cape of the inner membrane complex and serves to anchor the

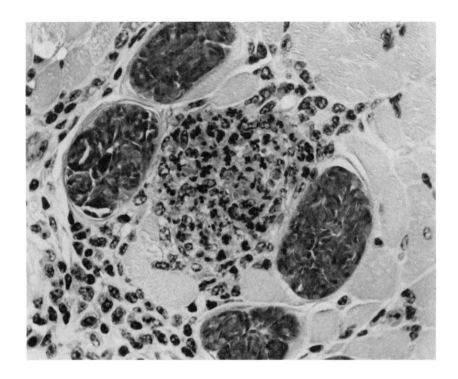

FIGURE 18. A degenerate (center) sarcocyst, surrounded by mononuclear cells and infiltrated by granulocytes, and four intact sarcocysts. (H. & E.; magnification × 400.) (From Dubey, J. P., *Am. J. Vet. Res.*, 43, 2147, 1982. With permission.)

ABBREVIATIONS FOR FIGURES 19 TO 62

Am	—	amylopectin granule
Ao	—	axoneme of microgamete flagellum
Ap	—	apical complex
Ar1,2	—	apical rings 1 and 2
Av	—	autophagic vacuole
Bb	—	basal body
Bd	—	bradyzoite
Bl	—	basal lamina
Bm	—	budding merozoite
Cb	—	crystalloid body
Cc	—	centrocone
Ce	—	centriole
Co	—	conoid
Db	—	dome-shaped base of sarcocyst wall
Dc	—	developing conoid
Dp	—	disk-shaped plaque in primary sarcocyst wall
Ds	—	distal segment of sarcocyst wall protrusion
Eb	—	electron-dense body
Ec	—	electron-dense collar of micropore or exocystosis pore
Eg	—	electron-dense granule
El	—	electron-dense layer of primary sarcocyst wall
En	—	endothelial cell
Er	—	endoplasmic reticulum
Fc	—	foot process of podocyte
Fl	—	flagellum
Gl	—	granular layer

ABBREVIATIONS FOR FIGURES 19 TO 62 (continued)

Go	—	Golgi complex
Gr	—	refractile granule
Hc	—	host cell cytoplasm
Hm	—	host cell mitochondrion
Hn	—	host cell nucleus
If	—	intermediate finger-like segment of sarcocyst wall protrusion
Im	—	inner membrane complex
Io	—	inner layer of oocyst wall
Ip	—	interposed strip in sporocyst wall
Is	—	inner layer of sporocyst wall
Lb	—	lipid body
Lu	—	lumen of capillary
Mc	—	metrocyte
Md	—	mucin droplet
Mf	—	microfilament
Mi	—	mitochondrion
Mn	—	microneme
Mp	—	micropore
Mr	—	merozoite
Ms	—	mesenchymal area in glomerulus
Mt	—	microtubule
Mv	—	multivesicular body
Mw	—	membranous whorls
Na	—	nucleus of macrogamont
Nd	—	electron-dense portion of microgamont nucleus
Ne	—	nucleus of endothelial cell
Ni	—	nucleus of microgamete
No	—	nucleolus
Np	—	nuclear pore
Nu	—	nucleus of parasite
Oo	—	outer layer of oocyst wall
Os	—	outer layer of sporocyst wall
Ow	—	oocyst wall
Pc	—	podocyte
Pe	—	pellicle
Pl	—	plasmalemma of parasite
Pm	—	parasitophorous vacuolar membrane
Pn	—	podocyte nucleus
Pr	—	polar rings
Pv	—	parasitophorus vacuole
Pw	—	primary sarcocyst wall
Rh	—	rhoptry
Sa	—	spindle apparatus
Sb	—	sharply conical base of sarcocyst wall
Sc	—	sarcocyst
Se	—	septa
Sm	—	subpellicular microtubule
Sp	—	sporozoite
St	—	spindle microtubule
Sw	—	sporocyst wall
Us	—	urinary space
Vp	—	villar protrusion
Vs	—	vesicle

subpellicular microtubules in those bradyzoites with a single polar ring.[563] In those species with two polar rings, polar ring 2 is located immediately beneath and slightly posterior to polar ring 1 and serves an an anchoring point for the subpellicular microtubules (Figure 28). Two electron-dense projections are attached to the inner surface of polar ring 1 (Figures 28 and 29). The anterior projection points anteriorly, the posterior projection posteriorly.

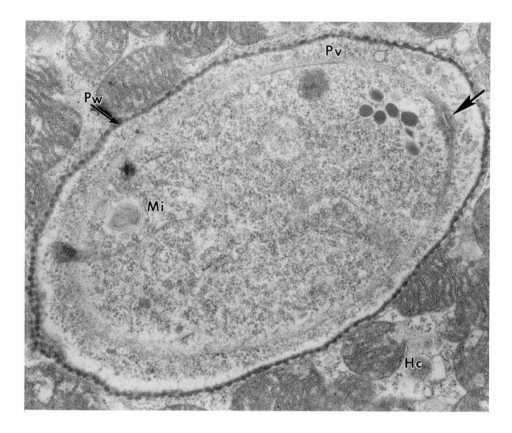

FIGURE 19. Transmission electron micrograph (TEM) of unizoite sarcocyst. The merozoite is in a late stage of transformation to a metrocyte. The infected myocyte has already formed a primary sarcocyst wall (Pw), and remnants of the apical complex are visible at the upper right (arrow) (*S. rauschorum*; 9 DPI). (Magnification × 25,000.) (Courtesy of R.J. Cawthorn, Atlantic Veterinary College, Prince Edward Island, Canada.)

4. Microtubules

Microtubules are found in virtually every stage of *Sarcocystis*. A few are cytoplasmic microtubules, whereas most other microtubules are components of specialized structures such as subpellicular microtubules, conoid, centrioles, flagellar axoneme, and the mitotic spindle apparatus. Zoites of *Sarcocystis* species have 22 subpellicular microtubules that originate at evenly spaced intervals around the polar ring and extend posteriorly immediately beneath the inner membrane complex for about 1/2 to 2/3 the length of the zoite (Figure 29).

In addition to existing as individual structures (cytoplasmic microtubules), microtubules are the basic building blocks of flagellar axonemes (found only in microgametes), centrioles, basal bodies, mitotic spindles, and the conoid. The subpellicular microtubules are similar to those found in a wide variety of plant and animal cells. They are cylinders approximately 25 nm in diameter. They usually contain 12 or 13 protofilament strands running longitudinally. They appear striated because the protofilaments appear to be composed of repeating subunits, probably α- and β–tubulin, arranged in transverse or helical rows. Subpellicular microtubules probably provide structural integrity to the overall shape of zoites and may be involved in motility (gliding and flexing), transport of cytoplasmic components near the margin of the zoite, or serve to anchor protein molecules in the inner membrane complex. Treatment with the antimicrotubular agents colcemid, colchicine, or vinblastine stops motility of bradyzoites of *Sarcocystis*.[129a]

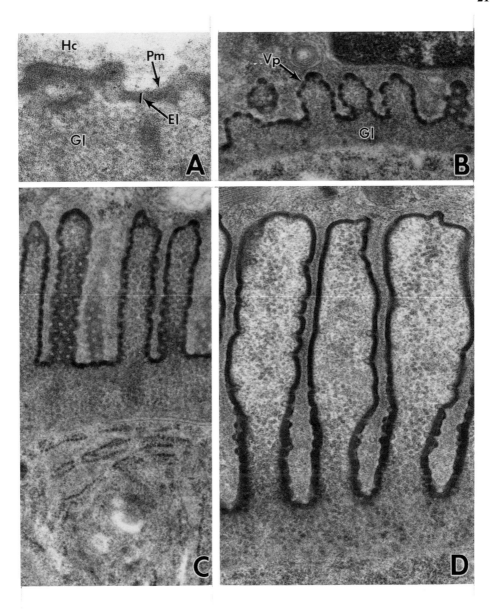

FIGURE 20. (A) Initial stage in formation of the sarcocyst wall (25 DPI). (B) The primary sarcocyst wall has folded to form short villar protrusions; the villar protrusions and the granular layer contain numerous small vesicles which appear to arise from pinocytotic-like indentations of the primary sarcocyst wall (30 DPI). (C) The villar protrusions are elongate with parallel sides and the granular layer is about twice as thick as in B (30 DPI). (D) The villar protrusions are longer and club shaped (30 DPI). (Magnification [A] × 100,000, [B to D] × 30,000.)

5. Micropore

Micropores are present in all stages of *Sarcocystis* species except for microgametes. They consist of an invagination of the parasite plasmalemma into the parasite cytoplasm (Figures 27 and 30). The invaginated membrane is encircled on the cytoplasmic side by an electron-dense membranous collar which the inner membrane interrupts and is juxtaposed at a 90° angle to it. Although most micropores appear to be inactive, some are observed actively ingesting particulate matter in developing or multiplying stages such as metrocytes (Figure 25).

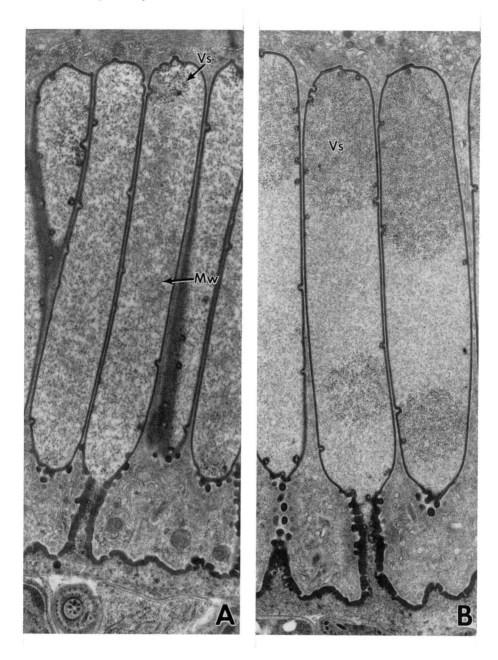

FIGURE 21. (A) The villar protrusions have become sausage shaped with a narrow stalk and contain small vesicles (Vs) and membranous whorls (Mw) that appear to arise from indentations in the primary sarcocyst wall (176 DPI). (B) The villar protrusions are slightly longer and wider and contain granules and aggregates of vesicles (176 DPI). (Magnification × 15,000.)

6. Rhoptries and Micronemes

Rhoptries and micronemes are electron-dense, membrane-bound, elongate structures occupying the anterior half of the zoite (Figures 27 and 30). These organelles are believed to constitute a single functional unit involved in secreting substances to aid penetration of cells. Although this appears likely, no direct evidence has been provided for their role in cell penetration. These organelles may also serve to secrete proteins that become antigenic or are inserted into the zoite plasmalemma just above the conoid.

The term rhoptry was introduced by Sénaud[617a] for the organelles that had been previously

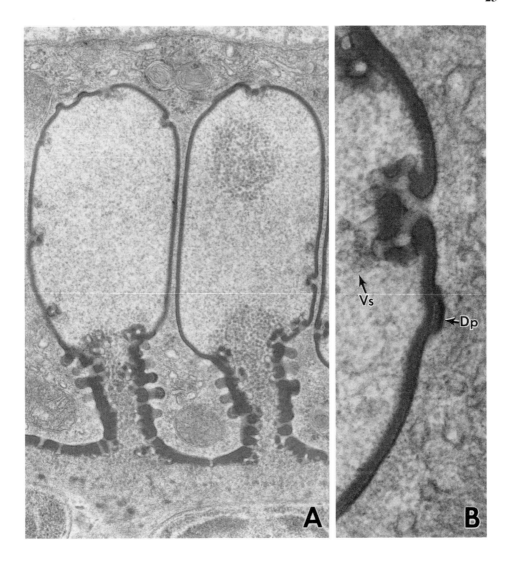

FIGURE 22. (A) The villar protrusions have decreased in size to about one fourth the length of those in Figure 21 (184 DPI). (B) High magnification of the apex of a villar protrusion showing disk-shaped plaques (Dp) and indentations in the primary sarcocyst wall that give rise to small vesicles (176 DPI). (Magnification [A] × 30,000, [B] × 100,000.)

called "paired organelles" or "club-shaped organelles". Rhoptries consist of a narrow, duct-like anterior region that originates within the core of the conoid and a posterior club- or sack-like region. Bradyzoites of *S. tenella* have only two rhoptries,[563] whereas bradyzoites of other species of *Sarcocystis* have several rhoptries. Bradyzoites of all species of *Sarcocystis* contain numerous micronemes. Micronemes are elongate, rod-like structures that usually appear as ovoid or round in cross-section. In some Apicomplexa, narrow interconnections have been found between micronemes and rhoptries,[609c,617a] but interconnections between the rhoptries and micronemes in bradyzoites of *S. tenella* are controversial.[563,609c] Micronemes appear to form within single membrane-bound vesicles in the apical region of bradyzoites and apparently give rise to rhoptries.[486c,609c]

7. Conoid

The conoid is a hollow, conical-shaped organelle located at the anterior tip of sporozoites, bradyzoites, and merozoites that consists of microtubular elements. The conoid of bradyzoites

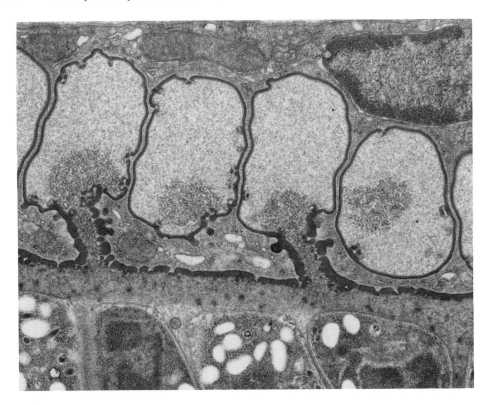

FIGURE 23. Aged sarcocyst wall. The villar protrusions have become even shorter, with the distal portion irregular in shape (184 DPI). (Magnification × 20,000.)

of *S. tenella* contains 20 microtubules that are arranged in a helical pattern.[563] Some investigators believe the conoid may be involved in penetration of host cells because when it is protruded the anterior tip becomes shaped like a long, narrow stylet that would facilitate active penetration into the cell. Within the core of the conoid are the ducts of rhoptries and one or two eccentrically located microtubules that extend for a short distance into the zoite cytoplasm (Figure 29).[563] These microtubules may be attached to the conoid and serve to protrude or retract the conoid.[563]

8. Golgi Complex

The Golgi complex is present but reduced in bradyzoites of most *Sarcocystis* spp. It is more developed in those stages that are metabolically more active, such as metrocytes and schizonts. The functions of the Golgi complex are probably similar to those described for other cells.

9. Endoplasmic Reticulum

Although somewhat reduced, both smooth and rough endoplasmic reticula are present in bradyzoites, especially in the area just posterior to the nucleus.

10. Ribosomes

Bradyzoites have an abundant amount of free ribosomes (Figure 27) which can be found from the conoid to the posterior tip, filling the bradyzoite cytoplasm not occupied by organelles or inclusion bodies.

11. Mitochondria

Mitochondria are found in bradyzoites as well as all other stages of *Sarcocystis* spp. Bradyzoites may have one to several mitochondria. This organelle is double membrane bound with tubular cristae, which are characteristic of most protozoans.

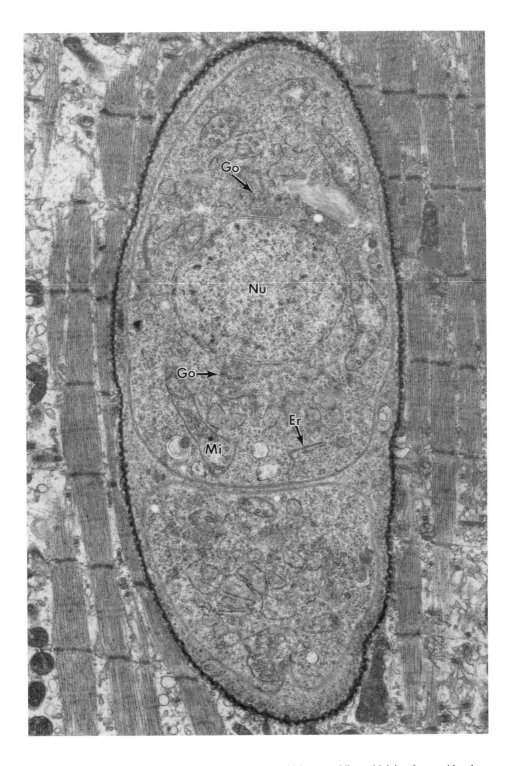

FIGURE 24. Young sarcocyst containing two metrocytes which are rapidly multiplying forms with a large nucleus, numerous mitochondria, Golgi complex, ribosomes and endoplasmic reticulum (*S. rauschorum*; 10 DPI). (Magnification × 15,000.) (Courtesy of R. J. Cawthorn, Atlantic Veterinary College, Prince Edward Island, Canada.)

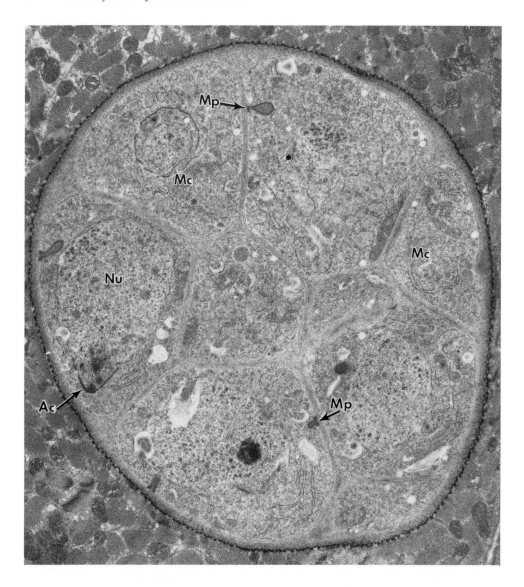

FIGURE 25. Young sarcocyst containing nine metrocytes; a developing apical complex (Ap) is visible in one metrocyte, and micropores (Mp) are present in several others (*S. rauschorum*; 15 DPI). (Magnification × 8800.) (Courtesy of R. J. Cawthorn, Atlantic Veterinary College, Prince Edward Island, Canada.)

12. Multivesicular Bodies

Multivesicular bodies (also called multimembrane-bound vesicles or Golgi adjuncts)[72,155,619,634] are peculiar, ill-defined membranous bodies found between the nucleus and anterior tip of bradyzoites. They may also occur in certain other stages such as metrocytes, merozoites, and schizonts. Although they vary considerably in composition and appearance, all are membrane bound by one, two, or several membranes, with the interior containing many or only a few membrane-bound vesicles. They may also vary considerably in function or have multiple functions. Because they are usually situated near the parasite nucleus, it is likely that multivesicular bodies function, at least in part, as a Golgi complex or as anlagen for organelles of the apical complex. In bradyzoites and merozoites of *S. tenella*, multivesicular bodies appear to give rise to micronemes and rhoptries either internally[486d] or externally near their outer margin.[155,634] In some species of *Sarcocystis*, the multivesicular body persists in fully formed bradyzoites (Figure 27) and merozoites.[155,634]

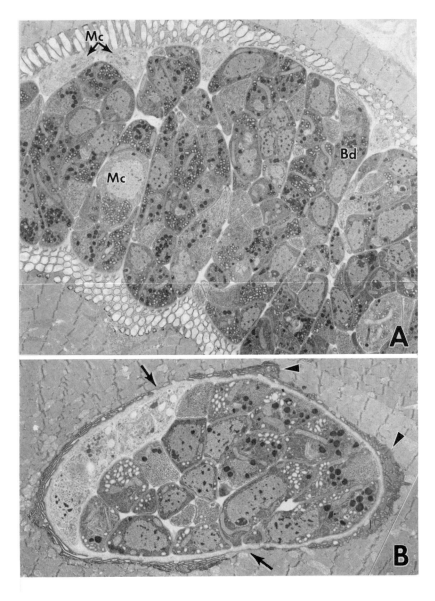

FIGURE 26. (A) Mature sarcocyst containing numerous bradyzoites (Bd) and a few metrocytes (Mc). The zoites are arranged in packets (*S. tenella*; 103 DPI). (B) *S. arieticanis* sarcocyst without septa. The sarcocyst wall is thin at some places (arrows) and thick at others (arrowheads) (103 DPI). (Magnification × 2800.)

13. Inclusion Bodies

Bodies that appear free in the cytoplasm, such as amylopectin, lipid, and electron-dense bodies, are generally considered to be inclusion bodies (Figure 27). It is likely that lipid and especially amylopectin represent energy reserves. Although electron-dense bodies are seen in nearly all stages, their function is still not known. They probably serve as a site for storage of energy reserves, protein, lipid, and perhaps enzymes.

D. Endodyogeny

Endodyogeny is a form of asexual multiplication in which two progeny develop within a parasite that is ultimately consumed in the process (Figures 31 and 32). (Figures 31 and 32 are a series of TEMs showing endodyogeny in metrocytes of *S. singaporensis*.) The process usually

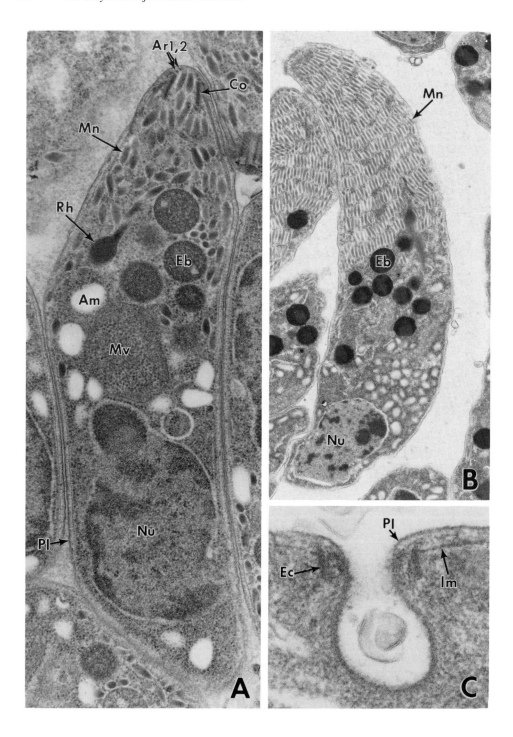

FIGURE 27. TEMs of organelles in bradyzoites of *S. singaporensis* (A and C) and *S. gigantea* (B). (A) Micronemes are randomly arranged (179 DPI). (B) Micronemes are numerous and arranged in parallel rows (391 DPI). (C) Micropore (176 DPI). (Magnification [A] × 30,000, [B] × 10,500, [C] × 100,000.)

begins with the transformation of the merozoite, metrocyte, or bradyzoite from a fusiform to an ovoid form. The parasite cytoplasm becomes electron lucent, and micronemes, rhoptries, amylopectin, and electron-dense bodies become localized or scattered and some disappear (Figure 30). The Golgi complex divides into two parts, and two centrioles appear at the anterior

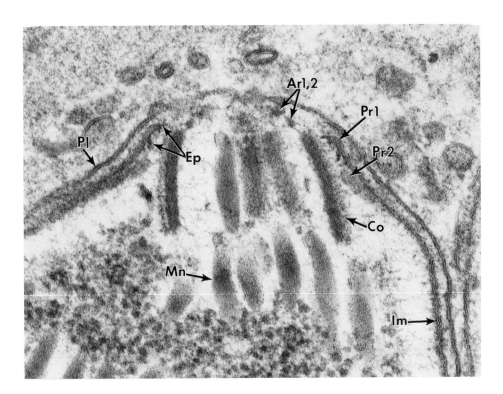

FIGURE 28. Apical complex of bradyzoite (*S. campestris*; 62 DPI). (Magnification × 100,000.)

margin of the large vesicular nucleus. The nucleus becomes bean-shaped with each of the two lobes containing a spindle apparatus consisting of nuclear spindle microtubules and a centrocone. The centrocone is a sharply conical structure from which the spindle microtubules originate. The spindle microtubules pass from the centrocone into the nucleus via perforations (not nuclear pores) in the nuclear envelope, which remains intact during nuclear division. Each of two centrioles is situated immediately above and slightly lateral to each centrocone. The inner membrane complex and subpellicular microtubules of each progeny form a conical-shaped structure above each spindle apparatus in the anterior part of the metrocyte. The apical rings and polar rings appear early with the appearance of the inner membrane complex and subpellicular microtubules. During this early stage of endodyogeny, a conoid is gradually formed at the tip of the developing complex. A centriole is often seen in close proximity to the developing conoid and may actually be responsible for its formation in much the same manner that centrioles give rise to a subsequent generation of centrioles.

As division progresses, the inner membrane complexes and subpellicular microtubules grow posteriorly, incorporating more of the dividing nucleus. The electron-dense areas present at the edges of the inner membrane and at the ends of the subpellicular microtubules probably represent areas in which components of these structures are assembled. As the inner membrane complex nears the posterior end of the metrocyte, the nucleus pinches into two nuclei, with each portion being incorporated into a progeny. At this point, each progeny usually has a completely formed conoid, some micronemes and rhoptries, and various inclusion bodies. The progenies increase in volume, acquiring more of the metrocyte cytoplasm and organelles until they occupy most of the metrocyte. As the progenies approach the surface, the inner membrane complex of the metrocyte disappears and the outer membrane becomes the outer membrane of the progenies. At this point, they separate into two progeny. During endodyogeny various spheroidal structures appear in the anterior end, which probably represent anlagen of rhoptries and micronemes of the developing progeny.

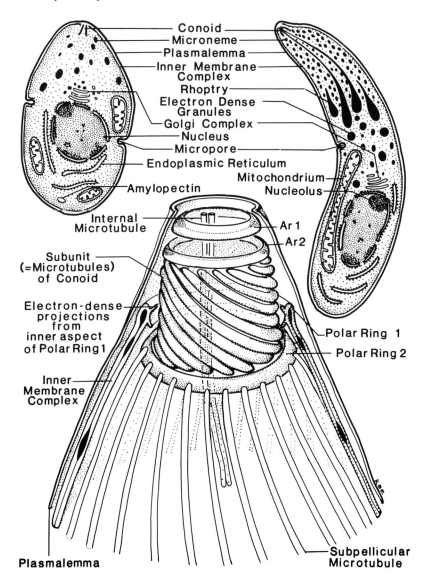

FIGURE 29. Schematic representation of typical bradyzoites, metrocytes, and apical complex of *Sarcocystis* spp.

E. Types of Sarcocyst Walls

Numerous ultrastructural reports have shown that the sarcocyst walls of *Sarcocystis* spp. vary between being relatively simple to highly complex. After examining published reports, there appear to be at least 24 distinct types of sarcocyst walls which herein are referred to as Types 1 to 24. It is likely that additional types will be discovered.

The primary sarcocyst wall consists of a parasitophorus vacuolar membrane (Pm) and an electron-dense layer immediately beneath the Pm. A granular layer is immediately beneath the primary sarcocyst wall. Septa which arise from the granular layer traverse the sarcocyst separating it into compartments (Figure 26A) which contain bradyzoites and metrocytes.

The various sarcocyst wall types are described as follows:

- Type 1 sarcocysts have a primary sarcocyst wall with minute undulations (Figures 33 and 34). Occasionally, the Pm invaginates into the granular layer, and the entire primary sarcocyst wall may invaginate superficially or deeply into the sarcocyst (Figure 34A).

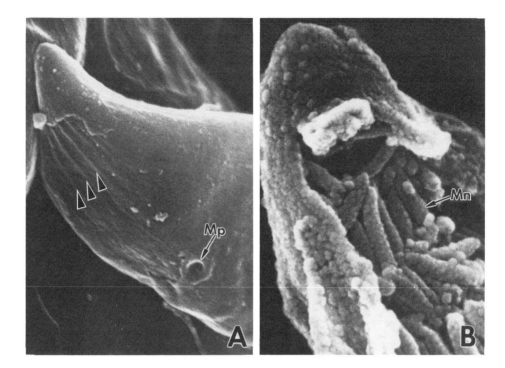

FIGURE 30. Scanning electron micrographs (SEMs) of the apical ends of *S. cruzi* bradyzoites. (A) Note opening of micropore (Mp) and ridges (arrowheads) in surface created by underlying subpellicular microtubules. (B) Portion of pellicle has been removed exposing the micronemes (Mn). (Magnification × 19,250.) (From Dubey, J. P., *J. Protozool.*, 29, 591, 1982. With permission.)

- Type 2 is similar to Type 1, except that thin, hair-like structures arise from some of the apices of the undulating primary cyst wall (Figure 34C).
- Type 3 has an undulating primary sarcocyst wall plus a flattened mushroom-like protrusion (Figure 35A).
- Type 4 has short, irregularly shaped protrusions with a finely granular core (Figure 35B). The primary sarcocyst wall is undulating at the base of the protrusions and may be smooth or undulating over the surface of the protrusions.
- Type 5 consists of an undulating primary sarcocyst wall that is highly folded to form hair-like protrusions with granular cores (Figure 35C). Type 5 may also have a relatively thin granular layer.
- Type 6 has finger-like protrusions with each arising from a dome-shaped base (Fig. 35D). The primary sarcocyst wall has minute undulations in the dome-shaped base, but is relatively smooth over the surface of the protrusions. The protrusions have a fine granular core and are oriented perpendicularly to the sarcocyst surface.
- Type 7 has protrusions with three distinct regions: a dome-shaped base, an intermediate finger-like segment, and a thin thread-like distal segment (Figure 36A). The intermediate and distal segments are often bent 90° and run parallel to the sarcocyst surface. In certain areas, however, the protrusions are arranged collectively in conical tufts. The core contains coarse granules in the base segment and fine granules in the intermediate and distal segments.
- Type 8 is similar to Type 7, except that the basal and intermediate segments appear as one, the protrusion narrows abruptly to form the distal segment, and the distal segment is branched (Figures 36B and C).
- Type 9 has widely spaced villar protrusions that are conical or tongue shaped, with a core containing electron-dense granules and microtubules which extend from the villar tips into

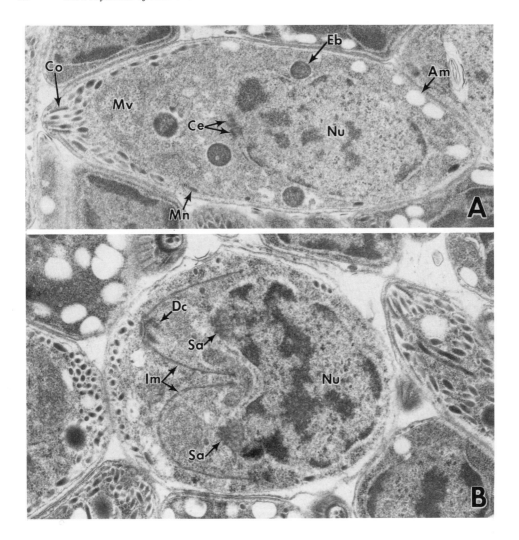

FIGURE 31. *S. singaporensis*: (A) TEM of a metrocyte in early stage of endodyogeny; note pair of centrioles (Ce) at anterior margin of nucleus and micronemes scattered around the margin of the metrocyte (176 DPI). (B) TEM of a metrocyte containing two developing zoites. The apical complex of each zoite has begun to form immediately above each lobe of the U-shaped nucleus (60 DPI). (Magnification × 18,000.)

the granular layer (Figures 37A and 38). The Pm has minute undulations that are present over the entire sarcocyst surface.
- Type 10 is similar to Type 9, but with tightly packed villar protrusions (Figure 37B).
- Type 11 is similar to Types 9 and 10, except that the microtubules extend from the villar tips to the plasmalemma of bradyzoites located immediately beneath the granular layer (Figure 37C).
- Type 12 has tightly packed, thin, finger-like protrusions with microtubules in the basal one third of the protrusion (Figure 39A).
- Type 13 consists of widely spaced, mushroom-like protrusions with microtubules extending from the core into the granular layer (Figure 39B). The primary cyst wall has irregularly spaced indentations.
- Type 14 has tightly packed, cylindrical-shaped protrusions, the tips of which contain disk-shaped plaques (Figures 40A and B). Minute undulations in the primary sarcocyst wall are present only at the base of the protrusions.

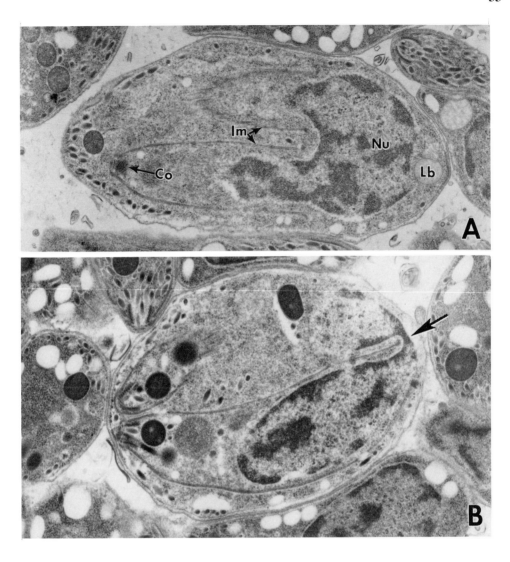

FIGURE 32. *S. singaporensis*: (A) TEM of intermediate stage of endodyogeny showing a completely formed conoid (Co); the inner membrane complex (Im) and subpellicular microtubules of each daughter zoite have extended further posteriorly, incorporating more of the parasite nucleus (176 DPI). (B) TEM of advanced stage in endodyogeny; note that the two lobes of the nucleus are connected by a narrow isthmus (arrow) (60 DPI). (Magnification × 18,000.)

- Type 15 consists of tombstone-like protrusions with a core containing scattered granules, scattered microfilaments, or microtubules (Figure 40C). The primary sarcocyst wall has minute undulations over most or all of the sarcocyst wall.
- Type 16 is similar to Type 15, except that the sarcocyst wall consists of alternating conical and club-shaped villar protrusions (Figure 40D).
- Type 17 has disk-shaped plaques at the apex and lateral margins of the villar protrusions, a highly branched primary sarcocyst wall at the base of the protrusions and a granular core (Figure 41A).
- Type 18 has irregularly shaped protrusions, which may appear T-shaped in appropriately cut sections, with a granular core (Figure 41B).
- Type 19 has club-shaped protrusions consisting of a cylindrical stalk and a sausage-shaped distal segment (Figure 41C). In older sarcocysts, the distal segment may be shorter and irregular in shape (see Figure 23). Invaginations of the primary sarcocyst wall are numerous

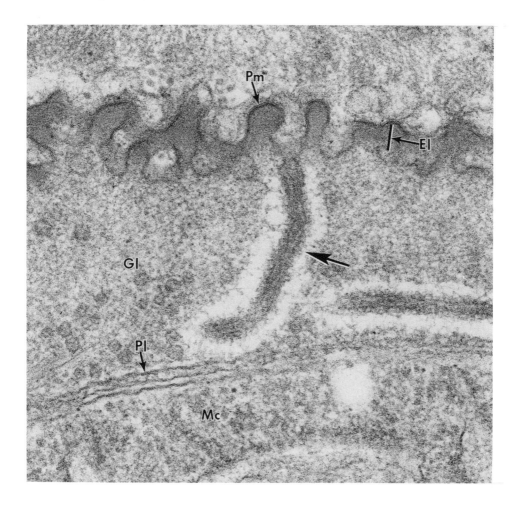

FIGURE 33. High magnification of sarcocyst wall of *S. rauschorum*. The primary sarcocyst wall consists of an undulating parasitophorous vacuolar membrane (Pm) and an electron-dense layer (El) immediately beneath the Pm. At certain points, the Pm invaginates (arrow) into the granular layer; the latter contains fine granules and small vesicles which arise from the Pm (60 DPI). (Magnification × 90,000.)

in the stalk and widely scattered in the distal segment. Small vesicles pinch-off by pinocytosis from the Pm at the invaginations. In young and intermediate sarcocysts, the core is filled with numerous small vesicles and membranous whorls, whereas in mature sarcocysts the core contains fine granules with localized areas of small vesicles.
- Type 20 has angular protrusions that may be pyramidal, trapezoidal, or mushroom like (Figures 42A and B). Snake-like projections arise from the surface of the protrusions as well as from the surface of the sarcocyst. The entire sarcocyst surface is covered by tiny bristles that project outward at 90° from the parasitophorous vacuolar membrane. The core contains tightly packed microtubules that probably account for the angular appearance of the protrusion.
- Type 21 has cauliflower-like protrusions with microtubules or microfilaments, fine granules, and coarse, electron-dense granules (Figure 42C).
- Type 22 has protrusions with a proximal, sharply conical base and a distal segment that is cocklebur-like with short, radiating projections. Cross-sections of the distal segment appear as cogwheels. Some of the projections from the distal segment connect with the margin of the sarcocyst wall (Figure 42D). The core contains numerous vesicles especially in the distal one half.

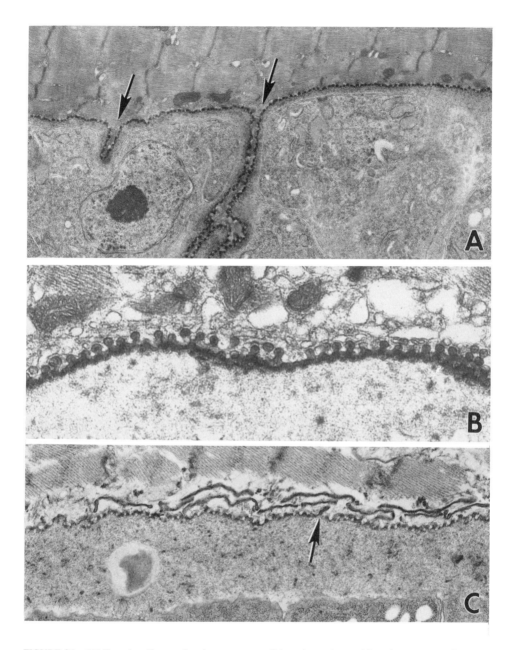

FIGURE 34. (A) Type 1 wall; note that the sarcocyst wall invaginates (arrows) into the sarcocyst (60 DPI; *S. rauschorum*). (Magnification × 3250.) (Courtesy of R. J. Cawthorn.) (B) Type 1 sarcocyst wall (*S. muris*). (Magnification × 29,575.) (C) Type 2 showing thin hair-like structures arising from the apex of the undulating primary sarcocyst wall (arrow) (*S. wapiti*). (Magnification × 20,000.) (From Speer, C. A. and Dubey, J. P., *Can. J. Zool.*, 60, 881, 1982. With permission.)

- Type 23 consists of anastomosing, cauliflower-like protrusions that contain fine granules and microfilaments (Figure 43A).
- Type 24 has numerous mushroom-like protrusions with a core of tightly packed microfilaments. Mushroom-like protrusions also arise from the surface of other protrusions (Figure 43B).

There are only a few reports that used scanning electron microscopy to study the ultrastruc-

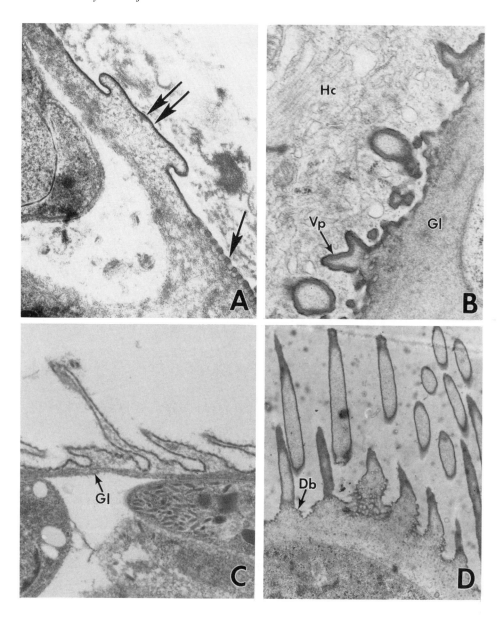

FIGURE 35. (A) Type 3 wall showing undulating primary sarcocyst wall (arrow) and a flattened mushroom-like protrusion (double arrows) (*S. ferovis*). (Magnification × 33,000.) (From Dubey, J. P., *Proc. Helminthol. Soc. Wash.*, 50, 153, 1983. With permission.) (B) Type 4 wall showing short, irregularly shaped protrusions of the primary sarcocyst wall (*S. sigmodontis* from cotton rat). (Magnification × 25,000.) (C) Type 5 wall consists of an undulating primary sarcocyst wall that is highly folded to form hair-like protrusions. This particular species of *Sarcocystis* also has a relatively thin granular layer (*S. sulawesiensis*). (Magnification × 25,000.) (From O'Donoghue, P. J., Watts, C. H. S., and Dixon, B. R., *J. Wildl. Dis.*, 23, 225, 1987. With permission.) (D) Type 6 wall showing finger-like protrusions arising from a dome-shaped base; the primary sarcocyst wall has minute undulations in the dome-shaped base, but is relatively smooth in the finger-like protrusions; the protrusions are oriented perpendicularly to the sarcocyst wall (*Sarcocystis* sp., red deer). (Magnification × 11,500.) (Courtesy of R. Entzeroth, University of Bonn, Federal Republic of Germany.)

ture of sarcocysts.[47,48,160,180,324,493] Compared to a TEM, a scanning electron micrograph (SEM) provides little information because it is limited to examining surfaces (Figures 44A to C). The sarcocyst surface of type 14 sarcocysts of *S. capracanis* have a honeycomb-like appearance, and the protrusions appear villus like (Figures 44A and B).[180]

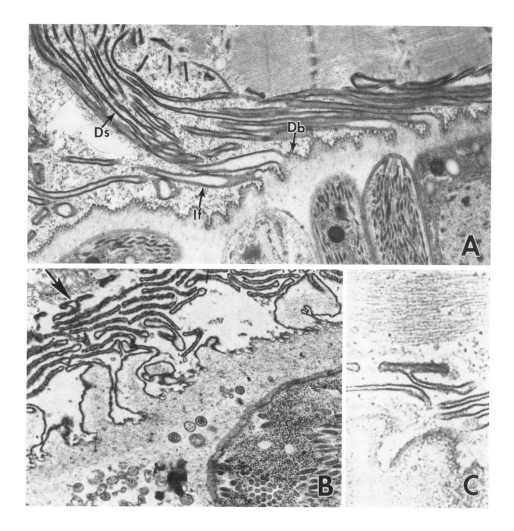

FIGURE 36. (A) Type 7 consists of a dome-shaped base with an undulating primary sarcocyst wall, an intermediate finger-like segment, and a thin, thread-like distal segment. The intermediate and distal segments are often bent 90° and run parallel to the sarcocyst surface, whereas in other areas, the protrusions are arranged collectively to form conical tufts (*S. arieticanis*). (Magnification × 13,200.) (B) Type 8 is similar to Type 7, except that the distal segment of the protrusion is branched (arrow) (*Sarcocystis* sp. from fallow deer). (Magnification × 20,700.) (From Entzeroth, R., Chobotar, B., Scholtyseck, E., and Nemeseri, L., *Z. Parasitenkd.*, 71, 33, 1985. With permission.) (C) Higher magnification of Type 8 showing branched distal segment (*S. sybillensis*). (Magnification × 40,000.)

F. Gametogenesis

After ingestion by the appropriate definitive host, bradyzoites escape from sarcocysts and penetrate enterocytes, usually goblet cells, in the small intestine (Figure 45A). Development from bradyzoite to micro- or macrogamonts is rapid, being completed in 6 to 18 h or less. Bradyzoites are situated within a PV and begin immediately to develop into either a micro- or macrogamont. How the sex is determined or regulated is not known.[108] Bradyzoites quickly transform from a fusiform to an ovoid or spheroid form, and most of the organelles of the apical complex undergo dissolution. Micronemes become scattered in the gamont cytoplasm and soon disappear. Simultaneously, the early gamont produces much endoplasmic reticulum, numerous ribosomes, mitochondria, and a vesicular nucleus with scattered heterochromatin and a nucleolus. Macrogamonts have two ultrastructural features that appear to be lacking in microgamonts. These are an exocytosis pore and electron-dense granules which are transported

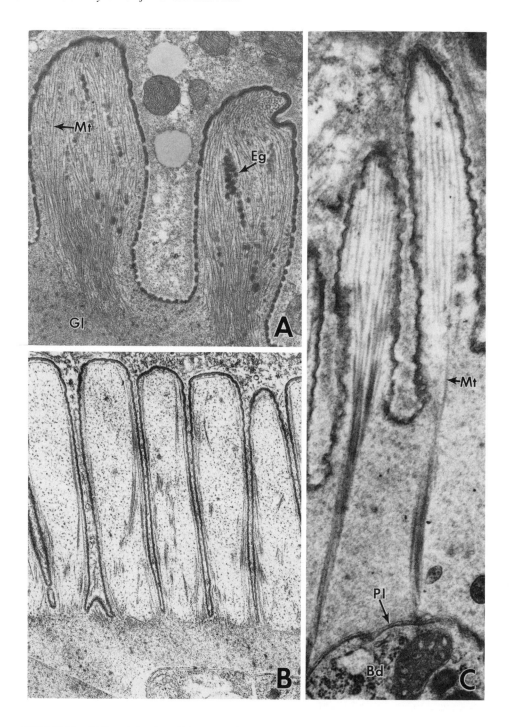

FIGURE 37. (A) Type 9 showing widely spaced villar protrusions; the parasitophorous vacuolar membrane has minute undulations over the entire sarcocyst surface and the villar protrusions contain prominent electron-dense granules (Eg) and microtubules (Mt) that extend from the villar tips into the granular layer (*S. campestris*). (Magnification × 9500.) (B) Type 10 is similar to Type 9, but with tightly packed villar protrusion (*S. odoi*). (Magnification × 12,450.) (C) Type 11 is similar to Types 9 and 10, except that the microtubules extend from the villar tips to the plasmalemma of bradyzoites (arrow) (*S. fayeri*). (Magnification × 38,500.) (From Tinling, S. P., Cardinet, G. H., III, Blythe, L. L., and Vonderfecht, S. L., *J. Parasitol.*, 66, 458, 1980. With permission.)

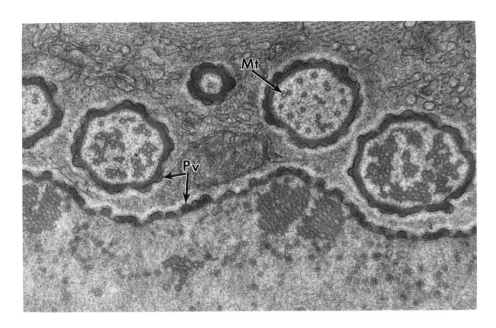

FIGURE 38. Type 11 showing villar protrusions and microtubules (Mt) in cross-section (*S. falcatula*). (Magnification × 41,000.) (From Box, E. D., Meier, J. L., and Smith, J. H., *J. Protozool.*, 31, 521, 1984. With permission.)

to a location near the surface of the gamont and discharged through the exocytosis pore (Figure 45B). In *S. cruzi*, the exocytosis pore is similar in size to micropores, but differs ultrastructurally (Figure 45B). The exocytosis pore protrudes slightly to produce a flattened dome on the surface of the gamont. The plasmalemma is invaginated at the apex of the dome, and the plasmalemma continues uninterrupted through the invagination. The single inner membrane is interrupted at the site of the micropore, allowing the invaginated plasmalemma to extend into the gamont cytoplasm. The margin of the interruption in the inner membrane is electron dense, thick, and projects outward at approximately a 45° angle instead of inwardly as in micropores. The membranes surrounding the electron-dense bodies fuse with the invaginated membrane of the exocytosis pore, which causes the electron-dense bodies to be discharged into the PV where it fuses to form a thin electron-dense layer (Figure 45B). This layer eventually becomes continuous to form the oocyst wall. Oocyst wall-forming bodies, typical of *Eimeria* spp. and certain other coccidia, are lacking in macrogamonts of *Sarcocystis* spp.

It has been reported that bradyzoites of *S. muris* have an exocytosis pore which discharges dense granules into the PV.[205a] The contents of the dense granules may stimulate the adhesion of the host cell endoplasmic reticulum to the PV to form a three-membrane complex or cause antigenic mimicry, preventing a host cell immune response or inflammation.[205a]

In most *Sarcocystis* species, macrogamonts begin development in the epithelium, but as development proceeds to an intermediate stage, the host cell degenerates and lyses (Figure 46), and the macrogamonts enter the lamina propria where they undergo oogony and sporogony. In addition to forming the oocyst wall, the contents of the dense granules may also contain enzymes that cause lysis of the enterocyte, enabling the parasites to enter the lamina propria. Since the oocyst wall is being formed during this time, it seems likely that the electron-dense granules may contain cytolytic substances that cause lysis of the host cells. Neighboring cells are apparently not harmed.

In contrast to macrogamonts, most microgamonts complete their development within the intestinal epithelium, usually a goblet cell (Figure 47). Several ultrastructural features distinguish microgamonts from macrogamonts. Early microgamonts increase in size and become

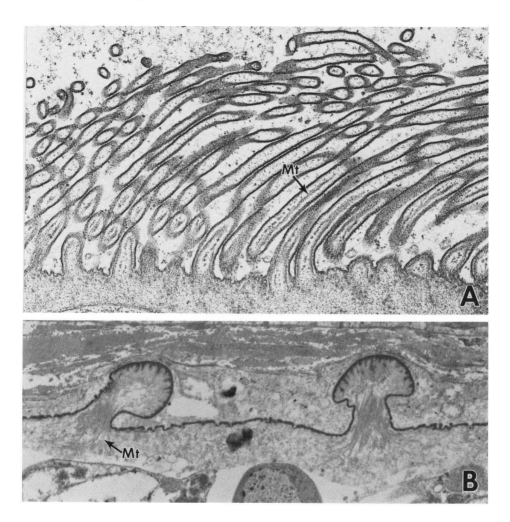

FIGURE 39. (A) Type 12 consists of tightly packed, thin, finger-like protrusions with microtubules (Mt) in the basal one third of the protrusion (*S. sybillensis*). (Magnification × 20,000.) (B) Type 13 has widely spaced, mushroom-like protrusions with microtubules that extend from the core of the protrusion into the granular layer (*S. mucosa*). (Magnification × 10,000.) (From O'Donoghue, P. J., Obendorf, D. L., O'Callaghan, M. G., Moore, E., and Dixon, B. R., *Parasitol. Res.*, 73, 113, 1987. With permission.)

ovoid (Figure 48A), have a well-developed endoplasmic reticulum, several mitochondria, numerous ribosomes, and a vesicular nucleus, but in contrast to macrogamonts, they lack electron-dense bodies and an exocytosis pore, but contain a mitotic spindle apparatus (Figure 47). The microgamont increases in size to about 10 μm in diameter, the inner membrane disappears, and the nucleus becomes lobulated with each lobe situated near the gamont surface. A single microgamete develops immediately above each nuclear lobe (Figure 48B). Two centrioles become basal bodies that form two flagella, each of which projects outward from the gamont. The chromatin becomes more compact and dense and eventually is situated in that portion of the nucleus nearest the gamont plasmalemma. The microgamete buds at the surface of the gamont (Figure 49), incorporating the dense portion of the nucleus and a mitochondrion, and eventually the microgamete plasmalemma separates from that of the gamont. Fully formed microgametes have a plasmalemma, two basal bodies associated with the base of each of two flagella, a dense nucleus, a mitochondrion, and several longitudinally oriented microtubules that extend from near the basal bodies to a point about midway along the nucleus (Figure 49).

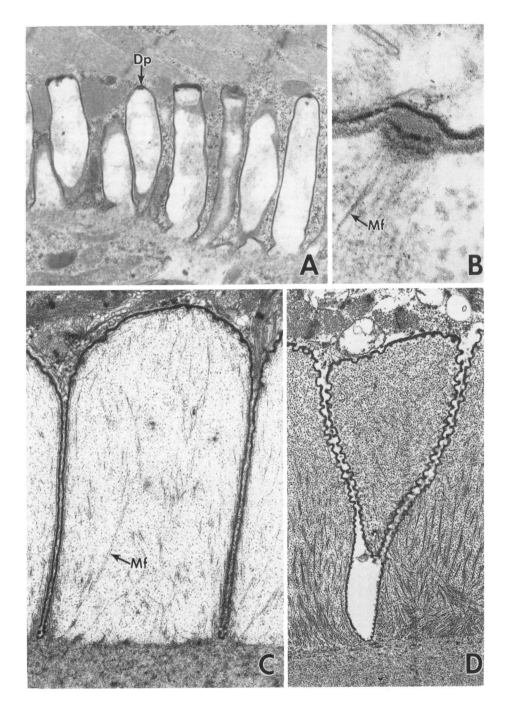

FIGURE 40. (A) Type 14 consists of tightly packed villar protrusions, the tips of which contain disk-shaped plaques (Dp) (*S. tenella*). Microtubules are absent. (Magnification × 13,980.) (B) High magnification of a disk-shaped plaque in the primary sarcocyst wall (*S. hemionilatrantis*). (Magnification × 99,000.) (From Speer, C. A. and Dubey, J. P., *J. Protozool.*, 33, 130, 1986. With permission.) (C) Type 15 consists of tombstone-like protrusions that contain widely scattered microfilaments or microtubules. In some species with Type 15, the villar protrusions have interdigitating lateral margins (*Sarcocystis* sp. from white-tailed deer). (Magnification × 12,450.) (D) Type 16 is similar to Type 15, except that the sarcocyst wall consists of alternating conical and club-shaped villar protrusions (*S. youngi*). (Magnification × 6600.) (From Dubey, J. P. and Speer, C. A., *J. Wildl. Dis.*, 21, 219, 1985. With permission.)

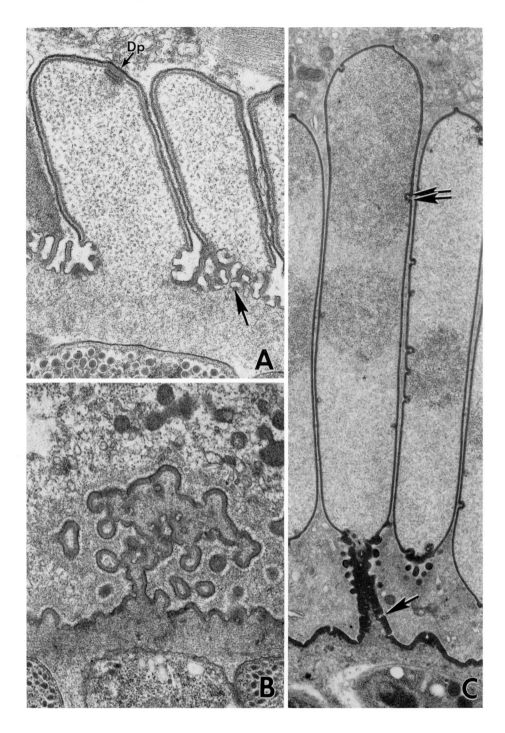

FIGURE 41. (A) Type 17 has disk-shaped plaques (Dp) at the apex and lateral margins of the villar protrusions and a highly branched primary sarcocyst wall at the base of the protrusions (arrow) (*S. odocoileocanis*). (Magnification × 30,000.) (B) Type 18 consists of an irregularly shaped protrusion which may appear T shaped in appropriately cut sections (*S. dispersa*). (Magnification × 24,000.) (From Sénaud, J. and Černá, Z., *Protistologica*, 14, 155, 1978. With permission.) (C) Type 19 has club-shaped protrusions consisting of a cylindrical stalk and a sausage-shaped distal segment. In older sarcocysts, the distal segment may be shorter and irregular in shape (see Figure 23). Invaginations of the primary sarcocyst wall are numerous in the stalk (single arrow) and widely scattered in the distal segment (double arrow) (*S. singaporensis*). (Magnification × 15,000.)

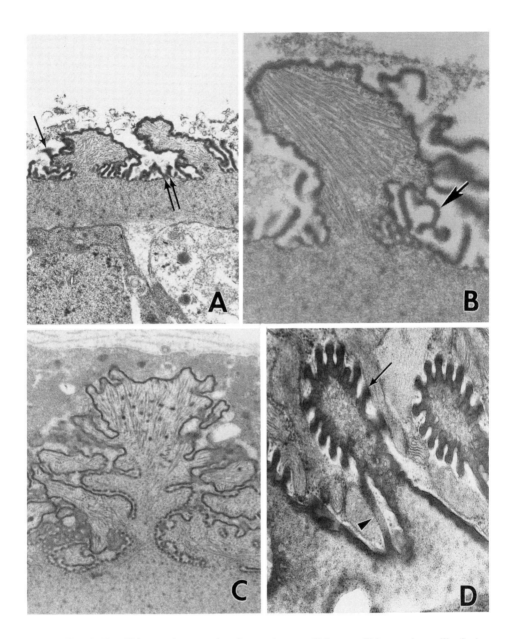

FIGURE 42. (A) Type 20 has angular protrusions that may be pyramidal, trapezoidal, or mushroom like. Snake-like projections arise from the surface of the protrusions (single arrow) as well as from the surface of the sarcocyst (double arrow) and some connect surface to protrusions (*S. medusiformis*). (Magnification × 16,218.) (B) Type 20 showing fine bristles (arrow) arising from the surface of the protrusions as well as from the surface of the snake-like projections. The villar core contains tightly packed microtubules (*S. medusiformis*). (Magnification × 30,000.) (C) Type 21 has cauliflower-like protrusions containing microtubules or microfilaments and fine or coarse granules (*S. gigantea*). (Magnification × 15,000.) (D) Type 22 has protrusions with a proximal, sharply conical base and a distal segment that is cocklebur like with short, radiating projections. The conical base is shown here in oblique section. Cross-sections of the distal segment appear as cogwheels (arrow). Some of the projections from the conical base connect with the surface of the sarcocyst wall (arrowhead). Vesicles are located within the protrusions, especially in the distal one half (*S. villivillosi*). (Magnification × 34,700.) (From Beaver, P. C. and Maleckar, J. R., *J. Parasitol.*, 67, 241, 1981. With permission.)

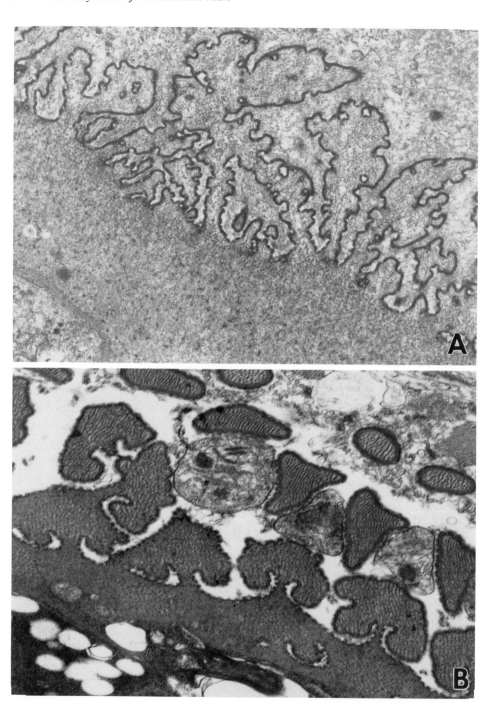

FIGURE 43. (A) Type 23 consists of anastomosing, cauliflower-like protrusions that contain fine granules and microfilaments (*S. rileyi*). (Courtesy of R. Cawthorn, Atlantic Veterinary College, Prince Edward Island, Canada.) (B) Type 24 has mushroom-like projections arising from mushroom-like protrusions (*Sarcocystis* sp. from mountain goat). (Magnification × 12,000.) (Courtesy of W. Foreyt, Washington State University, Pullman.)

G. Fertilization

Fertilization usually occurs within 18 h postinoculation.[494,624] Microgametes penetrate apparently through the oocyst wall by an unknown process and become closely associated with the macrogamont (Figure 50). The plasmalemma of the microgamete fuses with that of the

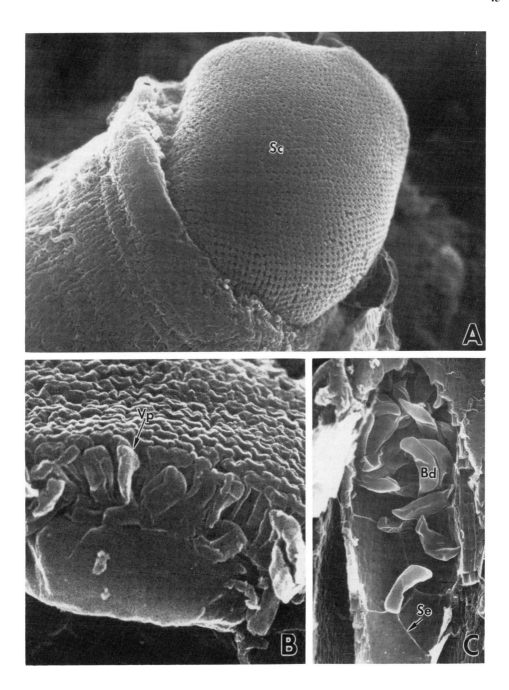

FIGURE 44. SEMs of sarcocysts. (A) Sarcocyst (Sc) protruding from myocyte. The villar protrusions give a honeycomb appearance to the surface of the sarcocyst (*S. capracanis*). (Magnification × 2630.) (B) Cross-sectional view of the margin of a sarcocyst showing villar protrusions (Vp) (*S. capracanis*). (Magnification × 6800.) (From Dubey, J. P., Speer, C. A., Epling, G. P., and Blixt, J. A., *Int. Goat Sheep Res.*, 2, 252, 1984. With permission.) (C) Ruptured sarcocyst showing bradyzoites and septa (*S. cruzi*). (Magnification × 2000.) (From Dubey, J. P., *J. Protozool.*, 29, 591, 1982. With permission.)

macrogamont to create a cytoplasmic bridge through which the microgamete nucleus enters the macrogamont (Figure 50). Evidently, soon thereafter the cytoplasmic bridge closes and the gamete nuclei fuse to form a zygote. The remnant of the microgamete becomes vesicular and eventually disappears *in situ* (Figure 51).

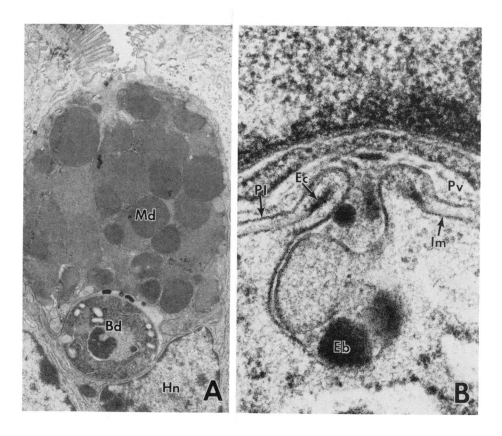

FIGURE 45. TEM of *Sarcocystis*. (A) Bradyzoite (Bd) in intestinal goblet cell (*S. cruzi*). (B) Exocytosis pore discharging electron-dense bodies into parasitophorous vacuole (*S. cruzi*). (Magnification [A] × 7590, [B] × 78,000.)

H. Oocyst

Unsporulated oocysts consist of a sporont surrounded by an oocyst wall. The oocyst wall has an electron-dense, finely granular outer layer and an inner layer of one to four membranes (Figure 51).[494] Electron-dense projections arise from the outer surface of the oocyst wall (Figure 51). The sporont is limited by a plasmalemma and a single inner membrane and contains a nucleus, endoplasmic reticulum, mitochondria, ribosomes, Golgi complex, amylopectin bodies, micropores, and an occasional electron-dense body (Figure 51).

I. Sporocyst

There are no reports on the ultrastructural changes that occur during sporulation of oocysts of *Sarcocystis* spp. The sporocyst walls are composed of a thin continuous outer layer and a thick inner layer consisting of four plates joined at sutures, similar to those in related coccidia.[52,87a,632c] The sporocyst wall of *S. cruzi* has a lamellated outer layer, approximately 50 nm thick, consisting of alternating electron-dense and electron-lucent layers, and an electron-dense inner layer, approximately 130 nm thick, that is separated from the outer layer by a single membrane (Figure 52A). The inner layer has alternating bands (≈5 nm thick) of electron-dense and -lucent material perpendicular to the surface of the sporocyst. A lip-like thickening (150 to 180 nm thick) is located at the margin of each plate. A thin strip of electron-dense material (≈20 to 30 nm thick) is interposed between the lip-like thickenings of two apposing plates (Figure 52B). In scanning electron micrographs, the sutures appear as ridges on the surface of the sporocyst (Figure 52). Treatment of sporocysts with NaOCl (Clorox® or Purex®) usually results in removal of the outer layer of the sporocyst wall (compare Figures 53A and B).

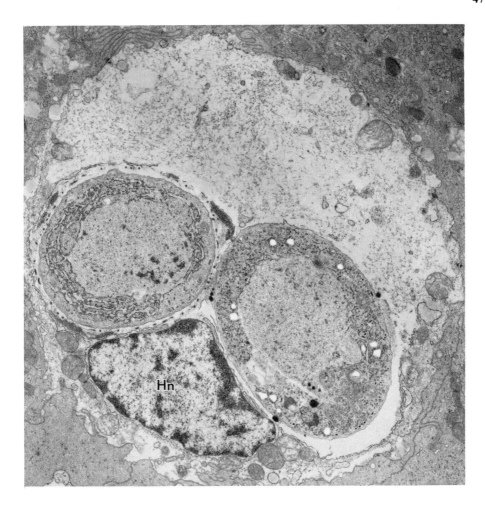

FIGURE 46. TEM of two macrogamonts in a degenerate goblet cell (*S. cruzi*). (Magnification × 8085.)

During excystation, bile salts and/or trypsin act upon the sutures, causing the plates to separate from the interposed strip, allowing the sporocyst to collapse, releasing the sporozoites randomly (Figures 52 and 53).

J. Sporozoites

Sporozoites of *S. cruzi* are banana shaped and have all the ultrastructural features described earlier for bradyzoites such as conoid, apical rings 1 and 2, polar rings 1 and 2, micronemes, rhoptries, micropore, nucleus, mitochondria, Golgi complex, endoplasmic reticulum, ribosomes, amylopectin bodies, pellicle, and subpellicular microtubules. They differ, however, from bradyzoites as well as merozoites by possessing a crystalloid body and structures similar to rhoptries and micronemes in the posterior one half of the sporozoite (Figure 54). The crystalloid body consists of electron-dense and -lucent granules (≈38 nm in diameter) and is usually located in the posterior one half of the sporozoite. Some sporozoites have several crystalloid bodies randomly scattered throughout the sporozoite with some bodies anterior to the sporozoite nucleus. Also, various other organelles and inclusion bodies such as rhoptries, micronemes, and amylopectin granules may be dispersed among the granules of the crystalloid body. This body is probably analogous to the homogenous refractile bodies of *Eimeria* spp. in which the body is believed to represent an energy or amino acid reserve.

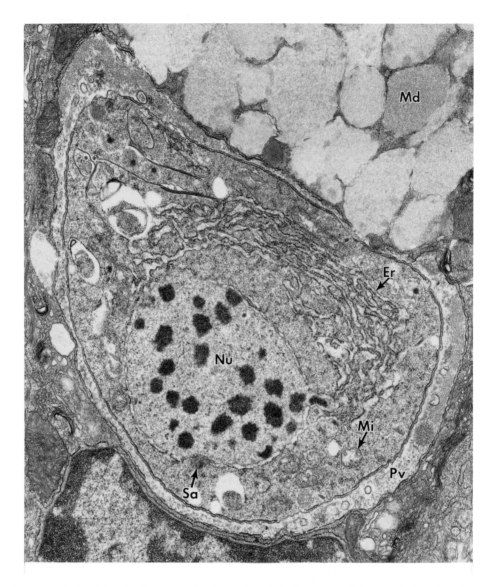

FIGURE 47. TEM of early microgamont showing portion of spindle apparatus (Sa) (*S. cruzi*). (Magnification × 15,840.)

K. Schizogony

Sarcocystis schizonts multiply by a process known as endopolygeny in which numerous merozoites begin development internally and later bud more or less simultaneously at the surface of the schizont.

Sporozoites and schizonts differ from other intracellular stages in the life cycle of *Sarcocystis* spp. in that they are located free within the cytoplasm of the host cell and are not surrounded by a parasitophorous vacuolar membrane (Figures 54 to 62). Soon after entering the appropriate host cell, the sporozoite transforms into a young ovoid schizont (≈10 × 8 μm) which contains a large nucleus with a single nucleolus, a double pellicular membrane, several mitochondria, endoplasmic reticulum, ribosomes, electron-dense bodies, Golgi complex, micropores, autophagic vacuoles, scattered micronemes and rhoptries, and lipid, amylopectin, electron-dense, and multivesicular bodies (Figure 55). Autophagic vacuoles contain parasite cytoplasm, ribosomes, and micronemes, indicating that this may be one of the sites for recycling the micronemes and

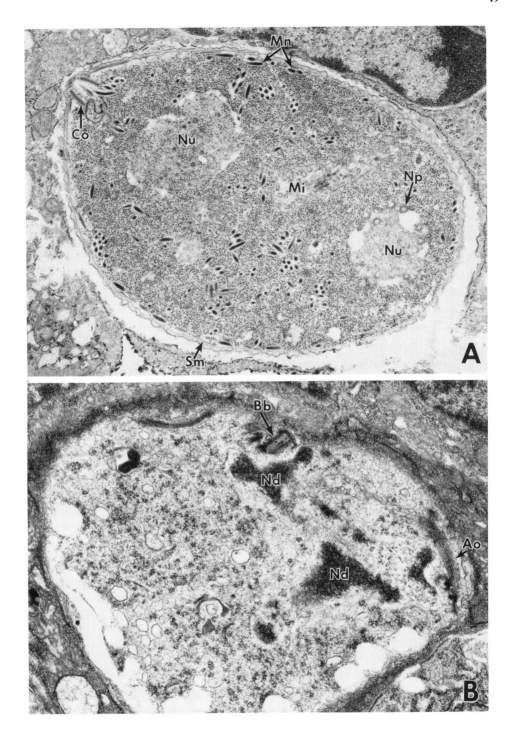

FIGURE 48. TEM of microgamonts. (A) Early microgamont showing two lobes of nucleus; note nuclear pores and scattered micronemes and conoid of original bradyzoite (*S. cruzi*). (B) Early stage of microgamete formation (*S. cruzi*). (Magnification [A] × 15,180, [B] × 18,675.)

other cytoplasmic components. In more advanced stages of the parasite, the nucleus becomes irregularly shaped with several nucleoli, and the micronemes are completely absent.

Intermediate schizonts have anlagen of developing merozoites and a highly lobulated

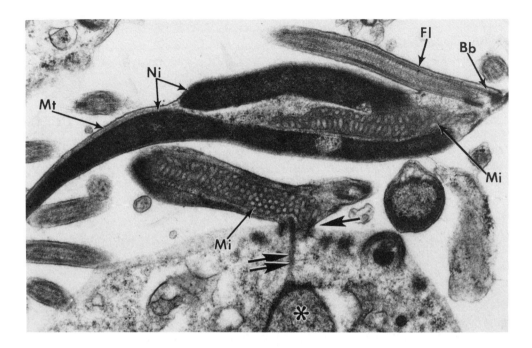

FIGURE 49. TEM of two microgametes in longitudinal section, one of which is still connected (single arrow) to the surface of the microgamont in which a nuclear bridge (double arrow) extends from residual nucleus (*) into microgamete (*S. suihominis*). (Magnification × 24,500.) (From Mehlhorn, H. and Heydorn, A. O., *Z. Parasitenkd.*, 58, 97, 1979. With permission.)

nucleus (Figures 56 and 57). A spindle apparatus appears in association with each lobe of the nucleus (Figure 57). The spindle consists of several microtubules that extend completely across the nucleus from one centrocone to the other, and of shorter diverging microtubules that cover an arc of ≈35°. Each of two centrioles are located immediately above and slightly lateral to the apex of each centrocone. In some species of *Sarcocystis*, one or two multivesicular bodies may be situated laterally along each centrocone, just outside the nuclear envelope.[155,634] Coincidental with the appearance of centrocones, the conoid, apical rings, a single-unit membrane, and subpellicular microtubules of a merozoite anlage form immediately above the centrioles (Figure 57). As each merozoite anlage elongates, the single-unit membrane and its associated subpellicular microtubules extend posteriorly, incorporating a part of the nucleus which eventually pinches off from the rest of the nucleus. Cytoplasm, ribosomes, endoplasmic reticulum, and one to three mitochondria are also incorporated into each merozoite anlage. Micronemes also appear at this stage in association with the Golgi complex or a multivesicular body between the apical end and the nucleus (Figures 58 to 60 and 63) and continue to be formed until the merozoite is complete. The double-membrane pellicle folds in around the developing merozoites until they appear to bud at the surface of the schizont (Figures 59 and 60). The double-membrane pellicle becomes closely associated with the single-unit membrane of each merozoite anlage to form the merozoite pellicle. Usually, one of the three membranes disappears so that the fully formed merozoite has a double membrane pellicle.[155,634]

Merozoites contain virtually all the organelles in bradyzoites, except rhoptries (Figure 62). Rhoptries are absent in merozoites of all *Sarcocystis* species. The merozoite is twisted slightly to form a helical pattern on its surface (Figure 63A). This helical pattern is probably due to a series of rectangular, contiguous strips (plaques) of the inner membrane aligned in longitudinal rows, as has been described for other coccidia.[193a] At the anterior tip, concentric ridges are visible by SEM which are formed by the apical rings and polar ring 1 which are located just beneath the merozoite plasmalemma (Figure 63).

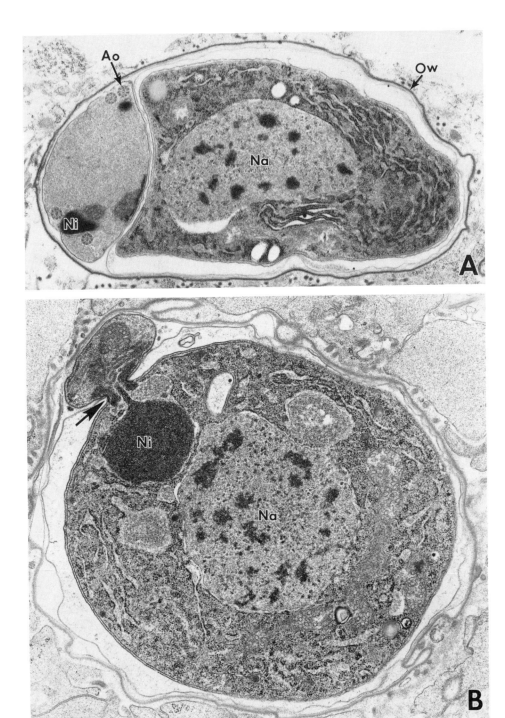

FIGURE 50. TEM of fertilization. (A) Early stage in fertilization of macrogamont by microgamete; note that the oocyst wall (Ow) has already formed (*S. cruzi*). (Magnification × 13,000.) (B) Intermediate stage in fertilization in which the microgamete nucleus (Ni) is entering the macrogamont through a cytoplasmic bridge (arrow) created by fusion of the plasmalemmae of the microgamete and macrogamont (*S. cruzi*). (Magnification × 15,770.) (From Sheffield, H. G. and Fayer, R., *Proc. Helminthol. Soc. Wash.*, 47, 118, 1980. With permission.)

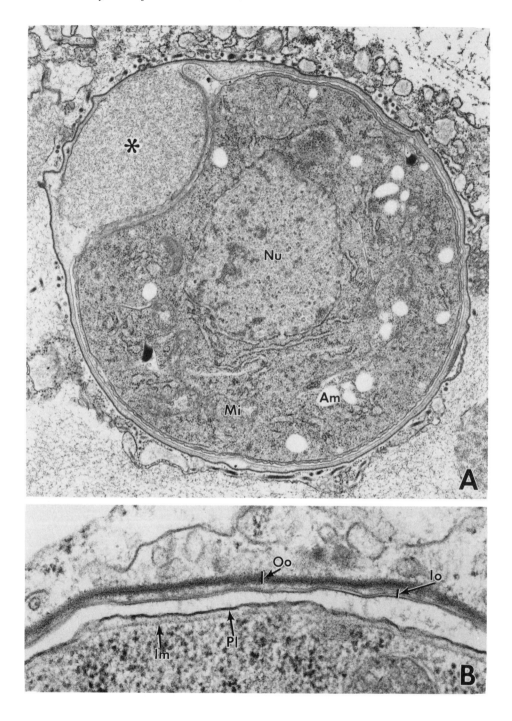

FIGURE 51. TEM of zygote. (A) Zygote soon after fertilization showing remnant of microgamete (*) (*S. cruzi*). (B) High magnification showing oocyst wall and pellicle of zygote; note that the oocyst wall consists of an inner layer (Io) of two membranes and an electron-dense outer layer (Oo) (*S. cruzi*). (Magnification [A] × 17,820, [B] × 60,000.)

In those species of *Sarcocystis* that undergo schizogony in the walls of blood vessels, the host cell appears to be an endothelial cell which has become displaced beneath the endothelium. In kidneys, the host endothelial cell is completely surrounded by a basal lamina, on one side of

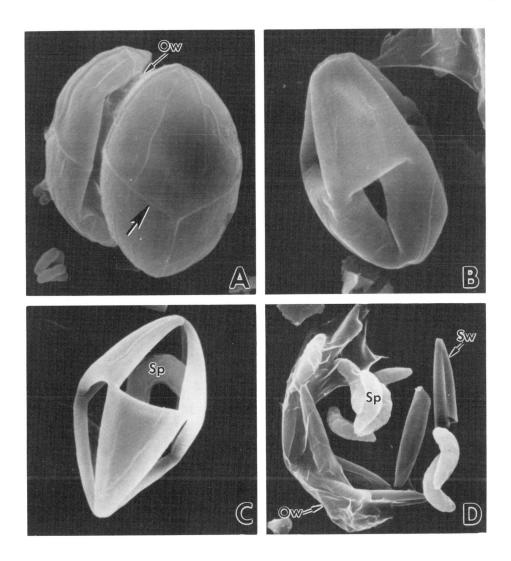

FIGURE 52. SEMs showing sporocysts and excystation of sporozoites. (A) Two sporocysts held together by a thin oocyst wall (Ow). Junctions between apposing plates appear as ridges (arrow) in the sporocyst wall. (B) Intermediate stage in excystation in which the apposing plates in the sporocyst wall have partially separated from each other in response to excysting fluid. (C) A single sporozoite is visible within a sporocyst in which the walls have almost completely separated. (D) Final stage in excystation in which two sporocysts have collapsed releasing sporozoites (*S. miescheriana*). (Magnification × 4300.) (From Strohlein, D. A. and Prestwood, A. K., *J. Parasitol.*, 72, 711, 1986. With permission.)

which is a typical capillary endothelial cell, and on the other side are foot processes of podocytes and the urinary space (Figure 61).

IV. TAXONOMIC CRITERIA

Before the discovery of the life cycle of *Sarcocystis* in 1972, the two major criteria for naming a new species of *Sarcocystis* were the structure of sarcocysts and the species of host. Because the age of sarcocysts and the method of fixation greatly influenced sarcocyst structure, some species are not clearly described. Life cycle studies have indicated that some structurally similar sarcocysts (e.g., *S. tenella* of sheep and *S. capracanis* of goats) are actually different species based on host specificity for the intermediate host. Furthermore, some species of *Sarcocystis*

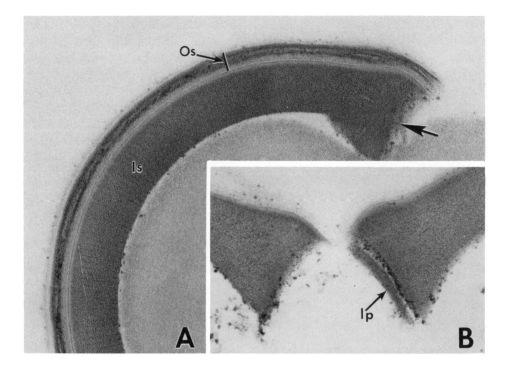

FIGURE 53. TEM of the sporocyst wall. (A) The outer layer is lamellated with alternating electron-lucent and -dense layers. The inner layer is cross-striated and has a lip-like thickening at its margin (arrow) (*S. cruzi*). (B) Site of apposition between two plates of the sporocyst wall after exposure to excysting fluid; note that the plates have separated and the interposed strip remains attached to the lip-like thickening of one plate; also note that treatment of the sporocyst with NaOCl prior to excysting fluid has removed the outer layer of the sporocyst wall (*S. cruzi*). (Magnification × 99,000.)

(e.g., *S. falcatula*) can infect numerous species of host (e.g., pigeons, budgerigars, canaries, and sparrows). Because it is often very difficult to complete the life cycle of *Sarcocystis* species of large animals under laboratory conditions, the relative value of different criteria that might be used to validate *Sarcocystis* species will be discussed.

A. Sarcocysts

The shape and size of the sarcocyst vary with the age of the sarcocyst, the type of host cell parasitized, and the techniques used for study. For example, sarcocysts of the same species in cardiac muscles and in the CNS are always smaller than those in skeletal muscles.[160,164,180] The size and shape of sarcocysts will also vary depending on fixation (they are smaller in fixed specimens than in live specimens) and possibly might vary with the type of fixative. Because they are often located in contractile muscles, the sarcocyst size will vary depending on whether the host cell was relaxed or contracted at the time of fixation.

Some *Sarcocystis* species continue to grow in size for several years after they have reached infectivity. For example, *S. bertrami* of the horse and *S. medusiformis* of sheep do not attain their maximum length (15 mm) in their hosts until they are approximately 4 years old, whereas *S. cruzi* attains a maximum size of 0.5 mm at approximately 4 months postinoculation. The shape of sarcocysts also varies with relation to the location. For example, *S. gigantea* sarcocysts in the esophagus are globular to pear shaped, whereas sarcocysts of the same species in the diaphragm are elongate and slender (see Chapter 4).

The structure of metrocytes is not a useful criterion for speciation because metrocytes are often irregularly shaped and their size is highly variable, depending on the stage of division. The

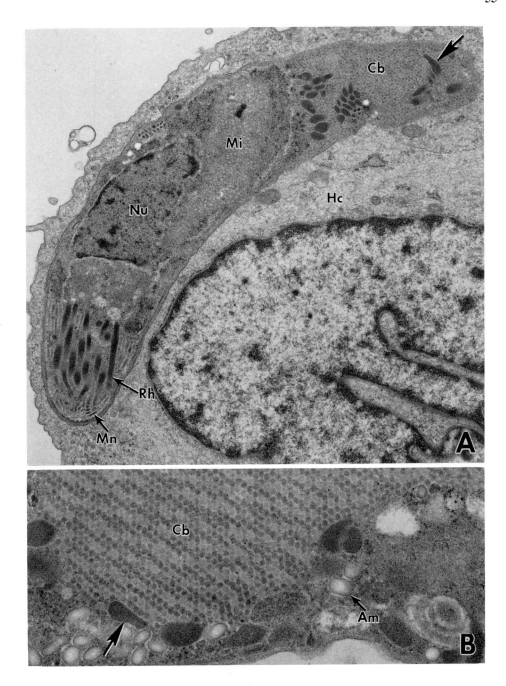

FIGURE 54. TEM of sporozoites. (A) Intracellular sporozoite free in the cytoplasm of a cultured bovine pulmonary artery endothelial cell; note absence of parasitophorous vacuole (*S. cruzi*). (B) High magnification of portion of sporozoite showing crystalloid body (Cb), amylopectin granules (Am) and rhoptry-like structures (arrows) (*S. cruzi*). (Magnification [A] × 15,000, [B] × 39,000.)

structure of bradyzoites also varies. Bradyzoites of some *Sarcocystis* species are densely packed within sarcocysts, whereas those of other species are sparsely found. Such conditions may affect size or shape. Because bradyzoites in most *Sarcocystis* species are banana shaped with great variation in curvature, it is difficult to measure them accurately. Measurements obtained from live preparations are unreliable because the size varies considerably depending on the pH and

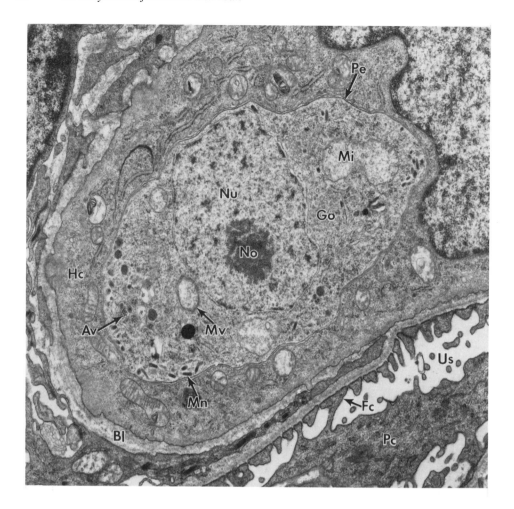

FIGURE 55. TEM of young schizont lying free in cytoplasm of glomerular capillary endothelial cell (*S. tenella*). (Magnification × 12,750.) (From Speer, C. A. and Dubey, J. P., *J. Protozool.*, 28, 424, 1981. With permission.)

osmolarity of the medium used, the weight of the cover glass, the volume of fluid under the coverslip, movement during measurement, and the effect of birefringence. The banana-shaped bradyzoites of *S. cruzi* become globular in acid-pepsin solution, used to digest away surrounding host tissue. Therefore, measurements of zoites fixed *in situ* within sarcocysts are more reliable than measurements of zoites liberated by digestion. The width of the bradyzoites at the level of the nucleus may be more reliable than the length of bradyzoites, because in histologic sections it is difficult to measure bradyzoites cut longitudinally along their entire length. Overall, unless the size differences of bradyzoites are obvious (Figure 4), this criterion should not be used for identification. The structure of organelles such as mitochondria, rhoptries, and micronemes is not a good taxonomic criterion.

The structure of the sarcocyst wall is a useful criterion for speciation. For example, all four species of ovine *Sarcocystis* (*S. gigantea*, *S. tenella*, *S. medusiformis*, and *S. arieticanis*) have sarcocyst walls characteristically different from one another, even though the sarcocyst wall of *S. tenella* is similar to that of *S. capracanis* of the domestic goat. In Figures 34 to 43, the sarcocyst walls of various species of *Sarcocystis* are illustrated and the information is summarized in Table 2 to facilitate species identification.

The structure of the sarcocyst wall seems to indicate a phylogenetic relationship. For example, Type 1 sarcocysts were found mainly in small, closely related mammals, Type 15 were only in closely related cervids, and Type 14 were found only in goats and sheep (Table 2). The

Table 2
CLASSIFICATION OF *SARCOCYSTIS* SPECIES BY THEIR SARCOCYST WALL STRUCTURE

Sarcocyst wall type	Structure — Light microscope	Structure — TEM Figure no.	*Sarcocystis* species and reference	Intermediate host
1	Thin, smooth	34A, B	*S. muris*[623]	House mouse (*Mus musculus*)
			S. crotali[218]	House mouse (*M. musculus*)
			S. booliati[397]	Moon rat (*Echinosorex gymnurus*)
			S. cymruensis[11]	Norway rat (*Rattus norvegicus*)
			S. sp.[212]	Red squirrel (*Tamiasciurus hudsonicus*)
			S. rauschorum[65]	Varying lemming (*Dicrostonyx richardsoni*)
			S. bozemanensis[173]	Richardson's ground squirrel (*S. richardsonii*)
			S. sebeki[655]	Vole (*Apodemus sylvaticus*)
			S. montanaensis[175]	Meadow vole (*Microtus pennsylvanicus*)
			S. sp.[213]	Eastern chipmunk (*Tamias striatus*)
			S. sp.[401]	Malaysian long-tailed monkey (*Macaca fascicularis*)
			S. sp.[491]	Baboon (*Papio cynocephalus*)
			S. sp.[491]	Rhesus monkey (*M. mulatta*)
2	Thin, smooth	34C	*S. wapiti*[637]	Wapiti (*Cervus elaphus*)
3	Thin, smooth	35A	*S. ferovis*[172]	Bighorn sheep (*Ovis canadensis*)
4	Thin, knotty	35B	*S. sigmodontis*[192a]	Cotton rat (*Sigmodon hispidus*)
			S. sp. (?)[410a]	Cat (*Felis catus*)
5	Thin, few hairy projections	35C	*S. sulawesiensis*[533]	Malaysian rats (*Bunomys* spp.)
6	Thin, hairy	35D	*S. capreoli*[207]	Roe deer (*Capreolus capreolus*)
7	Thin, smooth to hairy	36A	*S. cruzi*[488]	Cattle (*Bos taurus*, *B. indicus*, *Bison bison*)
			S. rangi[313]	Reindeer (*Rangifer tarandus*)
			S. arieticanis[191,368]	Sheep (*O. aries*)
			S. poephagicanis[679a]	Yak (*Poephagus grunniens*)
			S. hircicanis[9]	Goat (*Capra hircus*)
8	Thin	36B, C	*S. alceslatrans*[105a,617]	Moose (*Alces alces*)
			S. sp.[217]	Fallow deer (*Cervus dama*)
			S. grüneri[319]	Reindeer (*Rangifer tarandus*)
9	Thick, bristly to striated	37A	*S. campestris*[64]	Richardson's ground squirrel (*S. richardsonii*)
			S. microti[175]	Meadow vole (*M. pennsylvanicus*)
			S. putorii[655]	European vole (*M. arvalis*)
			S. sp.[611]	Opossum (*Didelphis virginiana*)
10	Thick, striated	37B	*S. gracilis*[216]	Roe deer (*C. capreolus*)
			S. tarandi[323]	Reindeer (*R. tarandus*)
			S. hemioni[181,186]	Mule deer (*Odocoileus hemionus*)
			S. odoi[178a]	White-tailed deer (*O. virginianus*)
			S. hirsuta[303a,489]	Cattle (*B. taurus*)
			S. poephagi[679a]	Yak (*P. grunniens*)

Table 2 (continued)
CLASSIFICATION OF *SARCOCYSTIS* SPECIES BY THEIR SARCOCYST WALL STRUCTURE

Sarcocyst wall type	Structure - Light microscope	Structure - TEM Figure no.	*Sarcocystis* species and reference	Intermediate host
			S. hominis[488a]	Cattle (*B. taurus*)
			S. suihominis[492]	Pig (*Sus scrofa*)
			S. cuniculi[201]	European rabbit (*Oryctolagus cuniculus*)
			S. leporum[201]	Cottontail rabbit (*Sylvilagus floridanus*)
11	Thick, striated	37C, 38	*S. fayeri*[667]	Horse (*Equus* sp.)
12	Thick, hairy	39A	*S. falcatula*[56]	Avians (several families)
13	Thin, small protrusions	39B	*S. sybillensis*[176]	Wapiti (*C. elaphus*)
			S. mucosa[533a]	Wallabies (*Macropus* sp. and others)
14	Thick, striated	40A, B	*S. capracanis*[9,180]	Goat (*C. hircus*)
			S. tenella[164,487]	Sheep (*O. aries*)
15	Thick, striated	40C	*S. sp.*[178a]	White-tailed deer (*O. virginianus*)
			S. rangiferi[322]	Reindeer (*R. tarandus*)
16	Thick, striated	40D	*S. youngi*[181,186]	Mule deer (*O. hemionus*)
17	Thick, striated to inverted T	41A	*S. odocoleocanis*[178a]	White-tailed deer (*O. virginianus*)
			S. taradivulpes[314]	Reindeer (*R. tarandus*)
			S. hemionilatrantis[181]	Mule deer (*O. hemionus*)
18	Thin—thick	41B	*S. tarandivulpis*[321]	Reindeer (*R. tarandus*)
			S. dispersa[619]	House mouse (*M. musculus*)
			S. zamani[24]	Norway rat (*R. norvegicus*)
19	Thick, striated	41C	*S. singaporensis*[24]	Norway rat (*R. norvegicus*)
20	Thin	42A, B	*S. medusiformis*[97]	Sheep (*O. aries*)
21	Thin	42C	*S. gigantea*[485a,523]	Sheep (*O. aries*)
22	Thick	42D	*S. villivillosi*[24]	Norway rat (*R. norvegicus*)
23	Thin—thick	43A	*S. rileyi*[63,69]	Ducks (*Anas* spp.)
24	Thin	43B	*S. sp.*	American mountain goat (*Oreamnos americanus*)

occurrence of Type 1 sarcocysts in primates might have resulted from accidental infections by species originating in small mammals.

It is important to stress that the structure of the sarcocyst wall should be used with caution in describing species of *Sarcocystis* because

1. The structure varies with fixative and the degenerative state of the host cell. For example, the smooth, thin wall of *S. cruzi* may appear thick and hirsute if not properly fixed or as a result of fixation with different fixatives.
2. In some species villar protrusions from the sarcocyst wall are embedded deep in the surrounding myocyte and thus may be impossible to detect with the light microscope, e.g., a hirsute sarcocyst wall could appear smooth (Figure 26B). Therefore, the ultrastructure of the sarcocyst wall is essential for describing a new species of *Sarcocystis*.

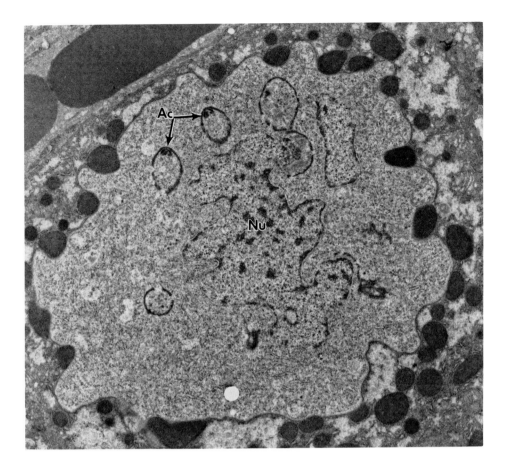

FIGURE 56. TEM of intermediate schizont with irregular nucleus and developing apical complexes (Ac) of merozoites (*S. rauschorum*). (Magnification × 6930.)

3. The shape and size of the villar protrusions on the sarcocyst wall vary in live vs. fixed preparations. Therefore, undue reliance should not be placed on their size.
4. As discussed earlier, the structure of sarcocyst walls may vary with age. Therefore, investigators should be cautious when studying specimens from naturally infected animals because there is no reliable way of determining the age of the sarcocyst.

B. Schizonts

The structure of the schizont is of limited taxonomic value because it varies greatly depending on the developmental cycle (e.g., first vs. second generation) and the host cell parasitized. For example, second-generation schizonts of *S. cruzi* and *S. tenella* are slender and elongated in skeletal muscles, but are globular in renal glomeruli.[160,180]

C. Oocysts and Sporocysts

The structure of the oocyst and sporocysts, traditionally used to determine species of other coccidia, is of little or no taxonomic value in *Sarcocystis* because

1. Except for minor variations in size, all *Sarcocystis* sporocysts and oocysts are structurally similar.
2. Numerous species of *Sarcocystis* in a given host overlap in dimensions. For example, dogs are definitive hosts for 22 species and cats are host for 12 species, and sporocysts of most

Table 3
ATTEMTPED TRANSMISSION OF *SARCOCYSTIS* SPECIES OF LARGE ANIMALS TO OTHER ANIMALS

Sarcocystis species	Intermediate host	Attempted transmission to	Results[a]	Ref.
S. arieticanis	Sheep	Goats	Neg.	367a
S. tenella	Sheep	Goat	Neg.	153
S. capracanis	Goat	Sheep	Neg.	543, 553
S. cruzi	Ox	Sheep, monkeys, rabbits, pigs, rats	Neg.	247
S. cruzi	Ox	Bison	Pos.	163
S. cruzi	Bison	Ox	Pos.	258
S. cruzi	Ox	Buffalo	Neg.	385
S. bertrami	Donkey	Horse	Pos.	467
S. ferovis	Bighorn sheep	Domestic sheep	Neg.	172
S. wapiti	Wapiti	Ox	Neg.	258
S. sybillensis				
S. gracilis	Roe deer	Sheep, mice	Neg.	216
S. odocoileocanis	White-tailed deer	Goats	Neg.	447
S. odocoileocanis	White-tailed deer	Sheep, ox	Pos.	115

[a] Sarcocysts in muscles of recipient animals.

canine species are about 15×10 μm and those of feline species are about 12×10 μm (see Chapter 11).

The prepatent and patent periods vary tremendously and are of no systematic value in *Sarcocystis*. For example, the prepatent period for *S. cruzi* is 7 to 33 d, and sporocysts are shed for many months.

D. Host Specificity

Species of *Sarcocystis* are generally more host specific for their intermediate hosts than for their definitive hosts. For example, for *S. cruzi*, ox and bison are the only intermediate hosts, whereas the dog, wolf, coyote, raccoon, jackal, and fox can act as definitive hosts. Coyotes and foxes also serve as efficient definitive hosts for other dog-transmitted species, e.g., *S. tenella* and *S. capracanis*. However, none of the species transmissible via dogs are transmissible by cats and vice versa. The only species of *Sarcocystis* transmitted via domestic cats that is known to be transmitted by other carnivores is *S. muris*, which is also transmitted by ferrets.[589]

The host specificity of *Sarcocystis* species for intermediate hosts varies (Tables 3 and 4). With the possible exception of *S. odocoileocanis* (transmitted from deer to sheep and cattle via dogs), no other *Sarcocystis* species of large animals is transmissible to unrelated hosts (Table 3). Less host specificity is found among *Sarcocystis* species of small mammals than those of large animals (Table 4). An extreme example of low host specificity is exhibited by *S. falcatula* of budgerigars. It is transmissible to canaries, cowbirds, pigeons, ducks, zebra finches, and grackles, but not to chickens and guinea fowl.[50,54]

E. Isoenzymes

Determination of isoenzyme electrophoretic patterns is useful in distinguishing species of *Sarcocystis*,[12,277,280,532] but the zoites must be obtained from experimentally infected hosts.

Table 4
ATTEMPTED TRANSMISSION OF *SARCOCYSTIS* SPECIES OF SMALL MAMMALS TO OTHER SMALL MAMMALS

Sarcocystis species	Intermediate host	Definitive host	Transmission attempted to	Results[a]	Ref.
S. rauschorum	Varying lemming (*Dicrostonyx richardsoni*)	Snowy owls (*Nyctea scandiaca*)	Norway rats (*Rattus norvegicus*)	Neg.	65
			House mouse (*Mus musculus*)	Neg.	
			Red-neck voles (*Clethrionomys gapperi*)	Neg.	
			White-footed mice (*Peromyscus leucopus*)	Neg.	
			Brown lemmings (*Lemmus sibiricus*)	Neg.	
S. dispersa	House mouse (*M. musculus*)	Barn owl (*Tyto alba*)	Voles (*Microtus arvalis*)	Neg.	71
S. muriviperae	House mouse (*M. musculus*)	Palestinian viper (*Vipera palaestinae*)	Guenther's voles (*M. guentheri*)	Neg.	472
			Jirds (*Meriones unguiculatus, M. tristrami*)	Neg.	
			Multimmate rats (*Mastomys natalensis*)	Neg.	
			European rabbits (*Oryctolagus cuniculus*)	Neg.	
S. clethrio-nomyelaphis	Voles (*Microtus* spp., *C. glareolus*)	Snakes (*Elaphe longissima*)	Voles		469
			M. oeconomus	Pos.	
			M. guentheri	Pos.	
			C. glareolus	Pos.	
			M. arvalis	Pos.	
			Lacerta agilis	Neg.	
			Podarcis muralis	Neg.	
			Apodemus silvaticus	Neg	
S. muriviperae			House mouse (*M. musculus*)	Neg.	
S. muris	House mouse (*M. musculus*)	Cat (*Felis catus*)	Norway rats (*R. norvegicus*)	Neg.	484a, 597 688
			Golden hamster (*Mesocricetus auratus*)	Neg.	
			Guinea pigs (*Cavia porcellus*)	Neg.	
			Meadow voles (*M. pennsylvanicus*)	Neg.	
			Jirds (*M. unguiculatus*)	Neg.	
S. singaporensis	Rat (*R. norvegicus*)	Snake (*Python reticulatus*)	Golden hamster (*M. auratus*)	Neg.	59, 694
			Pigeons (*Columba livia*)	Neg.	
			Chickens (*Gallus domesticus*)	Neg.	
			Multimammate rats (*M. natalensis*)	Neg.	
			Field voles (*Microtus* spp.)	Neg.	

Table 4 (continued)
ATTEMPTED TRANSMISSION OF *SARCOCYSTIS* SPECIES OF SMALL MAMMALS TO OTHER SMALL MAMMALS

Sarcocystis species	Intermediate host	Definitive host	Transmission attempted to	Results[a]	Ref.
S. idahoensis	Deer mouse (*Peromyscus maniculatus*)	Gopher snake (*Pitouphis melanoleucus*)	House mouse (*M. musculus*)	Neg.	36
			White-footed mouse (*Peromyscus leucopus*)	Neg.	

[a] Sarcocysts in muscles of recipient animals.

Table 5
LETHAL DOSES OF SPOROCYSTS OF PATHOGENIC SPECIES OF *SARCOCYSTIS*

Sarcocystis species	Intermediate host	Dose of sporocysts		Ref.
		LD_{50}	LD_{100}	
S. cruzi	Cattle	200,000	1,000,000	264
S. tenella	Sheep	100,000	1,000,000	164, 430, 509
S. arieticanis	Sheep	2,000,000	10,000,000	367a
S. capracanis	Goat	50,000	100,000	158
S. hircicanis	Goat	>2,000,000	10,000,000	367
S. miescheriana	Pig	3,000,000	>3,000,000	21, 225
S. suihominis	Pig	1,000,000	5,000,000	357
S. idahoensis	Deer mouse	15,000	—[a]	35
S. falcatula	Budgerigars	25,000	100,000	631a
S. sp.	Roe deer	100,000	>110,000	211, 216
S. campestris	Richardson's ground squirrel	9,000	—[a]	687
S. rauschorum	Lemming	501	501	65

[a] Not determined.

V. PATHOGENICITY

A. Intermediate Hosts

Not all species of *Sarcocystis* are pathogenic for intermediate hosts. Generally, species tramsmissible via canids are more pathogenic than those transmissible via other definitive hosts. For example, sheep have four species of *Sarcocystis*: *S. tenella*, *S. arieticanis*, *S. medusiformis*, and *S. gigantea*. Of these, *S. tenella* and *S. arieticanis*, transmitted via canids, are pathogenic, whereas *S. gigantea* and *S. medusiformis*, transmitted via cats, are nonpathogenic.

The severity of clinical sarcocystosis is dependent on dose, as shown in Table 5. The size or weight of the host does not appear to be relevant to resistance or susceptibility of clinical disease. Cows are as susceptible to a specific dose of sporocysts as calves. However, stress may play an important role in the severity of illness and the susceptibility to infection. Pregnancy, lactation, poor nutrition, weather, or other stresses may influence the severity of clinical sarcocystosis.

1. Clinicopathological Findings

Cattle usually do not develop acute sarcocystosis under experimental conditions unless

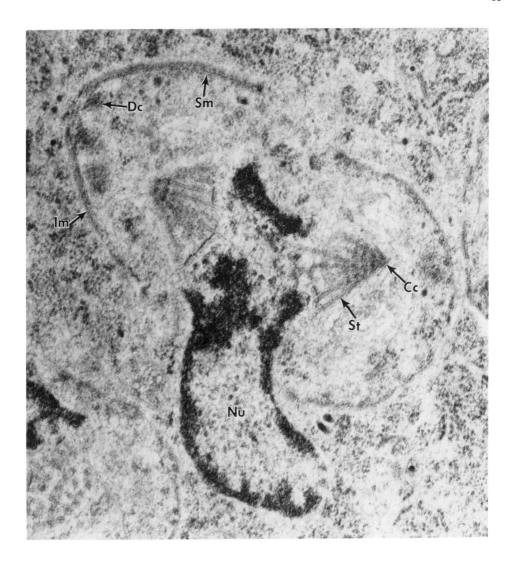

FIGURE 57. TEM of spindle apparatus and developing apical complexes of two merozoites (*S. tenella*). (Magnification × 45,500.) (From Speer, C. A. and Dubey, J. P., *J. Protozool.*, 28, 424, 1981. With permission.)

200,000 or more *S. cruzi* sporocysts have been ingested at a given time. Except for fever (≥40°C) between 15 and 19 DPI, clinical signs are not observed until 24 DPI.[165,239,264] Beginning the fourth week after inoculation, cattle may develop anorexia, diarrhea, weight loss, weakness, muscle twitching, prostration, and sometimes death. Pregnant animals may undergo a premature parturition, abortion, or produce a stillborn fetus.[248] Some or all of these clinical signs may last from a few days to several weeks.

Clinical laboratory findings indicate anemia, tissue damage, and clotting dysfunction. Anemia and packed cell volumes (PCV) below 20% are found in animals with moderate to severe infections.

Serum bilirubin, lactic dehydrogenase (LDH), alanine aminotransferase (AAT), sorbitol dehydrogenase (SBDH), and creatinine phosphokinase (CPK) are generally elevated for brief periods during the anemic phase.[257,286,453,454,566] Blood urea nitrogen (BUN) becomes elevated approaching terminal sarcocystosis.[165] Prothrombin time is longer, whereas platelet counts, clotting time, activated partial thromboplastin time, and thrombin time are generally not altered.[286,453,566] Acquired Factor VII deficiency and platelet dysfunction are found in some acutely infected cattle.[566]

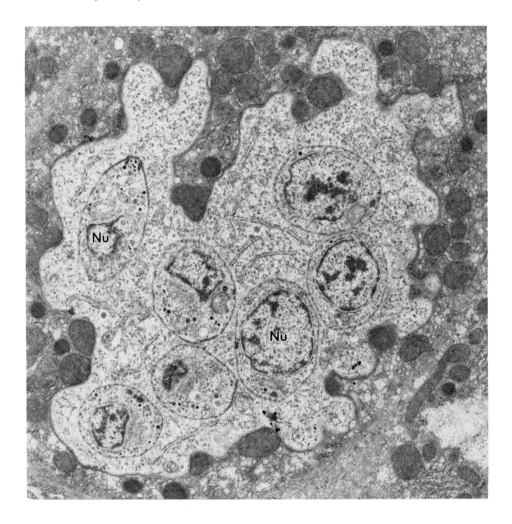

FIGURE 58. TEM of advanced intermediate stage in schizont in which each developing merozoite contains a nucleus but the merozoites are still located internally (*S. rauschorum*). (Magnification × 8580.)

As infections become chronic, other signs become apparent. Growth is adversely affected. Animals become hyperexcitable, they hypersalivate and they lose hair, especially on the neck, rump, and tail switch[264] (Figure 64). Some become emaciated. Some eventually develop CNS signs including recumbency, opisthotonous, nystagmus, cycling gait while laterally recumbent, and occasionally death. Significant laboratory findings have not been reported for these cattle. Clinical disease in relation to phases of the life cycle of the parasite is summarized in Figure 65.

With few exceptions, clinical signs similar to those seen in *S. cruzi*-infected cattle have been seen in goats,[100,158,366,367] sheep,[190,301,430,509] and pigs[21,225,357,606] infected with large doses of *S. capracanis*, *S. tenella*, and *S. miescheriana*, respectively. In sheep and goats, neural signs were more prominent than in cattle, but hypersalivation was not seen in sheep or goats.

2. Gross Lesions

Edema and focal necrosis in gut-associated lymph nodes are the first recognizable gross lesions; they are seen about 15 DPI.[165] The next gross lesions are seen at about 26 DPI. The most striking lesion seen at this time is hemorrhage (Plate 1*).[165,391] Hemorrhages are most evident on the serous surface of viscera, in cardiac and skeletal muscles, and in the sclera of the eyes.

* Plate 1 appears following page 70.

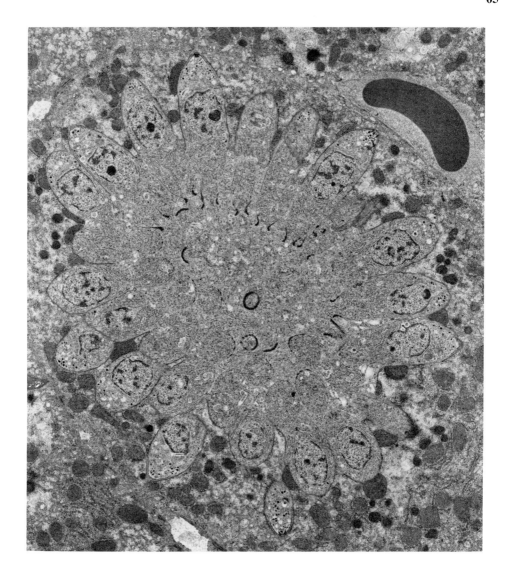

FIGURE 59. TEM of nearly mature schizont in which the merozoites are developing at the margin of the meront (*S. rauschorum*). (Magnification × 4700.)

Skeletal muscles mottled or striped with pale areas, interspersed with dark hemorrhagic areas are characteristic of acute sarcocystosis.[165,391] Hemorrhages vary from petechiae to ecchymoses several centimeters in diameter. Following acute infection, body fat becomes scanty and gelatinous. Body cavities contain straw-colored fluid, and organs become icteric. In chronically affected animals the most notable lesion is serous atrophy of fat, especially pericardial and perirenal fat, with white flecks of mineralization.

3. Microscopic Lesions

The earliest lesion observed is hypertrophy of endothelial cells associated with development of first-generation schizonts in arteries and arterioles[165,167] (Figure 66). This is followed by perivascular and interstitial infiltration of many organs with mononuclear cells (Figure 67). The mononuclear cells are seen as early as 11 DPI and may be present for several months. The predominant inflammatory cells are lymphocytes and macrophages, with few plasma cells and eosinophils.[165,391] Hemorrhage may be generalized and often not associated with inflammation. Necrosis may be found in many organs, especially in skeletal muscles, heart, and kidneys,

66 Sarcocystosis of Animals and Man

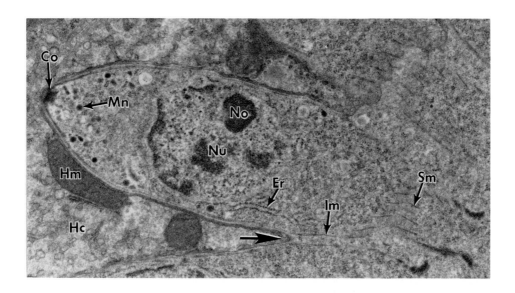

FIGURE 60. Higher magnification of portion of Figure 58 showing a budding merozoite; note that the merozoite plasmalemma (arrow) has invaginated just beyond the merozoite nucleus (*S. rauschorum*). (Magnification × 17,000.)

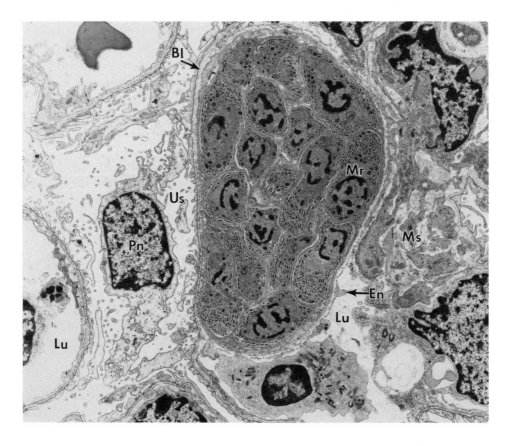

FIGURE 61. TEM of mature schizont in glomerular capillary endothelial cell (*S. tenella*). (Magnification × 7675.) (From Speer, C. A. and Dubey, J. P., *J. Protozool.*, 28, 424, 1981. With permission.)

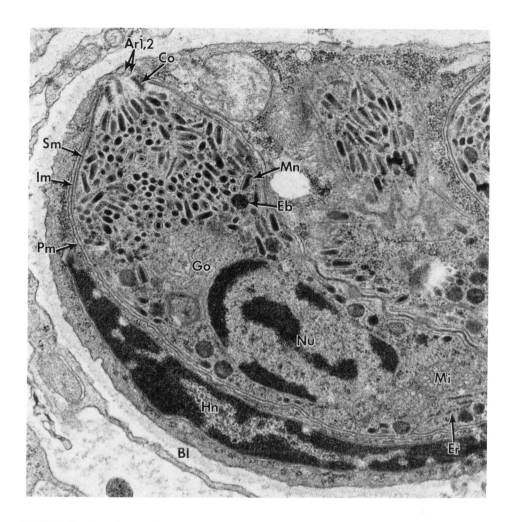

FIGURE 62. TEM of merozoite; note absence of rhoptries (*S. tenella*). (Magnification × 26,600.) (From Speer, C. A. and Dubey, J. P., *J. Protozool.*, 28, 424, 1981. With permission.)

probably associated with vasculitis[165] (Figures 67 to 72). The overall predominant lesion in sarcocystosis is inflammatory rather than degenerative.

Degenerating sarcocysts may be surrounded by mononuclear cells (Figure 73), neutrophils (Figure 74), eosinophils, giant cells, or a combination of these cells (Figure 75). In most livestock species, the cellular response is mainly mononuclear. How sarcocysts are removed from host tissue is not known. It is likely that some sarcocysts rupture spontaneously and inflammation follows. However, macrophages and plasma cells are seen around seemingly intact live sarcocysts.[542]

B. Definitive Hosts

Sarcocystis generally does not cause illness in definitive hosts. Dogs, cats, coyotes, foxes, and raccoons fed tissues infected with numerous species of *Sarcocystis* shed sporocysts, but are otherwise normal. A few dogs and coyotes vomited or were anorexic for 1 to 2 d following ingestion of meat, but such signs may have resulted from the change in diet from laboratory chow to raw meat. However, human volunteers who ingested beef and pork infected with *S. hominis* or *S. suihominis*, respectively, developed symptoms including vomiting, diarrhea, and respiratory distress. These symptoms were more pronounced in volunteers who ate infected pork than in those who ate infected beef (see Chapters 3 and 4).

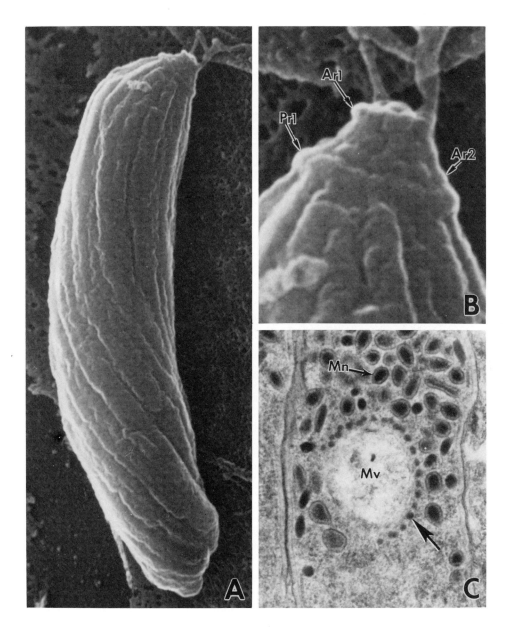

FIGURE 63. (A) SEM showing helical pattern on surface of merozoite; *S. cruzi* merozoite obtained at 57 DPI of sporozoites into cultured bovine pulmonary artery endothelial cells. (Magnification × 30,000.) (B) High magnification of anterior tip of merozoite in A. Note ridges created by underlying apical rings 1 and 2, and polar ring 1. (Magnification × 100,000.) (C) Anterior region of merozoite showing multivesicular body (Mv) which appears to give rise to micronemes (Mn) at its margin (arrow) (*S. tenella*). (Magnification × 52,000.) (From Speer, C. A. and Dubey, J. P., *J. Protozool.*, 28, 424, 1981. With permission.)

VI. PATHOGENESIS

A. Tissue Necrosis

Schizonts cause necrosis of cells and tissues depending on the species of *Sarcocystis*, location, and the multiplication potential.[165] For example, *S. falcatula* of birds multiplies extensively in endothelial cells and produces several generations of schizogony.[631a] The physical damage alone as a result of vasculitis might result in the death of birds with heavy infections with *S. falcatula*.[631b] The same hypothesis might apply to *S. idahoensis* infection in deer mice in which

FIGURE 64. An emaciated steer naturally infected with *S. cruzi*. Note loss of hair at the tip of tail. (From Giles, R. C., Tramontin, R., Kadel, W. L., Whitaker, K., Miksch, D., Bryant, D. W., and Fayer, R., *J. Am. Vet. Med. Assoc.*, 176, 543, 1980. With permission.)

hepatocytes are destroyed as a result of schizont multiplication.[35] However, localized tissue necrosis does not appear extensive enough to cause the severe illness or death seen in large animals (cattle, sheep, goats, pigs).

B. Inflammation

The perivascular mononuclear cell infiltration seen around *S. cruzi*-infected arteries at 7 to 11 DPI[165] indicates a host reaction to antigens liberated from sporozoites or immature schizonts or the expression of parasite antigens by host cells because merozoites are not liberated from these first-generation schizonts until 14 DPI or later. An intense inflammatory reaction is usually observed about the time when the second-generation schizonts mature and rupture, during the fourth to sixth week postinoculation. The myositis, during the penetration of myocytes by merozoites, may be related to products liberated from merozoites or myocytes.[165] The intense mononuclear cell infiltrations in the kidneys, liver, lungs, and other organs are probably stimulated by similar parasite antigens.[165,391]

C. Immune Regulation of Pathogenesis

Vascular endothelial cells have been recently found to possess MHC II antigens[632a] and to process antigen for presentation to lymphocytes.[481] Adherence of thymus-derived (T) lymphocytes to endothelial cells via the lymphocyte function-associated molecule, LFA-1, or other mechanisms[342a] is the first step toward emigration of these cells from the blood into the tissues. Lymphocytes also recognize specific MHC II antigen on vascular endothelial cells.[557a] In a rabbit model, a strong correlation was found between the magnitude of the *in vivo* immune response and the vascular endothelial expression of MHC II antigens.[632b]

Thus, during acute *S. cruzi* infections, endothelial cells may participate in eliciting an immune response leading to pathogenesis by processing and presenting antigens immunogenically, which results in sensitization of T lymphocytes.[643] These antigen-sensitive T lymphocytes

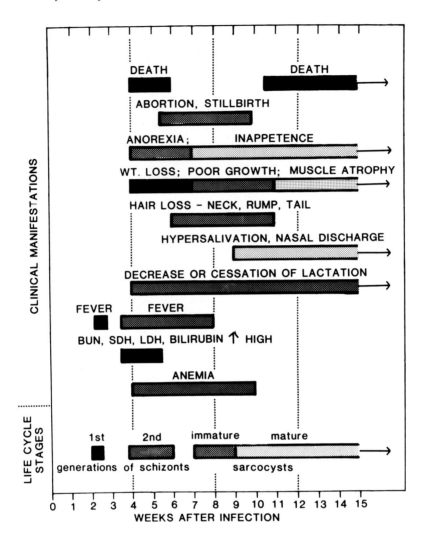

FIGURE 65. Relationship of life cycle stages of *S. cruzi* and weeks after infection to clinical sarcocystosis in cattle. (From Fayer, R. and Dubey, J. P., *Comp. Contin. Educ. Pract. Vet.*, 8, 130, 1986. With permission.)

might then recognize antigen plus MHC II[557a] on vascular endothelium and give rise to destructive immune responses such as local delayed hypersensitivity reactions involving mononuclear cell infiltration of vascular tissue.[165,632a,b]

D. Edema
Ascites and edema in tissues are probably related to hypoproteinemia and vasculitis.[165,631a]

E. Fever
The peaks of fever coincide with maturation of schizonts and release of merozoites into the bloodstream.[264] Fever is probably related to the release of pyrogen from mature rupturing schizonts directly on the hypothalamus or indirectly by stimulating the release of prostaglandins.

F. Anemia
Anemia is the most evident clinical finding of acute sarcocystosis in cattle, sheep, goats, and pigs, but the mechanism is unknown.[158,165,257,295,430,453,454,566] Although the anemia is regenerative, few or no reticulocytes are found. The anemia is normocytic, normochromic, and primarily

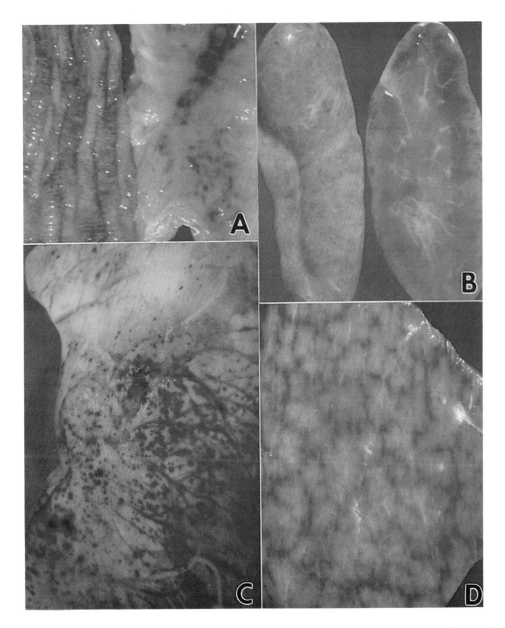

PLATE 1. Hemorrhages in tissues of calves experimentally infected with *S. cruzi*. (A) Serosal and mucosal surfaces of colon; (B) superficial cervical lymph node; (C) myocardium; (D) skeletal muscle. Hemorrhages vary in size and duration. (From Dubey, J. P., Speer, C. A., and Epling, G. P., *Am. J. Vet. Res.*, 43, 2147, 1982. With permission.)

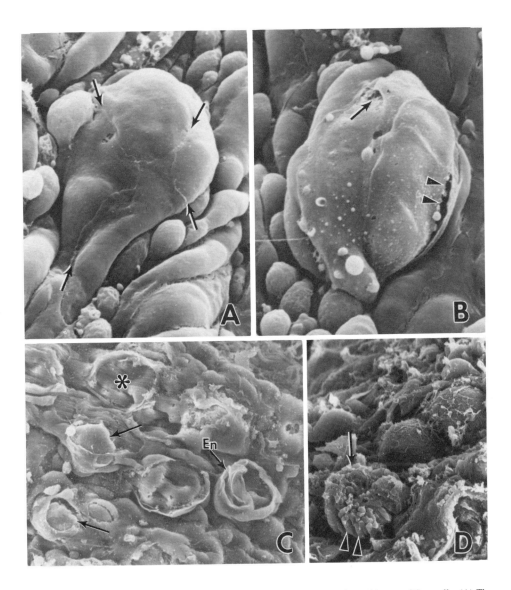

FIGURE 66. SEMs of ovine mesenteric arteries infected with first-generation schizonts of *S. tenella*. (A) The protuberance in the center includes four endothelial cells demarcated by intercellular junctions (arrows) covering a host cell containing a schizont (not visible). (Magnification × 1800.) (B) Perforation (arrow) and separation of two endothelial cells (arrowheads) covering schizont. (Magnification × 1970.) (C) Sloughing of endothelial cells (En) exposing internal elastic membrane (*) and host cells (arrows) harboring schizonts. (Magnification × 530.) (D) Most of the endothelium has sloughed exposing a schizont (arrow) and merozoites (arrowheads). (Magnification × 10,650.) (From Speer, C. A. and Dubey, J. P., *Can. J. Zool.*, 60, 203, 1982. With permission.)

hemolytic. Hemorrhage may account for part of the loss of blood cells. In addition, many red blood cells are removed from circulation and sequestered in the spleen, probably via immunologic mechanisms. It is possible that some unknown toxic factors or metabolite released from schizonts or infected host cells contributes further to the anemia. The increased prothrombin time and increased fibrin degradation product concentration in infected calves indicate intravascular coagulopathy, which may result in capillary extravasation.

G. Abortion

Abortion and fetal death can result when animals become infected with pathogenic species of *Sarcocystis* during pregnancy. Most cattle, sheep, swine, goats, and roe deer that developed

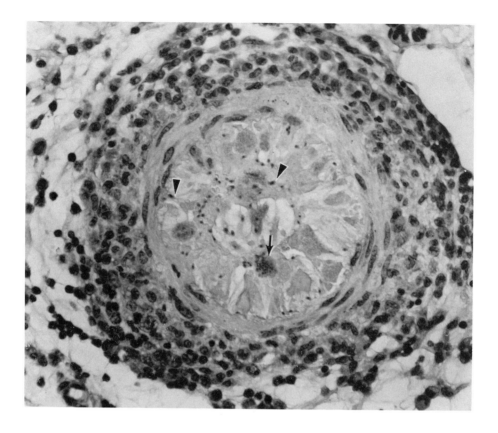

FIGURE 67. Necrosis and vasculitis in a mesenteric lymph node artery of sheep infected with *S. tenella*. The endothelial cells have been destroyed by rupture of second-generation schizonts (arrow), releasing merozoites (arrowheads) into the lumen. Mononuclear cells are in and around arterial walls. (H. & E.; magnification × 400.) (From Dubey, J. P., *Vet. Parasitol.*, 26, 237, 1988. With permission.)

clinical sarcocystosis from experimental infections induced in mid to late gestation aborted,[18,156,223,224,248,431,568,646] whereas most infected animals without signs of infection carried their fetus to term.[156,520] Lesions were not found in fetal tissues from any of these experimentally infected host species, with the exception of one lamb. Parasites (only 2) were found in only 2 of 18 bovine fetuses[18] and only 4 schizonts were found in fetal membranes of 3 sheep.[431] In contrast, parasites and lesions were found in maternal placentomes of cattle, sheep, and goats.[18,156,431] Thus, results from these experimentally infected animals indicate that although *Sarcocystis* is present in the maternal placenta, it rarely infects the fetus or fetal membranes. Unlike experimentally infected cattle, those with natural infections have had parasites, lesions, or both in the fetuses (Figure 76).[166,378,390,392,420,479,510,594,605] Unlike experimental infections, parasites as well as lesions were found in the placentas in several natural infections.[109,166,510] Furthermore, some naturally infected cows had no clinical signs of infection, although organisms were detected in the fetus or placenta.[166] It is not possible to confirm that the natural infections and the experimental infections were due to the same species of *Sarcocystis*, although the parasites were morphologically similar, parasitized the same type of host cells, and in some cases responded serologically to the same antigen. Thus, the question remains as to whether differences observed between experimentally infected animals and naturally infected animals were due to different species or strains of *Sarcocystis*.

There are many unanswered questions and observations that appear contradictory within the subject of how sarcocystosis affects fetal health. Some fetuses are infected and have lesions,

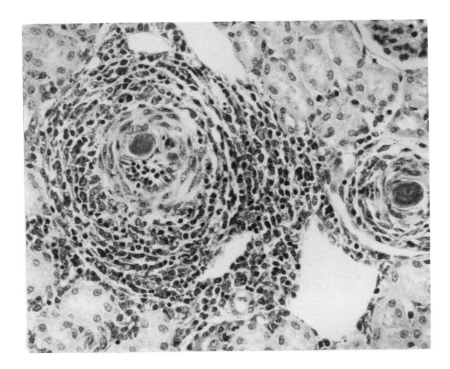

FIGURE 68. Severe mononuclear cell infiltrations in tunica adventia and tunica media in arteries in renal cortex of a calf infected with *S. cruzi*. The vascular lumen is almost occluded by hypertrophied endothelial cells and first generation schizonts. (Giemsa stain; magnification × 250.) (From Dubey, J. P., Speer, C. A., and Epling, G. P., *Am. J. Vet. Res.*, 43, 2147, 1982. With permission.)

others are unaffected. Some fetal placentas are infected and have lesions, others are unaffected. Some maternal placentomes are infected and have lesions, others are unaffected. Most pregnant animals have overt clinical illness at the time the fetus is affected, others do not. Several possible mechanisms have been hypothesized by which sarcocystosis may directly or indirectly affect fetal health.[267]

In naturally infected bovine fetuses, parasites were found in virtually all organs, but were most often found in the brain. Immature and mature schizonts and free merozoites were found, usually within endothelial cells of capillaries, but occasionally free in the lumen of a vessel or in neural tissue.[166,378,390,392,420] Brain lesions include nonsuppurative encephalitis or meningitis, with small foci of glial cells surrounding a central necrotic area throughout the gray and white matter of the cerebrum, cerebellum, or brainstem, with some perivascular mononuclear cell infiltration and occasional microthrombi in vessels in reaction foci. Other affected organs are similar to those of infected postnatal animals, including nonsuppurative myocarditis, pneumonitis, hepatitis, and renal glomerulitis accompanied by focal necrosis and hemorrhage.

One premature lamb born to an experimentally infected ewe had cerebral congestion and edema as well as areas of leukoencephalomalacia in the cerebrum and midbrain.[520] Another lamb, from a naturally infected ewe, was found with encephalitis characterized by mononuclear cell infiltration and gliosis.[188]

1. Placental Infection and Lesions

Free merozoites and schizonts, usually associated with lesions, have been observed in maternal or fetal placentas of some cattle, sheep, and goats. Immunofluorescent staining of placentomes from six experimentally infected cows revealed numerous merozoites and

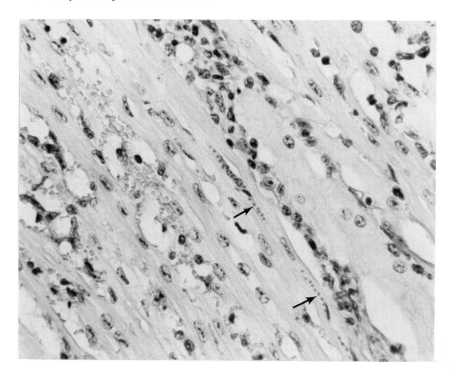

FIGURE 69. Hemorrhage and mild mononuclear cell infiltration in the myocardium of a calf experimentally infected with *S. cruzi*. Arrow points to merozoites and schizonts. (H. & E.; magnification × 250.) (From Dubey, J. P., Speer, C. A., and Epling, G. P., *Am. J. Vet. Res.*, 43, 2147, 1982. With permission.)

schizonts in the maternal but not the fetal placentome.[18] Placentomes of all six cows were atrophied. Maternal placentomes from naturally infected aborting cows were also found to contain numerous parasites, but were hemorrhagic and necrotic with calcification of placental villar epithelial cells and of the fibrous connective tissue.[109] In other natural infections resulting in abortion, fetal placental lesions were mild to severe with accompanying mononuclear cell infiltration.[166,510] Of 11 experimentally infected sheep, only 4 schizonts were found in fetal membranes from 3 ewes.[431] In experimentally infected goats, schizonts were found in the maternal placentome and endometrium, but not in fetal membranes.[156] Lesions were restricted to the maternal placentome and endometrium with heavy mononuclear cell infiltration and some focal necrosis.

2. Possible Mechanisms Causing Abortion or Fetal Death

Based on the foregoing descriptions of parasite locations and lesions, there appear to be several possible ways in which sarcocystosis might initiate abortion or fetal death (Figure 77).

One hypothetical sequence of events leading to labor might begin in the hypothalamus of the fetal brain, with transmission via the pituitary to the fetal adrenal where cortisol provides the fetal trigger to maternal endocrine changes.[83] In turn, there might follow a release of steroids and prostaglandins from the placenta, the placentome, and the fetal membranes which stimulate the myometrium to contract and the cervix to dilate. Under normal conditions it is thought there is a rise in concentration of plasma cortisol secreted from the fetal adrenal as it grows and develops, with the highest concentrations during the last few days of fetal life.[83] What mechanisms might result in high adrenocorticotrophic hormone (ACTH) or cortisol concentrations in the fetus leading to premature parturition or abortion? One sequence of events could begin with anemia in the dam, resulting in hypoxia in the fetus. Hypoxia could also result from insufficient blood

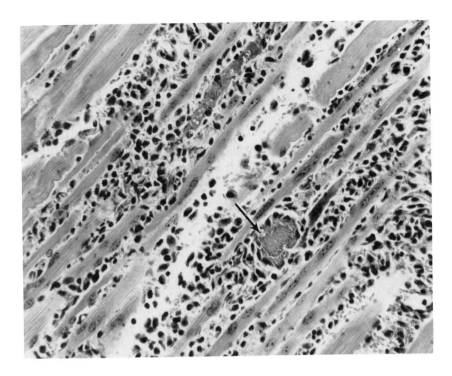

FIGURE 70. Severe necrosis, infiltration of mononuclear cells, and mineralization (arrow) in semitendinosus muscle of a calf infected with *S. cruzi*. (H. & E.; magnification × 250.) (From Dubey, J. P., Speer, C. A., and Epling, G. P., *Am. J. Vet. Res.*, 43, 2147, 1982. With permission.)

flow through the maternal placentome due to occlusion of capillaries by schizonts, localized hemorrhage, or intravascular coagulation. Evidence of hypoxia is the finding of leukoencephalomalacia in some fetuses. Vasopressin, known to stimulate ACTH release, rises severalfold with the approach of spontaneous parturition and in response to hypoxia.[83] Hypoxia itself has been shown to raise the concentration of immunoreactive ACTH in fetal plasma at all times during gestation when it was studied. Isolated fetal adrenal cells appear to be responsive and secrete large amounts of cortisol in response to ACTH, a form of GTP, or dibutyrylcyclic AMP at various times during gestation.[83]

Vascular damage associated with sarcocystosis in the dam might lead to changes in concentration of hormones associated with the maintenance of gestation. The plasma concentration of progesterone, E_1, E_2, PGFM, and α-fetoprotein was monitored in pregnant cows experimentally infected with *S cruzi*.[13,14]

Progesterone production after day 50 of gestation in sheep is primarily in the placenta. It acts by maintaining the myometrium refractory to stimulation by oxytocin or $PFG_{2\alpha}$. Decrease in placental progesterone at term follows and is caused by a rise in fetal cortisol levels. Progesterone withdrawal initiates increased $PGF_{2\alpha}$ secretion and parturition.

In late pregnancy, estrogens usually promote uterine contractility, increase responsiveness of the uterus to agonists such as oxytocin and $PGF_{2\alpha}$, inhibit progesterone production, and stimulate prostaglandin synthesis and release. Maternal estrogen increases in response to fetal adrenal activity.

Unfortunately, changes in progesterone, E_1, and E_2 concentration in the experimentally infected cows could not be interpreted as cause or effect. Although the changes preceded or were concurrent with abortion, similar changes were documented at the time of normal parturition.

Prostaglandins initiate labor by stimulating myometrial contractions and by softening the

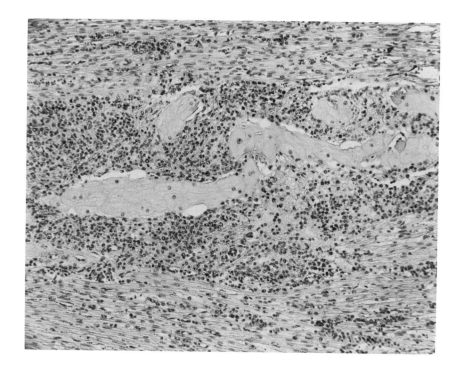

FIGURE 71. Severe myocarditis involving Purkinje fibers and myocardium of a calf experimentally infected with *S. cruzi*. (H. & E.; magnification × 100.) (From Dubey, J. P., Speer, C. A., and Epling, G. P., *Am. J. Vet. Res.*, 43, 2147, 1982. With permission.)

cervix. Two specific lines of observation suggest that events in acute sarcocystosis might result in increased concentration of prostaglandins in the peripheral circulation of the dam or in the maternal placentome and fetal placenta. First, release of second-generation merozoites is associated with damage of capillary endothelium, an event known to trigger the arachidonic acid cascade leading to $PGF_{2\alpha}$ release. Parasitic disruption of cell membranes and the action of inflammatory cells may also stimulate platelet activity including the release of $PGF_{2\alpha}$ by platelets and by damaged endothelium. Second, when second-generation merozoites mature, numerous mononuclear inflammatory cells are found perivascularly throughout the body and in the placenta. Many of these cells are macrophages, and *in vitro* studies indicate that macrophages can be stimulated by *Sarcocystis* lysates to release tumor necrosis factor (TNF).[267] One activity of TNF is the triggering of one arachidonic acid cascade leading to $PGF_{2\alpha}$ release. Because the findings in identifying those events leading to high maternal plasma $PGF_{2\alpha}$ concentration preceding bovine abortion were inconclusive; studies need to be conducted to clarify the relationship between sarcocystosis and prostaglandins.

Fever has been associated with premature parturition and sometimes regarded as the cause. However, it is not clear if fever itself is a cause or an effect. Pyrogens from infectious agents such as protozoa or from degenerating body tissues can cause fever. Pyrogens may act directly on the hypothalamus to raise the thermostatic control or may act indirectly by stimulating leukocytes to produce endogenous pyrogen. Endogenous pyrogen is thought to cause fever by inducing prostaglandin E_1 formation which elicits the fever reaction. Thus, it is not clear if fever directly affects the fetus or if the initiators of fever, in this case, *Sarcocystis* pyrogens or prostaglandins, induce lesions.

In addition to the foregoing hypotheses leading indirectly to death or premature parturition of the fetus, invasion of fetal tissues by the parasite might result in specific lesions with similar end results. Distribution of parasites throughout the brain with associated lesions might

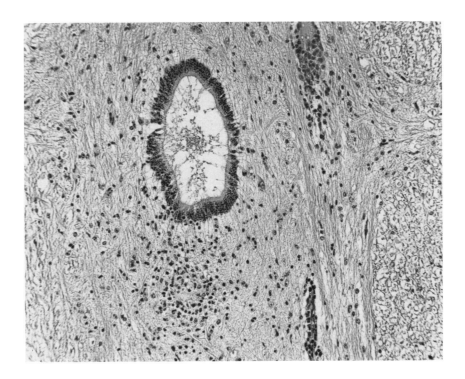

FIGURE 72. A glial nodule and perivascular infiltration of mononuclear cells in spinal cord of a calf infected with *S. cruzi*. (H. & E.; magnification × 160.) (From Dubey, J. P., Speer, C. A., and Epling, G. P., *Am. J. Vet. Res.*, 43, 2147, 1982. With permission.)

stimulate the fetal hypothalamic-pituitary-adrenal axis or destroy significant brain tissue leading to death. Evidence of fetal myocarditis, pneumonitis, and hepatitis indicate other possible avenues leading directly to fetal death.

H. Eosinophilic Myositis and Sarcocystosis

Eosinophilic myositis (EM) is a specific inflammatory condition of striated muscles, principally due to accumulations of eosinophils. It has been found mainly in cattle,[295,383,389,439,574,581] occasionally in sheep,[389] and rarely in pigs and horses (Figure 75). The affected animals may appear clinically normal.

Eosinophilic myositis is generally detected at the time the surface of the carcass is inspected or when it is cut into prime cuts or into quarters.[383] Sex and breed have no influence on the prevalence. The condition was reported with equal frequency in steers, cows, and heifers. Virtually all striated muscles, including skeletal muscles, the muscles of the eye, larynx, and heart may be affected.

In the U.S., federal meat inspection regulations require that parts, or in severe cases, the entire carcass, affected with EM be condemned. Generally, EM has been found in 1 of 100,000 slaughtered cattle, but in some feedlots up to 5% of the cattle were condemned.[383,389]

Two types of lesions are found associated with EM.[383,389] The most common lesion on the surface of muscles is multifocal, spindle-shaped to round, and 5 to 15 × 1 to 3 mm in size. The color of the lesion may vary. Green lesions result from accumulations of eosinophils, principally between myocytes. With the progression of EM, eosinophils and myocytes degenerate, resulting in granulomas with a central area of necrosis. Later, the central necrotic tissue becomes surrounded by zones of giant cells, epithelial cells, eosinophils, lymphocytes, and fibrocytes.[383]

A second type of lesion is less prevalent, but more conspicuous. It may be up to 15 cm long, bright green to pale yellow, and firm in consistency.

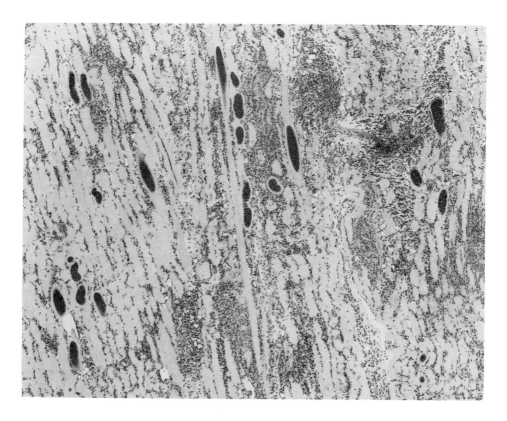

FIGURE 73. Severe mononuclear cell infiltration around blood vessels, myocytes, and perimysium, and several intact *S. capracanis* sarcocysts in diaphragm of a vaccinated-challenged goat. (H. & E.; magnification × 40.) (From Dubey, J. P., *Vet. Parasitol.*, 12, 23, 1983. With permission.)

The etiology of EM remains uncertain. Because sarcocysts are found in the same location as EM, *Sarcocystis* has been traditionally considered to be the cause of EM. Jensen et al.[389] considered that the large EM lesions were not due to *Sarcocystis*, but that the small lesions were associated with degenerating sarcocysts. The *Sarcocystis*-induced etiology of EM was supported by the finding of *Sarcocystis*-specific IgE antibody in the sera of eight cases of eosinophilic myocarditis, but not in two *Sarcocystis*-free calves.[332] Because of the small number of cases of EM compared with the high prevalence of intramuscular cysts, the confirmation of *Sarcocystis* as the cause and EM as the effect requires more substantial evidence. It is necessary to note that EM has never been seen in any experimentally infected ox, sheep, or other animals. In experimentally infected animals, degenerating sarcocysts are often surrounded by mononuclear cells and neutrophils. Because most of these observations were made in animals infected less than 1 year and with only one dose of sporocysts, repeated natural infections may result in a different reaction, perhaps involving other cell types such as eosinophils.

I. Chronic Sarcocystosis and Toxins

Little is known of the pathogenesis of chronic sarcocystosis. Anatomic lesions are insufficient, especially in cattle, to explain the etiology of CNS signs. Although sarcocysts in muscles or in the CNS are well adapted, usually without any host reaction, some sarcocysts probably rupture from time to time and may release toxic products. It has been known for about a century that an aqueous extract of bradyzoites is toxic (sarcotoxin) when inoculated into rabbits.[370,554,615,660] It is not clear how or if sarcotoxin might be released from intact sarcocysts and, if released, what role it might play in chronic sarcocystosis. Although lectins are also associated with bradyzoites,[79,450a,501a] their role, if any, in causing disease is unknown.

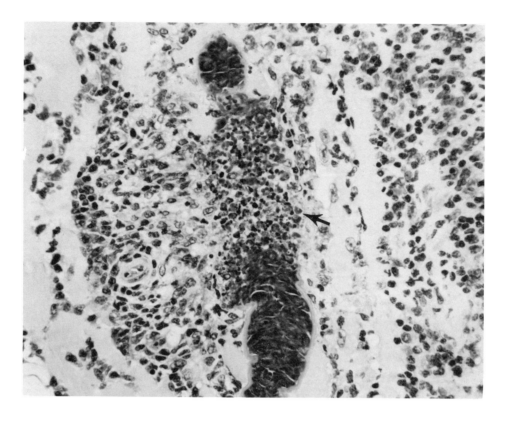

FIGURE 74. Part of a degenerating *S. capracanis* sarcocyst in diaphragm of a vaccinated-challenged goat being replaced by neutrophils (arrow). Mononuclear cells surround the sarcocyst. (H. & E.; magnification × 400.) (From Dubey, J. P., *Vet. Parasitol.*, 12, 23, 1983. With permission.)

It is possible that substances released from *Sarcocystis* stimulate production of tumor-necrosis factor (TNF).[257] TNF is known to be associated with wasting disease. Poor weight gain and low feed efficiency in sarcocystosis are regulated through growth-regulating hormones.[199,200] Any relationship between TNF and the growth-regulating hormones has not been established.

VII. IMMUNITY

Cellular and humoral immune responses in infected animals indicate that *Sarcocystis* species are immunogenic in intermediate hosts. Unfortunately, most information is available only from responses directed against antigens derived from bradyzoites. Because bradyzoites are obtained from the terminal stage (sarcocysts), only cross-reactive antigens are actually tested.

A. Antigenic Structure

Relatively large amounts of highly purified parasites are needed to study the proteins and antigens of *Sarcocystis* spp. Parasites can be obtained from experimentally infected animals or from *in vitro* cultivation. Parasites such as bradyzoites should be obtained from experimentally infected animals, and not from naturally infected ones, to insure accurate species identification. Whether parasites are obtained from *in vivo* or *in vitro* sources, biochemical analyses should also include controls of solubilized noninfected host cells or tissues.

Except for *S. cruzi*, there is little or no information concerning the proteins and antigens of *Sarcocystis* spp. Proteins and antigens of *S. cruzi* sporozoites, merozoites, and bradyzoites have been studied with sodium dodecyl sulfate-polyacrylamide gel electrophoresis (SDS-PAGE) and

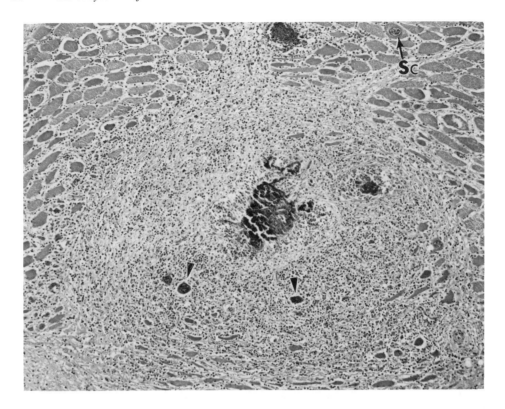

FIGURE 75. Eosinophilic granulomatous myositis in skeletal muscle of a naturally infected horse. The central mineralized area is surrounded by numerous layers of eosinophils and fibroblasts. A few giant cells (arrowheads) are present. Cross-section of a sarcocyst (Sc) is in the upper right area. (H. & E.; magnification × 75.) (Courtesy of W. J. Hartley, Wallaceville Animal Research Centre, Upper Hut, New Zealand.)

Western blotting with immune serum or monoclonal antibodies (Figures 78 to 81).[60b,639,643] Bradyzoites and merozoites had similar proteins with M_r 15.7 kDa, several between M_r 18 to 21.4, 29, and 110 kDa; proteins with M_r 15.7, 29, 49.9, and 54K kDa were common to sporozoites and merozoites. Only two proteins (M_r 15.7 and 16.5 kDa) were present in all three parasite stages.

Based on SDS-PAGE analysis of *S. cruzi* merozoite proteins, some appeared, disappeared, and then reappeared during *in vitro* development of merozoites. Merozoites harvested at 36 and 48 DPI (called 36D and 48D merozoites) from bovine pulmonary artery endothelial cells each had a unique protein of M_r 50 and 70 kDa, respectively (Figure 78).[639] Merozoites harvested at 36 and 60 DPI had many proteins with similar molecular weights as well as several unique proteins (Figure 79).[643] For example, 36D merozoites had a 43-kDa protein that was not present in 60D merozoites, and 60D merozoites had a prominent 63.1-kDa protein not present in 36D merozoites. A 63.1-kDa protein first appeared in 37 or 38D merozoites and was also present in 38, 40, 48, 52, and 60D merozoites, but was absent in 41 and 58D merozoites. Two other proteins also showed temporal variation in appearance. A 43-kDa protein was present in 31, 36, and 52D merozoites, whereas a 42.3-kDa protein was present in 29, 40, and 48D merozoites, and both proteins were present in 33, 38, 41, 58, and 60D merozoites.

In Western blot analyses, antibodies in bovine serum reacted with increasing numbers of antigen bands as the animals progressed through the course of infection.[643a] Proteins with M_r 21, 40, and 50 kDa might be potent antigens because serum reacted strongly with them beginning early in infection.

Three surface antigens of M_r 27, 43, and 90 kDa have been identified by ^{125}I labeling and immunoprecipitation of *S. muris* bradyzoites with specific rabbit or mouse antiserum.[1]

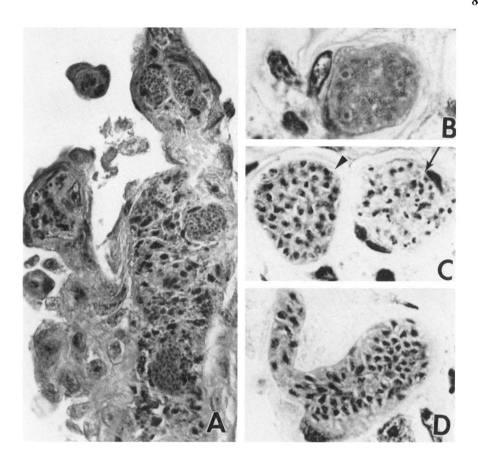

FIGURE 76. Necrosis and schizonts of *Sarcocystis* in fetal placental cotyledon of a naturally infected cow. (A) Necrosis, schizonts, and numerous merozoites; (B) early schizont with lobulated nucleus; (C) a degenerating (arrow) and normal (arrowhead) schizont; (D) an irregular-shaped schizont. (H. & E.; magnification [A] × 500, [B to D] × 1000.) (From Dubey, J. P. and Bergeron, J. A., *Vet. Pathol.*, 19, 315, 1982. With permission.)

Monoclonal antibodies (MAbs) have been generated against *S. cruzi*[60b,643] by immunizing female BALB/cBY mice 8 to 10 weeks of age with 10^6 merozoites or sporozoites per mouse. Merozoites were mixed 1:1 (v/v) in complete Freund's adjuvant and given by intraperitoneal injection. Mice were intravenously injected 1 month later with 8×10^5 merozoites or 5×10^5 sporozoites, respectively, and 3 d later the spleens were removed and fused to the myeloma cell line P3/X63-Ag8.6.5.3 and selected as previously described.[295a] Hybridoma culture supernatants were screened on acetone-fixed merozoites or sporozoites by an indirect immunofluorescence assay (IFA),[60a] cloned by limiting dilution, retested, and expanded in culture. MAbs were tested for their ability to react with whole parasites in a live IFA or with parasite proteins in Western blots.

A variety of reactivities occurred when either antisporozoite or antimerozoite MAbs were used to detect antigens of *S. cruzi* on Western blots of whole parasite (sporozoite, merozoite, or bradyzoite) antigens separated by SDS-PAGE.[60b,643] Several MAbs reacted in IFA with acetone-fixed sporozoites and merozoites and three MAbs reacted with the surface of live sporozoites or merozoites (Figure 80). In Western blots, the surface-reactive MAbs reacted with polypeptides ranging from M_r 20 to 74 kDa. Interestingly, only one of the MAbs analyzed reacted with blots of antigens prepared from homologous (eliciting) stages, although several did react with the surface of homologous stages in the live IFA. Thus, certain epitopes on the surface of both merozoites and sporozoites may have been irreversibly denatured in the blotting procedure, labile to certain proteases or nonprotein in composition (e.g., lipid, glycolipid). In addition,

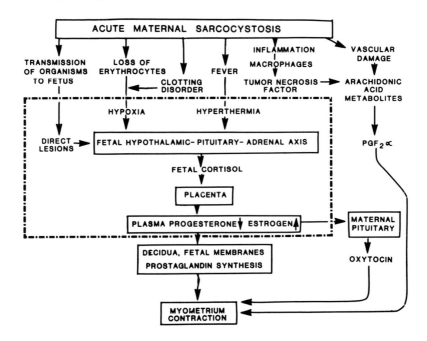

FIGURE 77. A hypothetical diagrammatic presentation of *Sarcocystis*-induced abortion. Fetal hormonal events are shown within the area marked by dotted lines. (From Fayer, R. and Dubey, J. P., in *Proc. NIH Conf. Transplacental Effects on Fetal Health*, 153, 1988.)

several epitopes on the surface of live merozoites appeared to be constituents of molecules within sporozoites. Such molecules may be precursors of surface components of merozoites or may simply share these epitopes on unrelated molecules. Antimerozoite MAbs also identified shared epitopes in sporozoites and bradyzoites that appeared to be located on molecules of both similar molecular weight (i.e., 40 kDa) and distinct molecular weight (i.e., 40 vs. 20 and 60 kDa). A MAb reactive with the surface of *S. cruzi* merozoites also identified epitopes on sporozoite and bradyzoite antigens, indicating that some of the 20- to 60-kDa sporozoite molecules may be precursors to internal or surface antigens of bradyzoites. Further investigations will be needed to elucidate the molecular relationships and functions of these molecules in the development of the major stages in the life cycle of *S. cruzi*.

Several MAbs generated against *S. cruzi* also revealed several patterns of fluorescence on infected bovine cardial pulmonary artery endothelial (CPA) cells (Figure 81). For example, all three surface-reactive MAbs reacted with single, intracellular or extracellular merozoites and with merozoites in mature schizonts, but did not react with schizonts in intermediate stages of development. Thus, the antigens identified by these three MAbs were expressed only by mature merozoites and not by intermediate schizonts. Results from the IFA indicated that intracellular merozoites shed the antigens into the host cell cytoplasm (Figure 81C). These antigens were then expressed on the surfaces of the host cell as well as on adjacent noninfected CPA cells (Figure 81D). Such observations indicate that active processing of antigen by neighboring CPA cells may occur after indirect acquisition of parasite antigens from adjacent infected CPA cells. Similar antigen processing and presentation may also occur in endothelial cells in infected animals.

B. Humoral Responses

Cattle,[248,252,260,283,299,309,450] pigs,[528,530,531,576,679,681,682] sheep,[299,516,533b] and mice[1,75,80,377,437a,575,662,663]

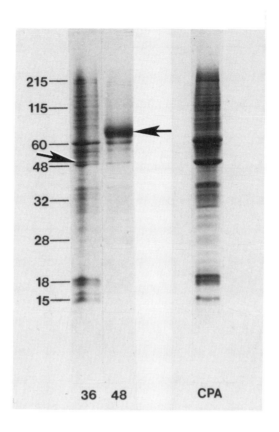

FIGURE 78. Sodium dodecyl sulfate-polyacrylamide gel (12.5%) after electrophoretic separation of proteins of solubilized merozoites of *S. cruzi* obtained from cultured bovine pulmonary artery endothelial cells (CPA) at 36 and 48 DPI of sporozoites. Approximate molecular weights (×10³) are shown at the left of the figure. The two merozoite populations share numerous proteins, but each also has one unique major protein (single arrows), a 50- and 70-kDa protein for 36- and 48-d merozoites, respectively. Proteins of CPA cells are shown at right side of figure. (From Speer, C. A., Whitmire, W. M., Reduker, D. W., and Dubey, J. P., *J. Parasitol.*, 72, 677, 1986. With permission.)

inoculated with *Sarcocystis* sporocysts developed IgG antibodies starting 3 to 5 weeks after inoculation with respective species of *Sarcocystis*. These antibodies were detectable by indirect hemagglutination (IHA),[80,450] enzyme-linked immunosorbent assay (ELISA) or dot-ELISA,[299,530,662,663] indirect fluorescent antibody (IFA),[75,662] or complement fixation (CF)[516] tests. The IgM antibodies appeared earlier than IgG antibodies, but were short lived, usually declining to low levels by the time sarcocysts matured. The IgG antibody concentration in serum peaked during the early period of sarcocyst formation and persisted at a relatively high concentration during the chronic infection. There was no anamnestic antibody response in calves orally challenged with sporocysts 3 to 5 months after primary infection.[299] IgA and IgG_2 antibodies were not found.[299]

The onset and persistence of *Sarcocystis* antibodies varied with the species of the host, species of parasite, source of antigen, and serological test. Although *Sarcocystis* species share antigens, the antibody titers were higher using antigen from homologous species of *Sarcocystis* than from heterologous species.[299] For example, titers in pigs inoculated with *S. miescheriana* were higher when the antigen was derived from *S. miescheriana* than from *S. muris* of the mouse or *S. gigantea* of sheep.[528]

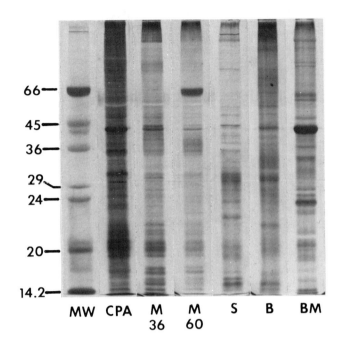

FIGURE 79. Sodium dodecyl sulfate-polyacrylamide gel after electrophoretic separation of proteins of solubilized merozoites (M; obtained at 36 and 60 DPI of CPA cell cultures with sporozoites), sporozoites (S), and bradyzoites (B) of *S. cruzi*, and of solubilized CPA cells and bovine skeletal muscle (BM). Approximate molecular weight standards are shown at the left.

C. Cellular Responses and Immunosuppression

As might be expected of an intracellular parasite, immune cells are mobilized during the *Sarcocystis* infection. The predominant cells infiltrating visceral and muscular tissue during *Sarcocystis* infection are lymphocytes and macrophages.[165,391] This mononuclear cell infiltration begins during the third week of infection and may last for several months, long after the parasite is no longer demonstrable in visceral tissues. Not only visceral tissues are affected, but lymphocytes from peripheral circulation also show a blastogenic response when stimulated with antigen-specific *Sarcocystis*. In calves and sheep inoculated with *Sarcocystis*, this blastogenic response was evident 2 weeks after infection, coincident with the release of first-generation merozoites in cattle and sheep.[299]

Whether these cellular events participate in the recovery of the host from sarcocystosis has not been established, and passive transfer of resistance via cells or antibodies has not been reported. In certain animals, sarcocystosis may depress immunity.[285] Goats with subclinical infection (1000 sporocysts) were more susceptible to intestinal coccidial infections than controls.[169]

The intense cellular response seen in immune animals that survive lethal challenge indicates cell-mediated immunity against *Sarcocystis*, (Figures 73 and 74). Cytotoxic antibodies or metabolites are known to destroy second-generation extracellular merozoites[635] (Figure 82).

D. Protective Immunity and Vaccination

Cattle, sheep, goats, and pigs inoculated orally with a dose of sporocysts that resulted in subclinical infection were protected against a challenge dose that normally would have been lethal. This protective immunity persisted for 80 d, but not 120 d in pigs,[681] at least 252 d in cattle,[263] 274 d in goats,[169] and at least 90 d in sheep.[276] The size of the immunizing dose is

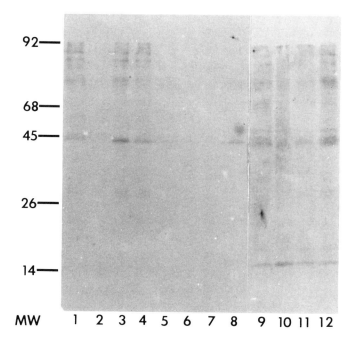

FIGURE 80. Western blot analysis of *S. cruzi* merozoite antigens with monoclonal antibodies. Lanes 1 to 8 were reacted with MAb 31.1B1.1; lanes 9 to 12 with MAb 15.5A5.1. Lanes 1 to 8 contained merozoites that were obtained from CPA cultures at 31, 33, 36, 38, 40, 41, 48, and 58 DPI of sporozoites, respectively; merozoites in lanes 9 to 12 were obtained at 40, 41, 48, and 58 DPI, respectively. In 31-d merozoites (lane 1), MAb 31.1B1.1 reacted with antigens with approximate molecular weights of 88 to 92, 85, 80, 76, 74, and 43 kDa; 33-d merozoites did not react (lane 2); reactions with 36- and 38-d merozoites (lanes 3 and 4) were similar to that with 31-d merozoites; only the 43-kDa antigen reacted in 40- and 58-d merozoites (lanes 5 and 8); 41- and 48-d merozoites did not react. MAb 15.5A5.1 reacted with the following molecular weight antigens in 40- and 58-d merozoites (lanes 9 and 12): 88 to 92, several between 64 to 72, 50, 45, 43, 40, 30, and 16 kDa; in 41-d merozoites (lane 10), 88 to 92, 80, 64 to 72, 50, 45, 43, 41.5, 20, and 16 kDa; in 48-d merozoites (lane 11), 88 to 92, 80, 43, 20, and 16 kDa.

important. Goats and pigs dosed with 100 or more sporocysts of *S. capracanis* and *S. miescheriana*, respectively, were protected; those dosed with 10 sporocysts were not protected. The degree of protection was better with 1000 sporocysts than with 100 sporocysts.[155a,697] Immunized animals had only mild illness after a lethal challenge, even though parasites from the challenge inocula underwent some development in the host.[155a,697] The protective immunity was induced only by homologous *Sarcocystis* species because cattle experimentally infected with *S. hirsuta* were not protected against challenge infection with *S. cruzi*.[263] The interruption of the second-generation schizogony by chemoprophylaxis did not affect the outcome of protective immunity. Cattle inoculated with lethal doses of sporocysts and treated with amprolium from day 21 to 35, survived acute sarcocystosis and developed protection against lethal challenge.[263]

The stage of the parasite responsible for protective immunity is not known, but is likely to be either the sporozoite or first-generation schizonts. Vaccination of pigs with live, killed, or fractions of *S. miescheriana* bradyzoites induced antibodies, but provided no protection.[531] Similar results were obtained in mice immunized with killed *S. dispersa* bradyzoites. Inoculation with *S. muris* sporocysts irradiated with 5 or 10 krad induced protective immunity in mice.[437a] Irradiated vaccines have not been tried in large animals.

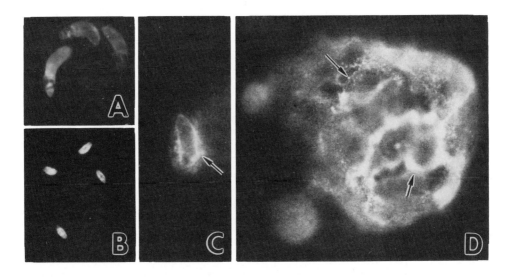

FIGURE 81. Photomicrographs of immunofluorescent antibody reactions of anti-*S. cruzi* monoclonal antibodies (MAb) with sporozoites, merozoites, and infected bovine pulmonary artery endothelial (CPA) cells. (A) Reaction of MAb 15.5A1.3 with acetone-fixed sporozoites. (Magnification × 1750.) (From Speer, C. A. and Burgess, D. E., *Parasitol. Today*, 4, 46, 1988. With permission.) (B) Reaction of MAb 31.1B1.1 with acetone-fixed merozoites. (Magnification × 500.) (From Burgess, D. E., Speer, C. A., and Reduker, D. W., *J. Parasitol.*, 74, 1988, in press. With permission.) (C) Intracellular merozoite binding MAb 15.5A1.3; note fluorescence in cytoplasm of host cell (arrow). (Magnification × 2000.) (From Speer, C. A. and Burgess, D. E., *Parasitol. Today*, 4, 46, 1988. With permission.) (D) Reaction of MAb 31.1B1.1 on surfaces (arrows) of uninfected CPA cells adhering to infected cells. (Magnification × 800.) (From Speer, C. A. and Burgess, D. E., *Parasitol. Today*, 4, 46, 1988. With permission.)

The administration of colostrum has no protective response in cattle and sheep, as these animals remain susceptible to clinical sarcocystosis from birth and throughout their life, despite repeated natural infection.[248,516]

The persistence of live sarcocysts is probably not essential for the maintenance of protective immunity. Goats were protected from challenge infection even when sarcocysts were no longer demonstrable or were markedly reduced in number.[169]

VIII. DIAGNOSIS

Diagnosis of acute sarcocystosis is difficult. The disease is generalized in nature with no specific signs; finding parasites in tissues of acutely infected cattle is not predictable. There is no commercially available serologic test, and virtually all cattle, regardless of clinical involvement, have some sarcocysts in muscles. A diagnosis of bovine sarcocystosis is based on the elimination of other causative agents, a good epidemiologic evaluation of the herd and its relationship to other animals (especially dogs), and clinical findings.

A presumptive diagnosis of acute sarcocystosis can be made if anemia, anorexia, fever, excessive salivation, abortion, loss of body hair (especially at the tip of the tail), increased levels of LDH, SBDH, CPK, BUN, and bilirubin, or lowered PCV are present. Finding *Sarcocystis* antibody or antigen in serum can aid diagnosis of acute sarcocystosis. There is a parasitemia during acute sarcocystosis. Although merozoites were found in buffy coat smears of cattle with acute sarcocystosis from 16 to 46 DPI,[161] this method is tedious, time consuming, and not practical for routine diagnosis. An ELISA to detect circulating antigen in mice and pigs has been reported.[528] This method may eventually prove useful in diagnosis of acute sarcocytosis in naturally infected animals. The detection of humoral antibodies can aid diagnosis, even though

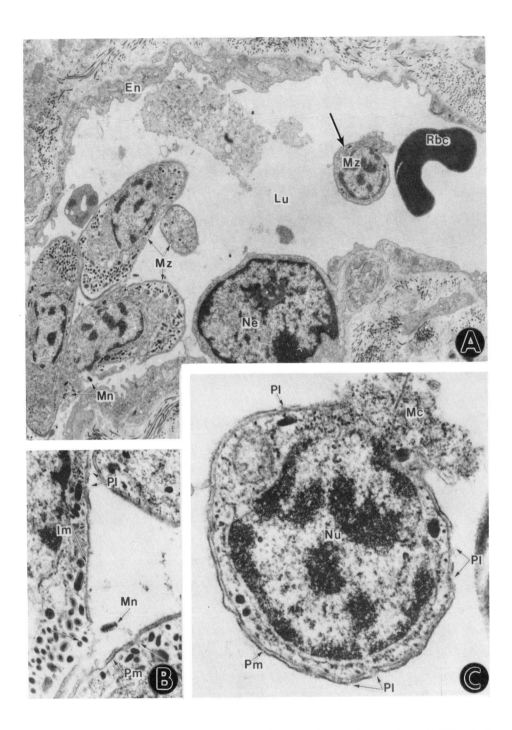

FIGURE 82. TEM of *in vivo* lysis of *S. cruzi* merozoites in venule of kidney of an experimentally infected calf. (A) Five merozoites (Mz) are present in the vascular lumen (Lu). Cytoplasm is extruding from one (arrow) merozoite lying next to a red blood cell (Rbc). (Magnification ×9225.) (B) Higher magnification of the merozoites in left corner of A, showing partial lysis of double pellicular membranes (Pm). Micronemes (Mn) are being released (arrowhead) from a pellicular lesion (Pl). (Magnification × 24,750.) (C) Higher magnification of a merozoite in A, showing cytoplasm (Mc) escaping from a lysed merozoite. (Magnification × 24,750.) En = endothelial cell, Ne = nucleus of endothelial cell, Nu = nucleus of merozoite. (From Speer, C. A. and Dubey, J. P., *J. Parasitol.*, 67, 961, 1981. With permission.)

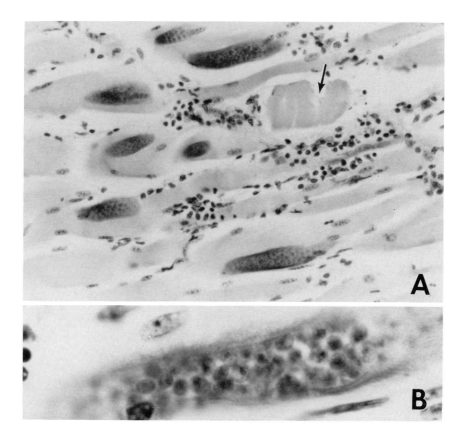

FIGURE 83. (A) Myocardial cell degeneration (arrow), infiltration by mononuclear cells, and seven immature *S. cruzi* sarcocysts in heart of a naturally infected steer. (H. & E.; magnification × 350.) (B) Higher magnification of a sarcocyst to show spheroid metrocytes. (H. & E.; magnification × 1000.) (From Giles, R. C., Tramontin, R., Kadel, W. L., Whitaker, K., Miksch, D., Bryant, D. W., and Fayer, R., *J. Am. Vet. Med. Assoc.*, 176, 543, 1980. With permission.)

titers sometimes do not correlate with the clinical state. Comparison of antibody titers in acutely ill animals with those in animals not showing clinical signs may be useful.[283,309] The more animals tested within an affected herd, the greater the ability to interpret trends in serologic data.

Sarcocystis antibodies have been detected by IHA, ELISA, dot-ELISA, and IFA tests. At present there is a lack of standardization of these tests. Because antigens are obtained from sarcocysts in the muscles of experimentally infected animals and consist of lysate of the bradyzoites,[450] variations in preparative methods yield antigens varying greatly from one batch to another. Although antigens obtained from merozoites may be more suitable for serologic diagnosis of acute sarcocystosis than antigens from bradyzoites, merozoite antigens have not been utilized for diagnostic purposes. Serologic cross-reactivity among *Sarcocystis* species is inconsistent. Although *S. muris* bradyzoite antigens from mice are more easily obtained than *S. cruzi* bradyzoite antigens from cattle, *S. muris* antigens are not suitable for the diagnosis of acute sarcocystosis in cattle. Antigens from *S. cruzi*, however, have been found suitable to detect antibody against *S. hirsuta* in cattle and *S. tenella* in sheep.[299]

Findings of vascular schizonts in biopsies of muscles and lymph nodes or in additional tissues post mortem may aid diagnosis. However, schizonts are often too few to be found in histologic sections, because they may disappear by the time clinical disease is obvious. An immunoperoxidase method to aid in detecting schizonts in tissues has recently been reported.[632] Finding large numbers of immature or mature sarcocysts at the same stage of development can also aid

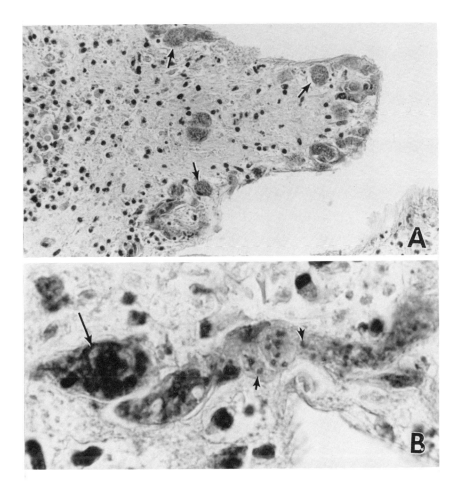

FIGURE 84. Placenta of an aborted cow naturally infected with *S. cruzi*. (A) Several schizonts (arrows) are located in the lamina propria of a villus exhibiting necrosis and infiltration of macrophages. (B) Higher magnification showing deeply stained immature schizont (arrow) and mature schizonts with merozoites (arrowheads). (H. & E.; magnification [A] × 300, [B] × 1000.)

diagnosis (Figure 83). Most clinical outbreaks are likely to occur between 5 and 11 weeks following the initial ingestion of sporocysts. Coupled with the history of feeding raw beef to local dogs (or the consumption of a carcass by dogs or wild carnivores), this time period might be a useful guide. Factored into this guide is the consideration that it takes 2 to 3 weeks for a carnivore to shed sporocysts after ingesting beef.

In histologic sections, *Sarcocystis* must be differentiated from *Toxoplasma gondii* and other closely related coccidians. *Sarcocystis* schizonts develop in endothelium of blood vessels. The immature schizonts are basophilic structures with or without differentiated nuclei (Figure 84). Ultrastructurally, *Sarcocystis* schizonts are free in the host cell cytoplasm, whereas all stages of *T. gondii* are separated from the host cell cytoplasm by a PV and can develop in virtually any cell in the body. Because each *T. gondii* organism divides into two, there is no immature stage in the intermediate host.

The diagnosis of *Sarcocystis*-induced abortion presents additional problems because the parasite is not consistently found in fetal tissues. Sarcocystosis produces gliosis and placental necrosis; therefore, brain and placenta should be examined (Figure 84). Within the placental cotyledons, schizonts have been seen in both the lamina propria and submucosa. The chances of diagnosis are improved if numerous fetal tissues are examined.

IX. ECONOMIC LOSSES

Millions of dollars are lost each year because of the condemnation or downgrading of meat containing grossly visible sarcocysts,[104,264] but economic losses from clinical and subclinical infections are difficult to calculate because (1) nearly 100% of cattle and sheep are infected, thus it is difficult to differentiate affected and nonaffected animals; (2) no dollar values are available for the losses due to poor feed efficiency, failure to grow, reduced milk and wool production, reproductive problems, and obvious clinical disease; and (3) clinical disease is difficult to diagnose and has been recognized in relatively few outbreaks involving cattle in natural conditions. The wide prevalence of pathogenic species of *Sarcocystis* in domestic animals leaves no doubt that clinical disease would be recognized when better diagnostic tests are available.

X. TRANSMISSION

Ingestion of sporocysts and oocysts in food or water is virtually the only major mode of transmission to the intermediate host. Although transplacental infection has been documented in cattle and sheep in nature by finding schizonts in some aborted fetuses, it is rare. It is possible that all transplacentally infected cattle and sheep die because *Sarcocystis* has never been found in newborn calves or in lambs. *Sarcocystis* has been found only once in a 3-day-old foal.[117] The possibility of lactogenic transmission via milk or colostrum was tested experimentally, but evidence of transmission was not found.[259]

The ingestion of mature sarcocysts (i.e., those containing bradyzoites) is the only known means by which the definitive host acquires *Sarcocystis* infection.

XI. EPIDEMIOLOGY

Sarcocystis infection is common in many species of animals worldwide. Virtually 100% of adult cattle in the U.S. are infected with this parasite. What conditions exist that permit such an unusually high prevalence?

1. A host may harbor any of several species of *Sarcocystis*. For example, sheep may become infected with as many as four species, and cattle may have as many as three species of *Sarcocystis*.
2. Many definitive hosts are involved in transmission. For example, cattle sarcocystosis is transmitted via felids, canids, and primates. *S. cruzi*, the most common species of *Sarcocystis* of cattle worldwide, is transmissible via dogs, coyotes, foxes, wolves, and raccoons. Wild carnivores, such as coyotes, may travel many miles for food each day and spread the parasite over long distances.
3. *Sarcocystis* oocysts and sporocysts develop in the lamina propria and are discharged over a period of many months.
4. Sporocysts or oocysts remain viable for many months in the environment. They may be further spread or protected by invertebrates.[459,631]
5. Large numbers of sporocysts may be shed. For example, dogs ingesting a relatively small amount of beef (250 g) shed 100 to 6000 sporocysts per gram of feces.[249] With an average fecal output of 250 to 350 g/d, a dog may shed from 250,000 to over 2 million sporocysts per day. Coyotes and foxes are even more efficient producers of *S. cruzi* sporocysts. As many as 200 million sporocysts were recovered from the intestinal scrapings of a coyote that was fed 1 kg of naturally infected beef.[153] *S. cruzi* is not the only species that produces large numbers of sporocysts. As many as 90 million *S. tenella* sporocysts were recovered from the feces of a dog fed naturally infected mutton.[274]

6. There is little or no immunity to reshedding of sporocysts.[249,349,582] Therefore, each meal of infected meat can initiate a new round of production of sporocysts.
7. Oocysts or sporocysts are resistant to freezing and, thus, can withstand winter on the pasture.[28,65,151,244,363,436] They are also resistant to disinfectants.[22] Apparently, sporocysts can be killed by drying and by 10 min exposure at 56°C.[363]
8. Unlike many other species of coccidia, *Sarcocystis* is passed in feces in the infective form and is not dependent on weather conditions for maturation and infectivity.

Not all species of *Sarcocystis* are highly prevalent. Most species of *Sarcocystis* transmissible via cats have been found less frequently than those transmissible via canids. There may be several reasons for this. One reason may be that cats are very poor producers of *Sarcocystis* sporocysts. Cats fed several grams of macroscopic sarcocysts of *S. muris* and *S. gigantea* containing millions of bradyzoites produced relatively few sporocysts. Another reason may be that sarcocysts of feline-transmitted species, such as *S. gigantea* and *S. medusiformis*, require several months or years to become infective. Such species are not able to complete the cycle in young animals. Still another reason may be that some host species are inherently more susceptible to infection with some agents than are others. For example, the pronghorn is rarely found infected with *Sarcocystis*. Whether breeds of livestock species vary in susceptibility to *Sarcocystis* is not known, but certain strains of mice appear more susceptible to infection than others.[591]

XII. CONTROL

There is no vaccine to protect livestock against clinical sarcocystosis. However, experimental studies indicate that cattle, sheep, goats, and pigs can be immunized by low dosages of live sporocysts. Thus, there is hope of developing a vaccine for sarcocystosis in the future. For the present, prevention is the only practical method of control.

Shedding of *Sarcocystis* in feces of definitive hosts is the key factor in the spread of *Sarcocystis* infection. Therefore, to interrupt this cycle

1. Carnivores should be excluded from animal houses and from feed, water, and bedding for livestock.
2. Uncooked meat or offal should never be fed to carnivores. Freezing can drastically reduce or eliminate infectious sarcocysts in meat.[243,302,331a,645] Exposure to heat at 55°C or higher temperatures for 20 min kills sarcocysts.[102,243,416]
3. Dead livestock should be buried or incinerated. This is particularly important where dead animals are left in the field for carnivores.
4. The prophylactic use of anticoccidials may be another practical method of controlling sarcocystosis in livestock.

XIII. CHEMOPROPHYLAXIS AND CHEMOTHERAPY

Several drugs routinely used as anticoccidials in poultry are effective against sarcocystosis in cattle, sheep, and goats when administered continuously for a month starting at or before inoculation with *Sarcocystis*.[245,365a,434,435] Amprolium (100 mg/kg body weight) administered from 0 to 30 DPI, prevented acute disease or death in cattle infected with *S. cruzi* and in sheep infected with *S. tenella*; 50 mg/kg body weight of amprolium was less effective in sheep than the larger dose.

Calves treated with 100 mg/kg of body weight of amprolium from 21 to 35 DPI with 100,000 sporocysts had only mild signs of acute sarcocystosis, indicating that second-generation schizonts and any subsequent asexual stages preceding sarcocyst development were affected.[263]

Salinomycin (1 or 2 mg/kg body weight), administered from 1 to 30 DPI with *S. tenella* also prevented deaths in sheep. Halofuginone (0.22 mg/kg body weight) similarly reduced or prevented acute sarcocystosis in goats infected with *S. capracanis* and in sheep infected with *S. tenella*; the drug was administered in feed from day 5 before inoculation and continued up to 36 DPI.[365a,677]

Lasalocid, decoquinate, and monensin, each incorporated in feed at 33 mg/kg of feed, had minimal or no effect on the course of acute sarcocystosis in cattle inoculated with 250,000 *S. cruzi* sporocysts; the medicated food was given from day 7 before inoculation through 80 DPI.[281]

For chemotherapy, only halofuginone has been found effective.[365a,677] Halofuginone (0.67 mg/kg body weight) given once or four times prevented death in goats and sheep acutely ill with sarcocystosis. Once the clinical signs of acute sarcocystosis appeared, the administration of salinomycin, sulfadoxin and trimethoprim, lasalocid, robenidin, or spiramycin did not prevent death in sheep and goats.[365a,435,677]

Although oxytetracycline (30 mg/kg body weight) given intravenously prevented death in two sheep with acute sarcocystosis, this drug may be too expensive.[365a]

Most anticoccidials affect *Sarcocystis* schizonts. Results with *S. muris* infections in mice provide further insight. Of the four phases of infections examined, sporozoite migration (2 to 10 DPI), schizogony (11 to 17 DPI), merozoite migration (18 to 27 DPI), and sarcocyst formation (28 to 50 DPI), anticoccidials were most effective against schizogony.[591] Of the 12 anticoccicidials tested in the *S. muris* model, including all four phases of *S. muris* infection, amprolium, monensin, arprinocid, and sulfaquinoxaline plus diaveridine had no antisarcocysticidal activity, and lasalocid, halofuginone, sulfadoxine plus trimethoprim, and sulfadimethoxine had only marginal anti-*S. muris* activity, whereas Zoalene®, primaquine diphosphate, sulfaquinoxalin plus pyrimethamine, and Bay® G 7183 had excellent anti-*S. muris* activity. Of particular interest is the activity of sulfaquinoxalin plus pyrimethamine against immature and mature sarcocysts because no other drugs are known to be effective against sarcocysts.[591,592] Results with the murine *S. muris* model also indicate that results with one species of *Sarcocystis* may not be directly applicable to other species of *Sarcocystis*. For example, amprolium was effective against *S. cruzi*, but not against *S. muris*.

Chapter 2

TECHNIQUES

I. EXPERIMENTAL INFECTION OF INTERMEDIATE AND DEFINITIVE HOSTS

Intermediate hosts are infected by feeding sporocysts. For this, sporocysts can be mixed in grain or dispersed in a small volume at the back of the tongue to avoid spillage. The age of the animal does not appear to affect infectivity; 1-d-old calves, goats, or lambs, as well as adults, are easily infected with *Sarcocystis*. Inoculated animals should be kept in isolation for 10 d, and all the waste and bedding should be incinerated to kill sporocysts that might pass through the gut unexcysted and be shed in the feces.[55,155a,523c] Sporocysts passively passed in the feces remain infectious for other animals.

The definitive hosts are infected by feeding intermediate host tissue containing mature sarcocysts; immature sarcocysts containing only metrocytes are not infectious. Time to maturity varies with each species of *Sarcocystis*. For example, in sheep, *S. tenella* and *S. arieticanis* become infectious within 3 months, whereas *S. gigantea* sarcocysts do not mature for more than 1 year or longer. The numbers of sarcocysts of some species (e.g., *S. miescheriana*, *S. capracanis*, *S. muris*) are reduced dramatically after 3 months postinoculation.

II. ISOLATION, PURIFICATION, AND PRESERVATION

A. Bradyzoites

Bradyzoites can be released from sarcocysts by incubation in digestive fluid (trypsin or acid pepsin). The sarcocyst wall is dissolved by digestive fluid, but the released bradyzoites survive in it.[383a] The following method[235,450] is useful for obtaining *S. cruzi* bradyzoites:

1. Select heavily infected tissue by microscopic examination. The number of sarcocysts in different tissues may vary among animals; heart and tongue are the most heavily infected organs.
2. Remove as much connective tissue and fat as possible and grind muscle in a meat grinder.
3. Suspend about 50 g of ground meat in 100 ml of warm (37°C) digestion fluid (pepsin, 2.5 g; concentrated HCl, 10 ml; water to make 1000 ml; dissolve pepsin in water, centrifuge to remove undissolved pepsin), and incubate at 37°C on a magnetic stirrer for approximately 10 min. The objective is to release bradyzoites from sarcocysts that were broken during grinding, and this method is best suited for heavily infected tissues. Lightly infected tissues may have to be digested longer in acid-pepsin solution, and in that case, the digestion fluid should be isotonic.[235]
4. Pour the homogenate through cheesecloth to remove large particles and centrifuge at 400 × g for 5 to 10 min using a floating-head centrifuge and 50-ml tubes.
5. Discard the supernatant fluid and suspend the sediment in Hanks' balanced salt solution (HBSS) at pH 7.4, preferably at 5 to 10°C.
6. Recentrifuge at 400 × g for 5 to 10 min and discard the supernatant. The pellet has two layers, the whitish bradyzoite layer at the bottom and brownish host tissue layer on the top. Remove host tissue by gently aspirating it from the whitish layer; this separation is not always possible.
7. To further clean zoites of host tissue, mix concentrated zoite suspension with isotonic Percoll® solution, one part of zoite suspension with two parts of Percoll®. To prepare isotonic Percoll®, mix Percoll® with 9% NaCl solution (nine parts Percoll® + one part NaCl).

8. Centrifuge Percoll®-zoite suspension at $400 \times g$ or more for 10 min. The bradyzoites settle at the bottom and host tissue floats in Percoll®.
9. Pour off Percoll® and wash bradyzoite pellet three times in HBSS by centrifugation.
10. Bradyzoites recovered by the above procedure remain viable.

B. Sporocysts

As previously stated, *Sarcocystis* sporocysts, after sporulating in the intestinal lamina propria, are normally released into the intestinal lumen over several months. To optimize the recovery of sporocysts,

1. Euthanatize the definitive host 3 to 7 d after oocysts or sporocysts are first passed in the feces.
2. Remove the small intestine, cut it lengthwise, and spread over paper towels with the luminal side up.
3. Lightly scrape the intestinal epithelium with a glass slide, removing only the tips of the villi, where most sporocysts are concentrated.
4. Suspend scrapings in water and then homogenize in a blender for about 2 min at top speed.
5. Centrifuge the homogenate at $400 \times g$ for 10 min.
6. Discard supernatant and suspend the sediment in water. Repeat the above process until most sporocysts are released from the host tissue.
7. Concentrate sporocysts by centrifugation, filter through a 25-µm stainless steel sieve, suspend in HBSS, and centrifuge.
8. Suspend the pellets in HBSS-antibiotic mixture (given below).
9. Although most sporocysts are concentrated towards the tips of the villus, some are located deep in the villus. To obtain these sporocysts, deeply scrape intestine and homogenize intestinal scrapings in water.
10. Emulsify homogenate in 5.25% sodium hypochlorite solution (1:1 ratio) in a cold bath for 30 min.
11. Centrifuge at $400 \times g$ for 10 min and discard supernatant.
12. Suspend the sediment in water, wash by centrifugation until the smell of chlorine is gone.
13. Suspend sporocysts in HBSS antibiotic mixture (penicillin, 10,000 U; streptomycin, 10 mg; Fungizone®, 0.05 mg; and Mycostatin®, 500 U/ml [mixture = PSFM] of HBSS PSFM mixture).
14. Sporocysts stored in PSFM solution at 4°C remain viable for 12 months or more. (One liter of antibiotic solution can be prepared and stored at –20°C in 100-ml aliquots. Each aliquot is thawed as needed).
15. Sporocysts can also be obtained from feces of definitive hosts by flotation methods used for other coccidia.

III. DIAGNOSTIC TECHNIQUES

A. Examination of Feces for Sporocysts

The number of sporocysts in feces is usually low; therefore, concentration methods are often necessary to detect *Sarcocystis* in feces. Either sugar or salt (NaCl or $ZnSO_4$) solutions of specific gravity 1.15 or more can be used to float sporocysts free of fecal debris. Sugar solution is less deleterious to sporocysts than salt solutions and has been used with the following method:

1. Mix 5 to 10 g of feces thoroughly in 50 to 100 ml of Sheather's sugar solution (sugar, 500 g; water, 320 ml; phenol, 6.5 g); to obtain a homogenous suspension, the feces should be mixed well in about 5 to 10 ml of water before adding sugar solution.
2. Filter through cheesecloth.
3. Centrifuge feces in 50-ml centrifuge tubes with a cap at about $400 \times g$ for 10 min.

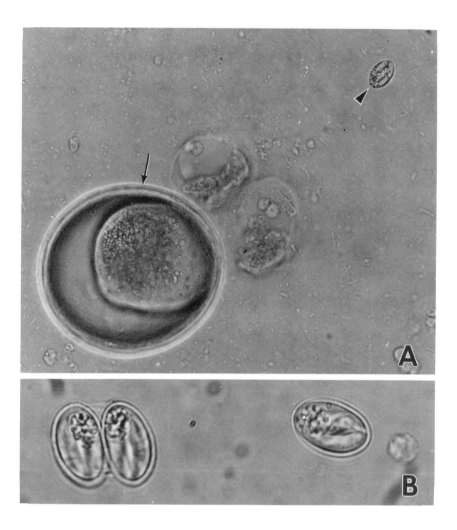

FIGURE 1. Float of a dog feces showing (A) a *Sarcocystis* sporocyst (arrowhead) and a roundworm egg, *Toxascaris leonina* (arrow); and (B) sporocyst on the right and an oocyst on the left. (Magnification × 500.) (From Streitel, R. H. and Dubey, J. P., *J. Am. Vet. Med. Assoc.*, 168, 423, 1976. With permission.)

4. With a pipette, remove several small drops from the very top, put drops on a glass slide and cover the drop with a coverslip.
5. Let the slide-coverslip lie flat for about 5 to 10 min so that the fecal particles can settle and the sporocysts can rise to the top before the slide is examined.
6. Examine at magnification of ×100 or more. *Sarcocystis* sporocysts are small (usually 15 × 10 µm), thin-walled, and of lower density than most other parasite eggs or oocysts; therefore, they lie just beneath the coverslip and are often missed (Figure 1).
7. To collect sporocysts from feces, aspirate the top 5 ml of the solution from the 50-ml centrifuge tube. Mix the aspirate with 45 ml of water, and centrifuge at $400 \times g$ for 10 min. Discard the supernatant fluid and resuspend the sediment in water, centrifuge, and repeat the process. Suspend the final sediment in PSFM solution.

B. Examination of Muscles for Sarcocysts

Sarcocysts can be detected in muscles by gross inspection, examination of unstained squash preparations, histologic examination, or by a digestion method. Each of these methods has its

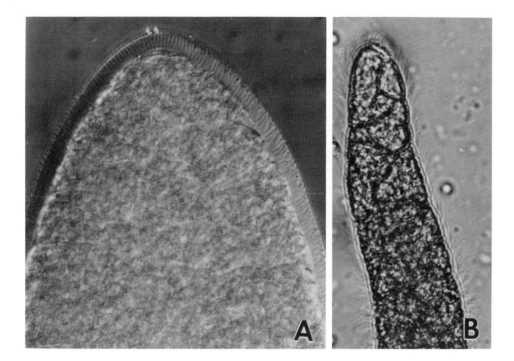

FIGURE 2. *S. tenella* (A) and *S. arieticanis* (B) in unstained wet preparations. Note wall thickness and villar protrusions. (Magnification × 1000.)

advantages and disadvantages. Only a few species of *Sarcocystis* form macroscopic sarcocysts. Examination of unstained squashes is useful to study sarcocyst structure (Figure 2); by this method one can examine relatively more tissue than by histologic examination. However, this method is unsuitable for fixed tissue. Histologic examination is necessary to study sarcocyst morphology, but this method is the least sensitive because only a small amount of tissue can be sampled.

Digestion of host tissue is the most sensitive method to detect light *Sarcocystis* infection. In the digestion procedure, host tissue (50 g) is incubated in ten volumes of either 1% trypsin solution at pH 7.4 or HCl-pepsin solution for 1 to 4 h.[51,101,191,383a,672a] The digest is examined for bradyzoites. By this method one may detect even a few sarcocysts in 50 g of tissue because several hundreds or thousands of bradyzoites are released from sarcocysts as the host tissue and sarcocysts are digested. Disadvantages of this method are (1) it is not possible to distinguish species of *Sarcocystis* because the sarcocyst wall is dissolved; (2) metrocytes are not resistant to digestion; therefore, immature sarcocysts may not be detected; and (3) it may not be possible to detect sarcocysts with relatively small bradyzoites (5 to 6 µm).

The following method is used to isolate individual sarcocysts from infected pigs (Heydorn, A. O., personal communication).

1. Remove fat and connective tissue from 100 g of heavily infected swine muscles (102 to 110 days postinoculation [DPI] after experimental infection).
2. Grind it in a meat grinder and suspend ground meat in 1 l of digestion fluid (0.3% trypsin [Merck], 0.05% ethylenediaminetetraacetic acid [EDTA] in phosphate buffer saline [PBS], pH 7.4 to 7.6).
3. Stir gently at room temperature with magnetic stirrer for 45 to 60 min.
4. Sieve through a coarse metallic sieve (or tea strainer) and save the filtrate.
5. To stop digestion, add 1 l of PBS with 15 ml of fetal calf serum.

6. Sediment sample in a tall glass cylinder for 15 to 20 min. Decant, and save 100 ml of the sediment. Refill with PBS, repeating sedimentation and decanting three to four times, using the PBS as the diluent.
7. Examine the sediment for intact and broken sarcocysts.

C. Serological Techniques

1. Preparation of Soluble Antigen from Bradyzoites

In IHA and ELISA tests, soluble antigen is prepared by the following method.[299]

1. Obtain bradyzoites as described and then freeze and thaw the suspension three times or more.
2. Suspend the bradyzoite suspension in Dulbecco's phosphate-buffered saline (DPBS).
3. Homogenize the bradyzoite suspension in a homogenizer (Polytron®, Brinkmann Instruments, Westbury, NY) at low speed until the zoites are disrupted. Let the bradyzoites stand at 4°C for 1 to 2 h to leach out soluble material.
4. Centrifuge at 16,000 to 20,000 × g for 30 min. Collect the supernatant fluid.
5. Dialyze the supernatant fluid overnight at 4°C against DPBS without added calcium and magnesium, using a dialysis membrane with a 6000- to 8000-M_r cut off.
6. Filter sterilize the dialysate through a 0.2-µm filter unit.
7. Estimate the protein concentration using a Bio-Rad® Coomassie blue dye binding assay or other method.
8. Store the antigen in aliquots at –70°C.

2. Method for ELISA Test

1. For antibody detection, coat vinyl 96-well assay plates overnight with 1 µg of *S. cruzi* antigen in 0.1 *M* carbonate buffer (pH 9.6).
2. The following day, coat the wells with 1% gelatin at 37°C for 1 h.
3. Rinse the plates three times with DPBS containing 0.05% Tween® 20.
4. Wash the plates, add 100 µl of diluted test serum, incubate for 1 h at 37°C. Rinse three times as in Step 3.
5. Add 100 µl of diluted rabbit anti-bovine IgM or IgG. Incubate for 1 h at 37°C. Rinse three times as in Step 3.
6. Add 100 µl (or optimal concentration) of goat anti-rabbit IgG, conjugated to alkaline phosphatase. Incubate at 37°C for 1 h. Rinse three times as in Step 3.
7. Add 100 µl of phosphate substrate at a concentration of 1 mg/ml in 0.1 *M* carbonate buffer (pH 9.6). Incubate for 1 h.
8. Read absorbance on an ELISA reader with a 405-nm absorbance filter.

IV. *IN VITRO* CULTIVATION

The first stages to be grown *in vitro* were sexual stages.[25,236,238,494] Only a few species of *Sarcocystis* have been grown *in vitro*. More recently, an *in vitro* system was developed to grow large numbers of merozoites of several species of *Sarcocystis* that infect livestock.[638-640] The development of merozoites to sarcocysts in cultured cells has not yet been reported.

Two factors determine the success of *in vitro* growth experiments. First, a large supply of sterile-stage-specific parasites (i.e., sporozoites, merozoites, or bradyzoites) must be readily available. Second, the selection of the cell line to be tested for its ability to support growth of the parasite should be based on the cell or tissue in which the parasite develops *in vivo*. For example, if schizonts develop *in vivo* in mouse hepatocytes, then a mouse epithelioid liver cell line should be tested. Because other coccidia have been shown to readily develop in some specific cell types, such as kidney cells, even though they do not normally develop in such cells *in vivo*,[637a] the literature on *in vitro* cultivation of coccidia should serve as a guide for possible cell types to test for *in vitro* growth of *Sarcocystis* spp.

A. Excystation and Cultivation of Schizonts

One of the most important preliminary steps in achieving *in vitro* development of sporozoites to merozoites is to obtain a relatively high rate of sporozoite excystation from sporocysts. After the sporocysts have been isolated from the definitive host, purified, and concentrated, they can be stored until used at 4°C in HBSS with antibiotics.[433] Sporocysts should not be stored in aqueous $K_2Cr_2O_7$, which is routinely used to store coccidian oocysts, because it has a deleterious effect on excystation.[68,433] The following method has proven useful for excysting sporozoites from sporocysts of *S. cruzi*, *S. capracanis*, and *S. tenella*.

1. Treat the sporocysts with 2.6% (v/v) NaOCl, at 4°C for 30 min (i.e., Clorox®, Purex®) to kill other microorganisms that would be certain to contaminate the *in vitro* cultures. A significantly greater excystation rate can be obtained by treating the sporocysts of *S. capracanis*, *S. cruzi*, and *S. tenella* with 2.6% NaOCl than with 1.3% or 5.3% NaOCl.[68] Treatment of the sporocysts with cysteine HCl for 18 h may also improve the rate of excystation.[433] Cysteine HCl and NaOCl evidently have similar beneficial effects on excystation, but NaOCl acts more rapidly (30 min vs. 18 h) and it must be used anyway to disinfect the sporocysts.
2. Wash the sporocysts several times (usually at least four times) by centrifugation and resuspension in cold saline or culture medium (i.e., RPMI 1640 without serum) until the odor of NaOCl can no longer be detected or until the pH color indicator in the culture medium remains unchanged in the presence of sporocysts.
3. Suspend the sporocysts in excysting fluid and incubate at 37°C for 1.5 to 4 h. The excysting fluid should consist of a balanced salt solution (i.e., HBSS, pH 7.2 to 7.4), containing 2% (w/v) trypsin or chymotrypsin (activity of 1:250 or 1:300) plus 5 to 10% bile or 2.5 to 5% (w/v) bile salt (i.e., sodium taurocholate obtained from the appropriate intermediate host such as bovine, ovine, etc.).
4. Remove the excysting fluid by washing the excysted sporozoites at least twice in culture medium without serum, and resuspend in 5 to 10 ml culture medium.
5. To remove debris and sporocyst walls, the sporozoites may be passed through a nylon wool column.[428a] Too much debris and sporocyst walls in the inoculum may have a deleterious effect on parasite development.[637a]
6. Count the number of sporozoites with a hemacytometer and resuspend them at the appropriate concentration in culture medium containing 5% (v/v) fetal bovine serum, 2 m*M* l-glutamine, 50 µg/ml dihydrostreptomycin, 50 U/ml penicillin G.
7. Inoculate 2 to 5×10^4 sporozoites per square centimeter of the monolayer of the cell line to be tested and incubate at 37 to 39°C in 5% CO_2, 95% air. Sporozoites of *S. cruzi*, *S. capracanis*, and *S. tenella* will develop in bovine pulmonary artery endothelial cells (CPA; American Type Culture Collection, Rockville, MD) or bovine monocytes (BM; see Reference 637b for details concerning the BM). The CPA cell line will produce approximately seven times more merozoites of *S. cruzi* than the BM line,[639] but the BM line is especially useful for investigating immunological parameters such as opsonizing antibodies and lymphokines.[637b]
8. Remove the culture medium at 1 d, and then at 2- to 7-d intervals after sporozoite inoculation, replace it with culture medium containing 1 or 2% fetal bovine serum (FBS).
9. The infected cell cultures might require maintenance for long periods to allow sufficient time for the parasites to undergo one or more generations of schizogony. Long cultivation periods can be achieved by varying the FBS concentration in the culture medium. As the cell cultures become sparsely populated due to parasite development (usually first

observed at 24 to 28 DPI), then treat them with cell culture medium (CM) containing 10 to 20% FBS, which can then be reduced to 1 to 2% FBS once the monolayer has reestablished itself.

10. Parasite development can be monitored and photographed at various intervals after sporozoite inoculation by direct microscopic observation with an inverted phase-contrast microscope or by examining double-coverslip preparations[546d] with a compound phase-contrast microscope.

Schizonts of *S. capracanis*, *S. cruzi*, and *S. tenella* develop similarly *in vitro*, except that *S. cruzi* produces considerably more merozoites per culture vessel than do *S. capracanis* or *S. tenella*.[638,639] *S. cruzi* and *S. tenella* develop to schizonts in BM and CPA cells, whereas *S. capracanis* develops in BM only. Soon after inoculation, sporozoites of all three species penetrate cells (Figure 3A) and gradually transform in approximately 2 to 3 weeks into ovoid or spheroidal schizonts with irregularly shaped nuclei with small nucleoli and small cytoplasmic granules (Figures 3B and C). More advanced stages appear to have numerous nuclei and nucleoli (ultrastructural studies have shown that the nucleus is highly lobulated, but by light microscopy the lobes appear as separate nuclei) (Figures 3D and E). Eventually, these schizonts form large schizonts in which merozoites bud radially from 8 to 12 residual bodies (Figure 3F). Large mature schizonts contain 150 to 350 merozoites (Figure 3G) and may be present at 2.5 to 10 weeks or more after sporozoite inoculation. Merozoites frequently glide rapidly through the host cell cytoplasm, escape through the host cell plasmalemma, and some of these immediately penetrate and glide through the cytoplasm of cells in the vicinity of the original host cell. Within a few minutes after being penetrated by one or more merozoites, these previously uninfected cells retract their cellular processes, become spheroidal, and some detach to float freely in the culture medium. Spheroidal host cells that remain attached transform within a few hours to flattened, epithelioid cells. Some of the merozoites that enter other cells or remain within the cytoplasm of the original host cell develop into small schizonts (Figures 4A to D).

Small schizonts contain 36 to 100 merozoites and may be present at 16 to 138 d or more after sporozoite inoculation, but are most numerous at 30 to 52 d. Several generations of small schizonts occur during this time, but the precise number of generations cannot be determined because schizogony is asynchronous. Merozoites bud radially from four to eight residual bodies (Figures 3A and B), are highly motile, easily exit the host cell, exhibit pivoting and gliding movements, and readily penetrate other cultured cells in the immediate vicinity. Noninfected CPA cells adhere to CPA cells infected with small schizonts to form aggregates of cells, and some of these aggregates reorganize into branching and occasionally anastomosing capillary-like structures which become centers of intense schizont activity and merozoite release. Cells that adhere to infected cells serve as host cells for additional schizogonous generations. Eventually, relatively large, elongated aggregates of cells (Figure 4D) are formed within which nearly all cells are infected with schizonts and merozoites (Figure 5). Production of *S. cruzi* merozoites cycles during the cultivation period, with peak merozoite numbers being present at 36 and 44 DPI and at 44 and 50 DPI for CPA and BM, respectively.[639]

The host cell type *in vivo* for large and small schizonts of *S. cruzi*, *S. capracanis*, and *S. tenella* has been considered, but not proven, to be vascular endothelial cells. The recent success in culturing merozoites *in vitro* indicates that the preferred host cell *in vivo* may indeed be an endothelial cell, since approximately 7 to 12 times more merozoites of *S. cruzi* and *S. tenella* were harvested from CPA than from BM.[638,639] In contrast, *S. capracanis* underwent schizogony in BM, but not CPA cells.[638]

Ultrastructural studies of infected BM and CPA cells revealed that sporozoites, schizonts, and merozoites of *S. cruzi* were located free in the host cell cytoplasm (i.e., parasites were not surrounded by a parasitophorous vacuole)[640,642] (Figures 5 and 6), which substantiates this finding in earlier studies of *in vivo* infections with *S. cruzi* as well as other species of

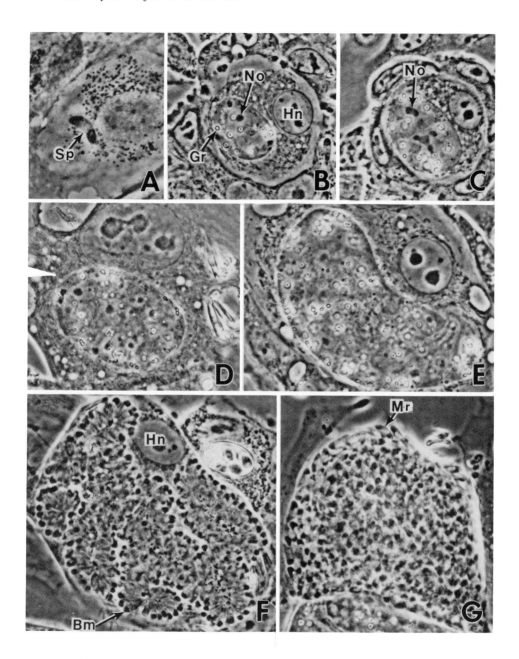

FIGURE 3. Phase-contrast photomicrographs of the development of sporozoites to large schizonts of *S. cruzi* in cultured bovine monocytes (A, F, G) and pulmonary artery endothelial cells (B to E). Day after sporozoite inoculation is given in parentheses. (A) Intracellular sporozoite (24); (B) young schizont (18); (C) young schizont with irregular nucleus and nucleoli (18); (D) small intermediate schizont (23); (E) large intermediate schizont (42); (F) large schizont with budding merozoites (35); (G) mature large schizont (35). (Magnification × 840.) (From Speer, C. A., Whitmire, W. M., Reduker, D. W., and Dubey, J. P., *J. Parasitol.*, 72, 677, 1986. With permission.)

Sarcocystis.[155,634] Certain endothelial cells are known to be part of the reticuloendothelial system and are capable of phagocytosing and digesting ingested matter and microorganisms. If such reticuloendothelial cells serve as host cells for schizonts of certain *Sarcocystis* species, being located free in the host cell cytoplasm may represent a means of escaping a potentially hostile host cell environment.

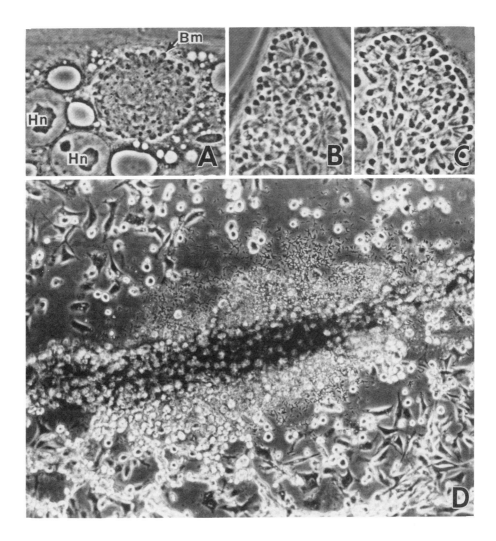

FIGURE 4. Phase-contrast photomicrographs of schizonts and merozoites of *S. cruzi* in cultured bovine pulmonary artery endothelial cells (CPA); day after inoculation of sporozoites given in parentheses. (A) Merozoites in early stage of budding at surface of schizont (42). (Magnification × 840.) (B) Schizont with merozoites budding from four residual bodies (48). (Magnification × 840.) (C) Mature schizont (42). (Magnification × 840.) (D) Large area of CPA culture showing numerous infected cells and extracellular merozoites (76). (Magnification × 50.) (From [A to C] Speer, C. A., Whitmire, W. M., Reduker, D. W., and Dubey, J. P., *J. Parasitol.*, 72, 677, 1986; and [D] Speer, C. A. and Burgess, D. E., *Parasitol. Today*, 4, 46, 1988. With permissions.)

This system for culturing merozoites of *Sarcocystis* species[638-640] will be particularly useful in elucidating the biochemistry, as well as the mechanisms of immunity and pathogenesis, especially since the parasite develops *in vitro*, as well as *in vivo*, in cell types (i.e., endothelial and mononuclear cells) that have been shown to participate in other immunologic and pathologic mechanisms.[557a,632a,632b] With minor modifications (i.e., selecting the appropriate cell line) this system may prove useful for culturing other species of *Sarcocystis* that are known to develop slowly within endothelial or endothelial-like cells in the intermediate host.

B. Cultivation of Gamonts

Fayer[236,238] was the first to describe the *in vitro* development of bradyzoites of *Sarcocystis* to microgamonts, macrogamonts, and oocysts (Figure 7C and D), which was also the first

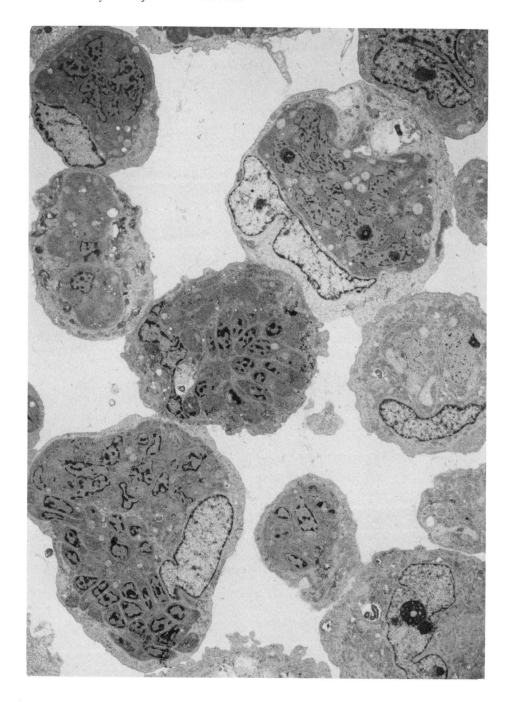

FIGURE 5. TEM of several CPA cells infected with schizonts of *S. cruzi* in various stages of development (78 DPI). (Magnification × 3000.)

description of such stages in the life cycle of *Sarcocystis*. Becker et al.[25] were able to achieve even greater *in vitro* development of this phase of the *Sarcocystis* life cycle, in which the oocysts of four species of *Sarcocystis* underwent sporulation in various mammalian cell lines. Gamonts of several other species have been grown in cell culture.[214,215,219,388b,494]

The ultrastructure of gametogony, oogony, and sporogony has been described for several species of *Sarcocystis* developing in cell cultures.[25,494,673a] The ultrastructural features of

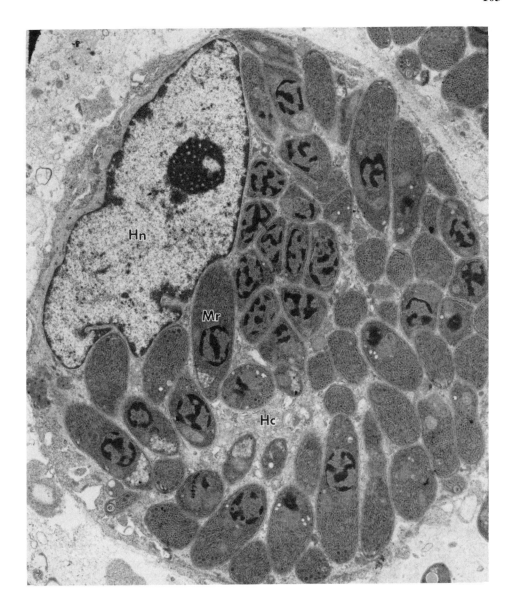

FIGURE 6. TEM of a mature schizont with merozoites of *S. cruzi* in a bovine CPA cell 35 d after sporozoite inoculation. (Magnification × 6200.) (From Speer, C. A. and Burgess, D. E., *Parasitol. Today*, 3, 2, 1987. With permission.)

gamonts and oocysts of five different species of *Sarcocystis* formed in cultured cells were essentially identical to those formed in the intestine of the natural hosts. These studies showed that bradyzoites developed directly into microgamonts and macrogamonts, usually within 18 h after inoculation. Each microgamont had a large, irregularly shaped nucleus which gave rise simultaneously to 20 to 30 microgametes. At 22 h after inoculation, oocysts began to undergo sporulation to form sporocysts.

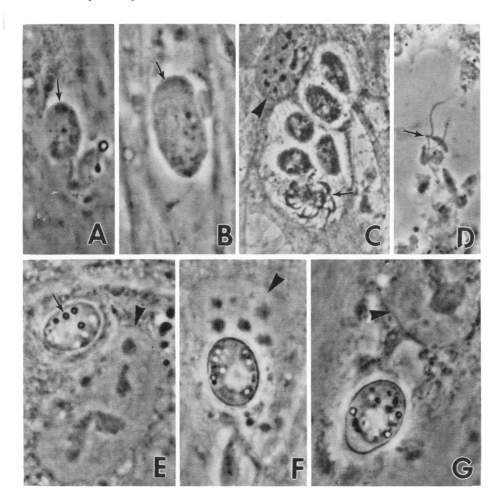

FIGURE 7. Phase-contrast photomicrographs of gamonts and oocysts in cultured cells. (A, B, E, F, G) Gamonts and oocysts of *S. muris* in mouse kidney cells. (C and D) Gamonts of *Sarcocystis* sp. from grackles in embryonic bovine trachea cells. (From Fayer, R., *Science,* 175, 65, 1972. With permission.) Arrowheads point to host cell nuclei. (A) Early microgamont, (B) advanced microgamont, (C) mature microgamont (arrow) and five macrogamonts, (D) microgamete (arrow) with two flagella, (E) macrogamont with four refractile granules (arrow), (F) unsporulated oocyst with oocyst wall (arrow), (G) unsporulated oocyst with slightly condensed sporont. (Magnification × 2000.)

Chapter 3

SARCOCYSTOSIS IN CATTLE (*BOS TAURUS*)

There are three species of *Sarcocystis* (Figure 1) in cattle: *S. cruzi*, *S. hirsuta*, and *S. hominis*, with canids, felids, and primates, respectively, as definitive hosts.

I. *SARCOCYSTIS CRUZI* (HASSELMANN, 1926) WENYON, 1926

Synonym — *S. bovicanis*, Heydorn et al., 1975.
Distribution — Worldwide.
Intermediate hosts — Cattle (*Bos taurus*) and bison (*Bison bison*).
Definitive hosts — Dog (*Canis familiaris*), coyote (*Canis latrans*), red fox (*Vulpes vulpes*), raccoon (*Procyon lotor*), and wolf (*Canis lupus*).

A. Structure and Life Cycle

Sarcocysts are microscopic, <500 μm in length, and are formed in virtually all striated muscles, Purkinje fibers of the heart, and in the CNS.[160] The sarcocyst wall is thin (<1 μm) and its surface is covered with long, narrow, ribbon-like protrusions.[48,160,488,545] The protrusions are up to 3.5 μm long and up to 0.3 μm wide at the base.[293] They taper towards the tip, which is blunt. There are no supporting tubules or microtubules in the villar core; therefore, the villi are often folded (Figure 2A). The sarcocyst wall surface has numerous small invaginations which are hexagonal when viewed from the surface.[293]

The life cycle which was used as an example in Chapter 1, is summarized in Table 1.

B. Pathogenicity

Sarcocystis cruzi is the most pathogenic species found in cattle.[165,239,245,248,260,393,526] It can cause fever, anorexia, anemia, loss of weight, hair loss, weakness, muscle twitching, prostration, abortion, reduced milk yield, hypersalivation, neurologic signs, and death, depending on the isolate and the number of sporocysts ingested. Ingestion of 1000 or fewer sporocysts under experimental conditions causes no clinical signs. Doses of 5 million or more are uniformly fatal. Cattle fed 200,000 sporocysts become clinically ill, some die of acute sarcocystosis, and some survivors do not grow to their full potential.[264] In experimental infections cattle show no clinical signs until the fourth week, irrespective of dose. Beginning as early as 24 d after inoculation they develop persistent fever, anorexia, weight loss, weakness, muscle twitching, prostration, and some die. Pregnant cows may abort, and milking cows may have reduced yield. Cattle that do not recover lapse into chronic sarcocystosis in which there is continued weight loss, hypersalivation, hyperexcitability, and loss of hair, especially on the neck, rump, and tail switch (Figure 64, Chapter 1). Some months after infection, a few animals develop signs of CNS involvement including nystagmus, opisthotonus, and lateral recumbency with a running gait. All cattle that develop CNS disorders eventually die.

Gross and microscopic lesions were described in Chapter 1.

C. Natural Outbreaks

In 1961, a febrile illness was noticed in a dairy herd.[109] Within 8 weeks a total of 25 animals was affected and 17 died. Acutely ill cows had intermittent anorexia, drop in milk yield, diarrhea, transient fever, nasal discharge, hypersalivation, and hemorrhagic vaginitis. Of 17 cows, 10 aborted in the last trimester of pregnancy and only 1 of the 10 survived. Erosions were found on the tongue and buccal mucosa of the salivating cows. Chronic cases were marked by emaciation, pale or icteric mucous membranes, mandibular edema, exophthalmos, cessation of lactation,

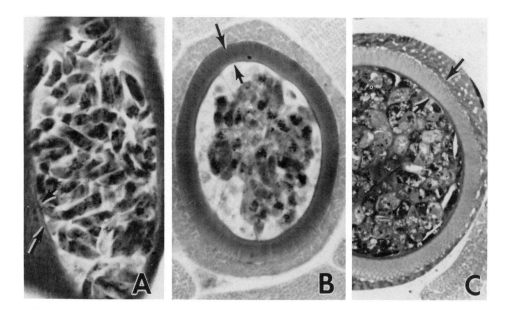

FIGURE 1. Sarcocysts of *S. cruzi* (A), *S. hirsuta* (B), and *S. hominis* (C) in skeletal muscles of cattle. Note outer (large arrow) and inner (small arrow) layers of sarcocyst walls. ([A and C] Toluidene blue, [B] PAS; magnification × 1000.)

and sloughing of the tail switch. Cachectic animals became recumbent, with muscle tremors over the body resembling those of hypocalcemia. Gross and histologic lesions resembled those described for acute experimental sarcocystosis. Numerous schizonts were present in the endothelium of numerous organs of 11 of 16 animals. Because neither the stage nor clinical signs of sarcocystosis had been described, this outbreak was called Dalmeny disease, after the small Canadian town in which it occurred. In retrospect, however, it was an outbreak of acute sarcocystosis.[109]

Subsequently, cases of natural clinical sarcocystosis in cattle have been reported in Canada,[485] England,[89,269,270,424] Ireland,[91] Norway,[427] Australia,[62] and the U.S.[283,284,309,505] Unlike Dalmeny disease, these were infections in yearlings or younger calves. A well-documented outbreak was reported in Kentucky.[309] Approximately 30 of 41 heifers in a 1-ha lot had diarrhea, were losing weight and consuming less feed, and had rough coats. The condition worsened to include severe laminitis, hypersalivation, extreme nervousness, and marked hair loss from the ears, lower limbs, and the tail switch, giving a "rattail" appearance. Within a month, more than half of the animals were severely debilitated and eight died. In the next 5 weeks, two thirds died or were euthanatized, only a third were salvaged for market.

Macroscopic erosions were found in several tissues. The skin of the carpus and tarsus just above the coronary band was denuded and abraded; interdigital erosions were present, and the hoof wall was cracked, necrotic, and about to separate from the coronary band; the cornea had erosions and opacities; erosions and shallow ulcers were found in the muzzle, lips, tongue, hard palate, esophageal mucosa, and omasum; the lungs were congested. Histologic examination revealed intense mononuclear cell infiltrates in the muscles and muscular organs with accompanying myodegeneration and numerous immature sarcocysts (Figure 83, Chapter 1). The *Sarcocystis* IHA antibody titers in affected heifers were fourfold greater than in unaffected cows on the same farm.[309] Hay contaminated with feces from farm dogs, was implicated as the source of infection.[283,309]

The association between sarcocystosis and ulcerations in the mouth and foot seen in naturally infected cattle has not been explained. Neither buccal nor pedal lesions were seen in over 100

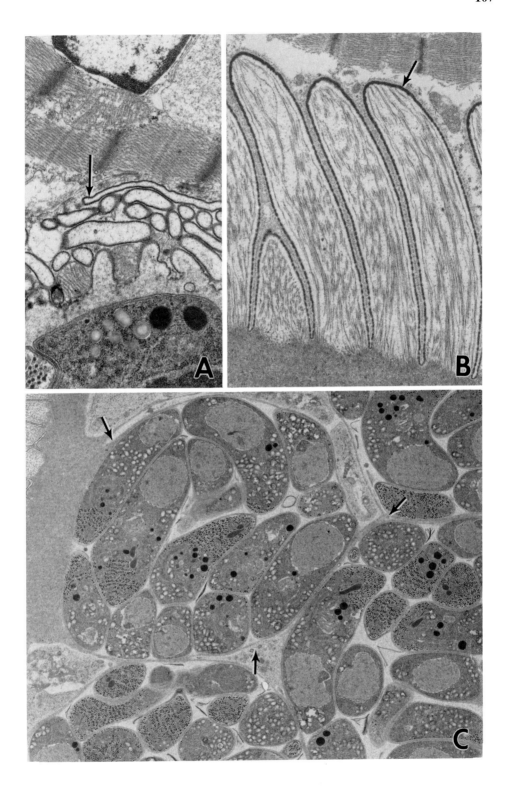

FIGURE 2. TEM of sarcocyst walls of *S. cruzi* (A) and *S. hominis* (B) in experimentally infected calves. (A) Filamentous villar protrusions (arrow), without microtubules, folded over the primary sarcocyst wall. (Magnification × 16,000.) (B) Finger-like villar protrusions with microtubules. (Magnification × 12,000.) (C) *S. hominis* bradyzoites arranged in packets separated by septa (arrows). (Magnification × 3900.)

Table 1
COMPARISON OF DEVELOPMENTAL STAGES OF *S. CRUZI* AND *S. HIRSUTA* IN CATTLE

	S. cruzi[a]	*S. hirsuta*[b]
First-generation schizonts		
Location	Arterioles of several organs	Mesenteric and intestinal arterioles
Duration (DPI)[c]	7—26	7—23
Peak development (DPI)	15	15
Size of meronts (μm)	41.0 × 17.5	37.2 × 22.3
No. of merozoites	>100	>100
Size of merozoites (μm)	6.3 × 1.5	5.1 × 1.2
Second-generation schizonts		
Location	Capillaries of several organs	Capillaries of striated muscles, heart
Duration (DPI)	19—46	15—23
Peak development (DPI)	24—28	16
Size of meronts (μm)	19.6 × 11.0	13.9 × 6.5
No. of merozoites	4—37	3—35
Size of merozoites (μm)	7.9 × 1.5	4 × 1.5
Parasitemia		
Duration (DPI)	24—46	11
Intraleukocytic multiplication	Yes	No
Sarcocysts		
1st seen (DPI)	45	30
Maturation time (DPI)	86	75
Wall (μm)	Thin (<1.0)	Thick (3—6)
Location		
CNS	Yes	No
Heart	Yes	Yes

[a] Data from Reference 160.
[b] Data frm Reference 159.
[c] DPI — days postinoculation.

experimentally inoculated cattle. Naturally infected cattle may be more susceptible to secondary infections.

Sporadic cases of bovine abortion and neonatal mortality have been reported in New Zealand,[510,594,674] Australia,[390,479,506,510] Canada,[109] the U.S.,[166,378,420,546c,605] and Cuba[81] in which fetal lesions and or protozoa were found. Fetal encephalitis, myocarditis, and hepatitis were often associated with the outbreaks. Intravascular schizonts resembling those of *Sarcocystis* were found in some of the fetuses in all of the reported outbreaks. Apart from focal placentitis and hypersalivation in a few cows, no clinical abnormalities were identified.

D. Protective Immunity

Cattle are susceptible to clinical sarcocystosis from birth and throughout their life, despite repeated natural infections. However, it has been possible to establish a protective immunity under experimental conditions.[263] Calves infected with 50,000 or 100,000 *S. cruzi* sporocysts were protected from illness and death that would have resulted from challenge infections with large numbers of *S. cruzi* sporocysts given 70 to 252 d later. Calves treated prophylactically with amprolium from 21 to 35 d after infection to reduce second-generation schizonts were also protected.[263] Infection with the nonpathogenic *S. hirsuta* did not protect calves against *S. cruzi* infection.

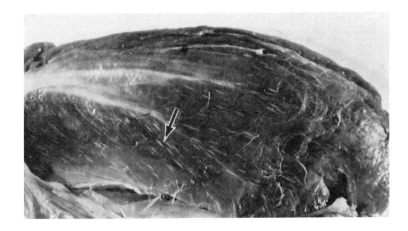

FIGURE 3. Macroscopic sarcocysts (arrow), probably *S. hirsuta*, in beef. (From Collins, G. H., *Monogr. Massey Univ. N. Z.*, 1, 1980. With permission.)

II. *SARCOCYSTIS HIRSUTA* MOULÉ, 1888

Synonym — *S. bovifelis* Heydorn et al., 1975.
Distribution — Probably worldwide.
Definitive host — Cat (*Felis catus*).

A. Structure and Life Cycle

Sarcocysts as large as 8 mm long and 1 mm wide (Figure 3) have been found in naturally infected cattle,[47] and as large as 800 μm long and 80 μm wide in experimentally infected cattle.[159] The sarcocyst wall may be as thick as 7 μm and appears radially striated or hirsute.[159,303a] The villar protrusions are up to 7.0 μm long and 1.5 μm wide and contain numerous microtubules.[47,48,303a] They may be closely packed (0.03 μm apart) or widely spaced (1 μm).[48]

Information on life cycle and structure is summarized in Table 1. There are two known generations of schizogony. First-generation schizonts have been found only in small arteries of the intestine and mesentery, and second-generation schizonts have been found only in capillaries of the heart and skeletal muscles.

B. Pathogenicity

Sarcocystis hirsuta is mildly pathogenic.[167] Calves fed 100,000 or more sporocysts became febrile, had diarrhea, and were mildly anemic; none died. At necropsy, hemorrhagic lesions were not seen.[167]

III. *SARCOCYSTIS HOMINIS* (RAILLIET AND LUCET, 1891) DUBEY, 1976

Synonym — *S. bovihominis* Heydorn et al., 1975.
Distribution — Europe.
Definitive hosts — Humans (*Homo sapiens*), rhesus monkey (*Macaca mulatta*), baboon (*Papio cynocephalus*), and possibly chimpanzee (*Pan troglodytes*).

A. Structure and Life Cycle

The sarcocyst wall is up to 6 μm thick and appears radially striated because of numerous villar protrusions which are up to 7 μm long and 0.7 μm wide.[488a] (Figure 1C). The villar core contains numerous microtubules that run from the base to the apex.[192,488a] Bradyzoites are 7 to 9 μm long and are arranged in packets.[488a]

Table 2
PREVALENCE OF THREE SPECIES OF *SARCOCYSTIS* IN NATURALLY INFECTED CATTLE

Country	Number examined	Percent positive	Percent positives with			Ref.
			S. cruzi	S. hominis	S. hirsuta	
Austria	320	87[a]	70	51	44	374
Austria	1337	94	56	48	42	408
Brazil	168	96	69	4	8	331
Federal Republic of Germany	1020	99.7[b]	66	63	34	37,44
	46	83[a]	30	4	41	613
New Zealand	500	100	98	?[c]	80	47

[a] Mixed infections 5 and 6%, respectively.
[b] By digestion.
[c] ? — unknown.

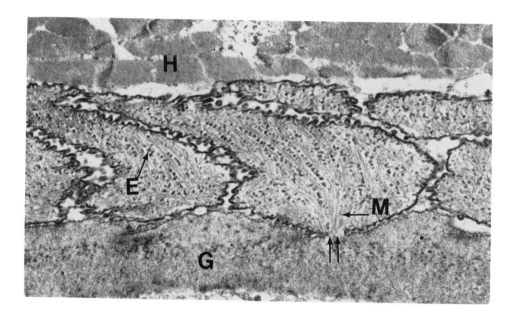

FIGURE 4. TEM of the sarcocyst wall of *S. hirsuta*. The villar protrusions are conical shaped, constricted at the base (double arrow), and contain electron-dense granules (E) and microfilaments (M); (G) granular layer; (H) host cell cytoplasm. Minute undulations of the primary cyst wall cover the entire surface of the villar protrusions. (Magnification × 9900.)

S. hominis is only mildly pathogenic.[192] Calves inoculated with 1 million sporocysts develop mild anemia, but survive.[192]

IV. PREVALENCE OF SARCOCYSTS IN CATTLE

Based on examination of tissues obtained at abattoirs, most cattle in the U.S. and throughout the world are infected with *Sarcocystis*. Recent surveys indicate that *S. cruzi* is the most prevalent species (Table 2), and this species is easily recognized in histologic sections, whereas *S. hirsuta* is difficult to distinguish from *S. hominis* microscopically. However, the sarcocyst

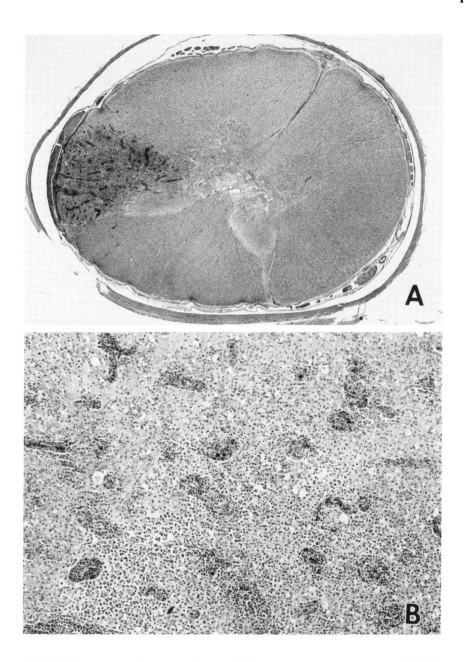

FIGURE 5. Encephalomyelitis due to a *Sarcocystis*-like organism in naturally infected calf. (A) An area of necrosis (arrow) in the spinal cord. (B) High-power magnification of A, showing severe vasculitis and infiltration of macrophages. (H. & E.; magnification × 75.)

walls of *S. hirsuta* and *S. hominis* can be differentiated ultrastructurally. The villar protrusions of the sarcocyst wall of *S. hirsuta* are approximately 8 μm long, constricted at their base, expanded laterally in the midregion, and tapered distally. The villar core contains numerous filaments and rows of electron-dense granules oriented parallel to the longitudinal axis of the protrusion. The sarcocyst wall appears only approximately 4μm thick because the distal region of each protrusion is folded over neighboring protrusions (see Figure 4).The villar protrusions of *S. hominis* measured 5.7 × 0.6 and 7 × 1.4 μm at 111 and 222 DPI, respectively, are cylinder shaped and contain scattered granules and filaments that extend from the base to the tip of the projection.[192]

V. SARCOCYSTOSIS-LIKE ENCEPHALITIS IN CATTLE

Isolated cases of fatal encephalitis were reported from Canada,[189] the Federal Republic of Germany,[661] and the U.S. (Dubey, unpublished). In both published instances, affected cattle developed neural signs before death. The case from Canada was in an 18-month-old steer. The steer became ataxic, recumbent, and blind. Grossly, a 1-cm-diameter, gray discolored area was seen in the cerebellum. Microscopically, the grossly visible lesion consisted of focal malacia, vasculitis, and infiltration of mononuclear cells, mainly macrophages (Figure 5). Numerous immature and mature schizonts and extracellular merozoites were seen; these stages resembled *S. cruzi* in structure. However, such lesions have never been seen in experimentally infected cattle.

BIBLIOGRAPHY

Information pertinent to the subject matter of this chapter may be found in References 13, 14, 18, 37, 46a to 48, 57, 62, 81, 89, 91 to 94, 109, 127, 146, 147, 151, 153, 155, 157, 160, 161, 163, 165 to 168, 171, 178, 189, 192, 199, 200, 239 to 245, 247 to 250, 252 to 254, 256 to 260, 263, 264, 266 to 271, 280, 282, 283 to 286, 293, 295, 299, 302 to 303a, 309, 327, 331, 332, 342, 342b, 349 to 354a, 356, 363, 365, 369, 374, 378, 382 to 383a, 385, 385a, 387a, 387b, 389 to 393, 408, 424, 427, 432, 433, 439, 450, 453, 454, 457, 479, 485, 488 to 490, 504, 506, 510, 526, 544, 545, 546c, 561, 565, 566, 568, 574, 576, 579, 581, 583, 584, 590, 594, 600a, 600b, 605, 613, 620, 624, 625, 627a, 635, 639, 640, 646, 661, 672, 672a, 673, 674, and 679b.

Chapter 4

SARCOCYSTOSIS IN SHEEP (*OVIS ARIES*)

There are four species of *Sarcocystis* in sheep (Figure 1). *Sarcocystis tenella* and *S. arieticanis* are transmitted via canids, whereas *S. gigantea* and *S. medusiformis* are transmitted via cats.

I. *SARCOCYSTIS TENELLA* (RAILLIET, 1886) MOULÉ, 1886

Synonym — *S. ovicanis* Heydorn et al., 1975.
Distribution — Worldwide.
Definitive hosts — Dog (*Canis familiaris*), coyote (*Canis latrans*), and red fox (*Vulpes vulpes*).

A. Structure and Life Cycle

Sarcocysts are microscopic, up to 700 μm long, and are found in striated muscles, including heart and tongue, and in the CNS.[164] The sarcocyst wall is 1 to 3 μm thick and has villar protrusions up to 3.5 μm long and up to 0.5 μm wide (Figure 1A).[164,487] There are no microtubules in the villi, but plaques are present (Figure 2A).

Information on the life cycle is summarized in Table 1. There are two well-documented generations of schizogony in the blood vessels; a possible third generation was reported in only two lambs at 36 days postinoculation (DPI) in macrophages in visceral lymph nodes and Kupffer cells.[529] These were small (7.4 × 5.1 μm) and contained six to nine merozoites. Differences in size, parasitemia, and number of phases of schizogony reported by various authors[164,355,367b,430,431,509,529] may reflect the techniques used, the number of stages measured, or the possibility that *S. arieticanis* might have been present as a contaminant.

Sarcocysts first seen at 35 DPI had a 1-μm-wide sarcocyst wall that appeared nonstriated until over 52 DPI when the wall was up to 2 μm thick and appeared cross-striated. At 75 DPI, sarcocysts were up to 700 μm long and contained both metrocytes and bradyzoites.[164]

B. Pathogenicity

S. tenella can cause anorexia, weight loss, fever, anemia, loss of wool, abortion, premature birth, nervous signs, myositis, and death, depending on the number of sporocysts ingested.[190,226,302,509,516,520,523a,524,555] Doses of 1000 to 100,000 sporocysts have been tested for their effect on lambs. Data using doses lower than 50,000 sporocysts will be reviewed.

1. Effect on Clinical Disease

At low doses of 1000 to 5000 sporocysts, 288 2- to 4-week-old *Sarcocystis*-free lambs were examined at 1 and 4 months postinoculation.[516,524] As few as 1000 sporocysts reduced hematocrits at 1 month postinoculation, but both hematocrits and weights at 4 months were the same as in uninoculated lambs (Table 2).

Ingestion of 10,000 sporocysts significantly depressed wool growth in lambs at 1 month of age, but not at 5 months of age.[523a]

Ingestion of 25,000 sporocysts affected the health and growth of lambs.[226] The mean carcass weight and mean daily weight of 40 infected lambs in 4 groups of 10 lambs each, based on identical weights, was compared with those of 40 noninfected control lambs; all lambs were reared and managed on a commercial farm. Three infected lambs died and one lamb was condemned at slaughter because of carcass appearance. The carcass weights of the four groups of the inoculated lambs were 27, 15.9, 18.5, and 6.5% less than the weights of the uninfected controls.[226]

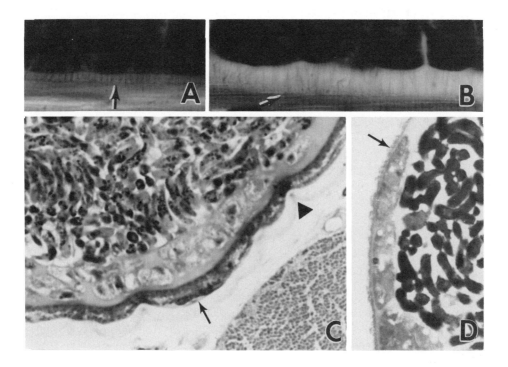

FIGURE 1. A comparison of sarcocyst walls of the four species of *Sarcocystis* in sheep. Arrows point to the tips of villi or outer margin of sarcocyst wall. (A) *S. tenella* with thick striated wall. (B) *S. arieticanis* with thin wall, but with long, hair-like protrusions. These protrusions are not always visible unless stained optimally. (C) *S. gigantea* with thin sarcocyst wall (arrow). A connective tissue capsule (arrowhead) surrounds the sarcocyst wall. (D) *S. medusiformis* with thin sarcocyst wall with small protrusions. ([A and B] H. & E., [D] toluidene blue; magnification × 1000.)

2. Effect on Parturition and Reproductivity

S. tenella can cause abortion in sheep.[431] Of 12 ewes mated 83 to 85 d earlier, 4 were each fed with 50,000, 100,000, or 500,000 sporocysts apiece. Of those fed 50,000 sporocysts, 1 was not pregnant, 1 aborted 33 DPI, and 2 lambed normally; all survived the acute phase of sarcocystosis. Of those fed 100,000 sporocysts, 1 aborted 40 DPI, 1 lambed normally, and 2 ewes carrying dead fetuses became moribund and were euthanatized. Of those fed 500,000 sporocysts, 3 died 27 and 39 DPI, and 1 became moribund and was euthanatized 39 DPI; 2 of these ewes aborted 28 and 30 DPI and 2 others had dead fetuses in utero. *Sarcocystis* was not found in tissue sections of any fetus or lamb.[431]

In another study to determine the effects of sporocyst dosage on reproduction, 20 *Sarcocystis*-free ewes were allotted to 5 groups, A to E (Table 3).[520] Ewes in Group A (6 ewes) were grazed on a 0.23-ha field 1 d after it had been sprayed with 1 million *S. tenella* sporocysts. In Group B, 3 ewes were fed 60,000 sporocysts each; in Group C, 3 ewes were fed 10,000 sporocysts each; and in Group D, 4 ewes were dosed with 2500 sporocysts each. In Group E, four ewes were not given sporocysts. All 3 ewes fed 60,000 sporocysts became severely ill and were euthanatized 48, 50, and 84 DPI. No other ewe became ill or died. All lambs were euthanatized soon after birth and their tissues were examined histologically. Two of the three lambs born to ewes in Group B were very weak and were estimated to be 1 to 3 weeks premature. One lamb from a ewe in Group A was stunted and also appeared premature; focal areas of leukoencephalomalacia were found in the cerebrum and midbrain; the accompanying placenta had small foci of necrosis. Lesions were not seen in other lambs. *Sarcocystis* was not seen in tissues of any lambs.[520]

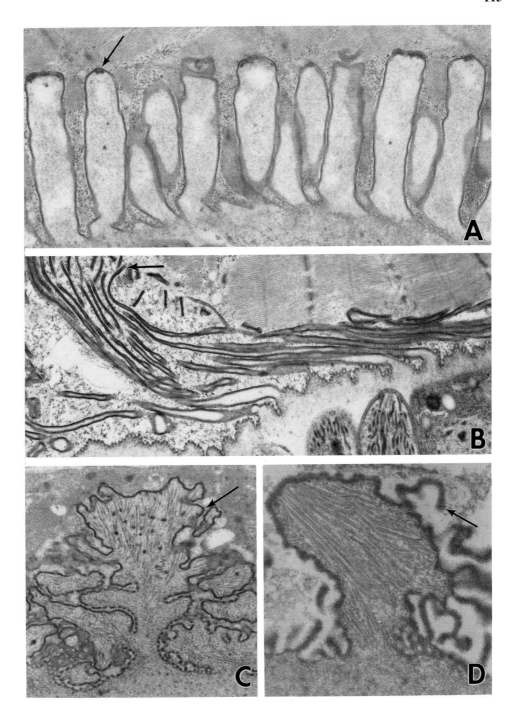

FIGURE 2. TEM showing a comparison of villar protrusions on the sarcocyst wall of the four species of *Sarcocystis* in sheep. (A) *S. tenella* with finger-like villi without microtubules. Unusual plaque structures are visible (arrow) at the tip of the villi. (B) *S. arieticanis* with long filamentous protrusions tapering at the distal end (arrow). (C) *S. gigantea* with cauliflower-like villi (arrow). (D) *S. medusiformis* with short, broad villi and numerous convoluted filaments (arrow) arising from the main villus. (Magnification [A to C] × 15,000, [D] × 30,000.)

Table 1
COMPARISON OF DEVELOPMENTAL STAGES OF *S. ARIETICANIS* AND *S. TENELLA* IN SHEEP

Species: Ref.:	*S. arieticanis* 367a	*S. tenella* 164[a]	*S. tenella* 529[b]
First-generation meronts			
Location	Mesenteric and mesenteric lymph node arteries	Arterioles and arteries throughout the body	Arterioles
Duration (DPI)	14—19	9—21	6—19
Peak development (DPI)	?[c]	16	?
Size of meronts (μm)	45—80 × 35—50	29—45 × 24—32	13.4—35.5 × 12.2—15.1
No. of merozoites	100	Up to 168	18—28
Size of merozoites (μm)	6.8—7.5 × 2.3—3.0	7 × 1.5	?
Second-generation meronts			
Location	Capillaries	Capillaries	Capillaries
Duration (DPI)	26—31	16—40	21—34
Peak development (DPI)	?	25	?
Size of meronts (μm)	?	10.5—42 × 7—17.5	15.2 × 10.6 (12.0—27.4 × 7.5—15.3)
No. of merozoites	?	Up to 54	18—38
Size of merozoites (μm)	6—7.5 × 2.3-3	6 × 15	?
Parasitemia			
Duration (DPI)	?	14—16, 25—32	?
Intraleukocytic multiplication	Yes	Yes	?
Sarcocysts	Microscopic	Microscopic	Microscopic
First seen (DPI)	31	35	41
Maturation time (DPI)	70	75	?
Location			
CNS	No	Yes	?
Heart	Yes	Yes	Yes

[a] In conventionally reared lambs.
[b] In SPF lambs.
[c] ? — unknown.

C. Immunity and Protection

Oral inoculation of low numbers of *S. tenella* sporocysts induces protective immunity in sheep.[276] In 20 *Sarcocystis*-free 1-year-old sheep divided into 5 groups of 4 sheep each, sheep in each group were fed 0, 5, 50, 500, or 5000 sporocysts apiece and were challenged with 50,000 homologous sporocysts apiece 13 weeks later.[276] The sheep were euthanatized 10 weeks after the challenge infection. In the sheep fed 50 or 500 sporocysts there were fewer sarcocysts, lower body temperatures, and higher hematocrits than in the sheep fed 0 or 5 sporocysts. The sheep fed 5000 sporocysts became ill; thus, this dose was not practical for immunization.[276]

Prophylactically administered anticoccidial drugs reduce clinical sarcocystosis.[434,435] Amprolium or salinomycin was administered to sheep from 1 d before until 29 d after 100,000 or 1 million *S. tenella* sporocysts were fed. Amprolium premix (50 or 100 mg/kg body weight) and salinomycin (1 or 2 mg/kg) were mixed in the daily grain ration. Both drugs reduced clinical sarcocystosis, but did not prevent the completion of the life cycle of some of the parasites. Then, 63 d after sporocysts were fed, lambs given salinomycin were challenged with 1 million *S. tenella* sporocysts and were found to have developed a protective immunity.[435]

D. Clinical Sarcocystosis in Naturally Infected Sheep

Because nearly 100% of adult sheep are infected with *Sarcocystis*, it is difficult to document

Table 2
COMPARISON OF HEMATOCRITS AND WEIGHTS OF LAMBS INFECTED WITH 1000 to 10,000 *S. TENELLA* SPOROCYSTS AND OF NONINOCULATED CONTROLS[a]

			Hematocrits (%)		Weights (kg)		
Expt. No.	No. of lambs	Dose	1 month PI	4 months PI	0 Day	1 month PI	4 months PI
1[b]	43	0	36	37	9	22	32
	38	5,000	28	35	11	22	28
	37	10,000	25	34	10	20	27
2[c]	32	0	36	38	17	25	30
	32	1,000	32	35	18	27	31
3[c]	34	0	37	37	13	20	28
	35	2,500	31	35	12	19	27
	37	5,000	29	35	12	19	26

[a] Data of Munday.[516,524]
[b] Number of sarcocysts percubic millimeter were <1, 117, and 102, respectively, in 3 groups.
[c] Sarcocysts were not found in control lambs, but were found in inoculated lambs.

Table 3
EFFECT OF *S. TENELLA* SPOROCYSTS ON LAMBING WEIGHTS[a]

Group	No. of sporocysts	No. of ewes	No. of sarcocysts/mm^3 [b]	Lamb weights (kg) and (number of lambs)	
				Singles	Twins
A	Grazed infected paddock	6	333	2.25(2)	3.21(8)
B	60,000	3	1,250	2.38(3)	None known
C	10,000	3	530	3.95(2)	2.5(2)
D	2,500	4	315	4.0(3)	2.38(2)
E	None	4	6	3.6(3)	3.25(2)

[a] Data of Munday.[520]
[b] Biopsied 1 d before and 90 DPI. No sarcocysts were seen 1 d before inoculation.

the adverse effects of *Sarcocystis* in naturally infected sheep. Scott[615,616] spent his career studying sarcocystosis in sheep in Wyoming, and although his efforts were hampered because the life cycle was unknown, he established strong circumstantial evidence that heavy infections retard the growth of sheep.

Debilitation and lymphadenopathy have been found in association with a *Sarcocystis*-like organism.[426] A 3-year-old ewe had been debilitated for 2 months, and for 2 weeks had enlarged prescapular lymph nodes. Microscopically, lymph node enlargement was due to proliferation of connective tissue. Multiple foci of degeneration and calcified myofibers were seen in the heart, along with sarcocysts and schizonts. Schizonts were also seen in endothelial cells of lymphatics and blood vessels in lymph nodes. There was mild arteritis and schizonts were seen in the adventitia. The schizonts were PAS negative and contained merozoites 5 to 6 × 1.5 to 2 μm. Although it was recorded that merozoites had rhoptries, none were visible in two illustrations. Rhoptries are absent from *Sarcocystis* merozoites (see Chapter 1).

Neonatal sarcocystosis was found in a 3-week-old lamb with generalized gliosis, and immature and mature schizonts were seen in the brain.[188]

II. *SARCOCYSTIS ARIETICANIS* HEYDORN, 1985

Distribution — Europe,[28a,39,229,367a] Australia,[532] New Zealand,[559] the U.S.,[191] and probably other countries.
Definitive host — Dog (*Canis familiaris*).

A. Structure and Life Cycle
Sarcocysts are found in striated muscles, but not in the CNS. They are up to 900 μm long, thin (<1 μm) walled, and have hair-like projections 5 to 9 μm long (Figure 1B).[191,368] The villar protrusions lack microtubules and thus are folded over the sarcocyst wall (Figure 2B). Septa were not seen.[191]
Salient life cycle features are summarized in Table 1.

B. Pathogenicity
S. arieticanis is less pathogenic than *S. tenella*.[367a] A group of 5 sheep fed 50,000, 1 million, 10 million or 30 million sporocysts developed 2 peaks of fever. The first peak of fever occurred 14 to 16 DPI, and the second peak was 26 to 31 DPI. Sheep fed 50,000 or 1 million sporocysts survived the acute phase of sarcocystosis.[367a] Sheep fed 2, 10, or 30 million sporocysts died between 16 and 31 DPI.

III. *SARCOCYSTIS GIGANTEA* (RAILLIET, 1886) ASHFORD, 1977

Synonym — *S. ovifelis* Heydorn et al., 1975.
Distribution — Worldwide.
Definitive host — Domestic cat (*Felis catus*).

A. Structure and Life Cycle
Sarcocystis gigantea sarcocysts are found primarily in the muscles of the esophagus, larynx (Figure 2B, Chapter 1), and tongue, and to a lesser extent in the diaphragm and the rest of the carcass[57,183,327a,523,523b,616] (Figure 3). Sarcocysts have not been found in the heart or the CNS. Macroscopic sarcocysts are found mainly in old sheep. Such sarcocysts are up to 1 cm long, dull white, round, oval, or pear shaped, sometimes resembling rice grains (Figure 3). The sarcocyst wall is thin (<2 μm), smooth, and is often surrounded by a PAS-positive connective tissue secondary cyst wall (Figure 1C). Ultrastructurally, the sarcocyst wall has cauliflower-like protrusions (Figure 2C).[523] The septa are thin (<1 μm wide). In older sarcocysts, live bradyzoites are located peripherally, and the centers of the sarcocysts are often empty.

Little is known about the endogenous development of *S. gigantea* in sheep because sporocysts are not very infectious, sarcocysts grow for 4 years or more, and bradyzoites are not highly infectious to cats.[90,183,278,279,303,523b] Only one schizogony is known.[534] First-generation schizonts were found 7 and 14 DPI, but not 21, 28, and 35 DPI. They were 21 to 44 × 5 to 12 μm and were located in capillaries and arterioles in the lung, kidney, and brain.[534] A sarcocyst was first seen 40 DPI; it measured 71.5 × 8 μm and contained only metrocytes. Bradyzoites were first seen 119 DPI.[523] Between 10 and 14 months postinoculation sarcocysts grew up to 1 mm.[183,523] At 47 months postinoculation sarcocysts were up to 7.5 × 5 mm. Sarcocysts first became infectious between 230 and 265 DPI, but few sporocysts were shed.[183]

B. Pathogenicity
Sarcocystis gigantea is mildly pathogenic for sheep. Other than fever, lambs inoculated with 1 million sporocysts did not develop clinical signs.[183,523b]

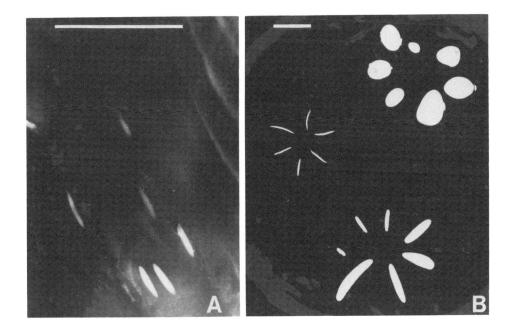

FIGURE 3. A comparison of macroscopic sarcocysts. (A) *S. medusiformis* in abdomen. (B) Sarcocysts in the upper right corner are *S. gigantea* fat sarcocysts from esophagus, thin sarcocysts towards the left corner are *S. medusiformis* from diaphragm, and the sarcocysts towards the bottom are *S. gigantea* from diaphragm. (Bar = 1 cm.) (Courtesy of S. Moore, Wallaceville Animal Disease Centre, Upper Hut, New Zealand.)

IV. *SARCOCYSTIS MEDUSIFORMIS* COLLINS, ATKINSON, CHARLESTON, 1979

Distribution — Australia[97,535] and New Zealand.[12,97]
Definitive host — Cat (*Felis catus*).

A. Structure and Life Cycle

Sarcocystis medusiformis sarcocysts are up to 8 mm long and 0.2 mm wide and are found primarily in the diaphragm, abdominal muscles, and the carcass (Figure 3). The sarcocyst wall is thin (<2 μm), and there is no secondary wall.[502] Ultrastructurally, the villar protrusions are trapezoidal. Snake-like filaments arise from the villar surfaces[97] (Figure 2D).

The endogenous development up to 188 DPI is not known. Immature sarcocysts were found in sheep at between 188 and 1132 DPI.[535] Sarcocysts at 188 and 260 DPI were microscopic; at 443 DPI they were 2 to 3.5 mm long; at 765 and 1132 DPI they were 4 to 5 mm long. Cats fed these sarcocysts did not shed sporocysts, therefore, maturation time is unknown.

V. *SARCOCYSTIS* INFECTION AND CARCASS CONDEMNATION

Whole carcasses or parts of adult sheep are condemned because of the appearance of macroscopic sarcocysts of *S. gigantea* and *S. medusiformis*.[104] Carcasses are also condemned for eosinophilic myositis (EM) in lambs or in adult sheep. In one report, EM was found in 7 of 8,179,104 sheep slaughtered.[389]

VI. *SARCOCYSTIS*-LIKE ENCEPHALOMYELITIS IN SHEEP

Ovine protozoal encephalomyelitis was first reported in an adult Hampshire ewe from the flock at the Veterinary Experimental Station, Cornell University, NY.[539] After 10 d of neural signs the ewe was euthanatized. Microscopic lesions, confined to the CNS, were more severe in the spinal cord than in the brain. Although the entire spinal cord was affected, more lesions were found in white matter than in gray matter. The predominant finding was nonsuppurative encephalomyelitis characterized by necrosis, perivascular cuffing, meningitis, and gliosis with *Toxoplasma gondii*-like organisms at the periphery of the lesion. Similar cases were subsequently reported from Ireland,[480] Scotland,[699] France,[336] Australia, New Zealand, and England.[341,388c,684] All cases were considered to be toxoplasmosis until Hartley and Blakemore[341] clearly described the structure of the parasite in paretic sheep and distinguished it from *T. gondii*. They reported encephalomyelitis in a 6-month-old Suffolk ram from Tasmania with incoordination, periodic head jerking, and apparent blindness, as well as in a 5-month-old Dorset Horn ewe lamb from New South Wales with rear limb incoordination.

The predominant lesion in both was necrosis of the white matter of the spinal cord (malacia). Lesions in the brain were mainly inflammatory and minor compared with those in the spinal cord. Cyst-like structures up to 40 µm in diameter and containing as many as 25 PAS-negative merozoites were seen in neural cells. Immature schizonts were also seen. Merozoites were 5 to 7×2 to 3 µm and were often arranged radially. Ultrastructurally, they were located in the host cell cytoplasm, without a parasitophorous vacuole, and apparently there were no rhoptries.

Recently, sarcocystosis-like ovine encephalomyelitis was reported from England.[388c,503,648] Only a few sheep were affected. In one instance, 2 of 60 Clun Forest-cross gimmer lambs developed progressive weakness of the front legs.[503] The brains in both sheep had microscopic generalized nonsuppurative meningoencephalitis, but no necrosis. Schizonts, similar to those reported by Hartley and Blakemore,[341] were found in neural cells and apparently not in blood vessels. The spinal cord was not examined.

In another instance, 5 of 140 lambs had a rapid onset of neural signs.[648] The lambs became ataxic with paresis of the rear limbs. The paraplegic lambs either lay in sternal recumbency with their hind limbs stretched out to one side or adopted a "sitting dog" posture. Four lambs were euthanatized and necropsied. There was a multifocal encephalomyelitis, characterized by gliosis, lymphocytic perivascular cuffing, and focal malacia in the spinal cord. A few schizonts were found in the spinal cord of one lamb and mature sarcocysts were seen in the brains of three other lambs. Numerous microscopic sarcocysts were seen in sections of skeletal muscles.

Whether these naturally occurring cases of encephalomyelitis in adult sheep were due to *S. tenella* or *S. arieticanis* is uncertain. The structure of the organism, however, leaves no doubt that the parasite in sheep is closely related to *Sarcocystis* and is not *T. gondii*.

BIBLIOGRAPHY

Information pertinent to the subject matter of this chapter may be found in References 10, 12, 28a, 39, 58, 90, 97, 102, 104, 147, 151, 153, 164, 168, 178, 183, 191, 226, 229, 232, 274 to 279, 280a, 301, 303, 327a, 336, 341, 355, 359, 365a, 367a to 368, 370, 382, 383a, 388c, 389, 426, 430, 431, 434, 435, 480, 484b and c, 485a to d, 486a to c, 487, 502, 503, 507 to 509, 513, 515, 516, 520, 521a, 523a and b, 524, 529, 530, 532, 532a, 533b to 535, 539, 541, 542, 550, 553, 555, 559, 576, 578, 582, 584, 614 to 616, 632, 636, 638, 648, 672b, and 684.

Chapter 5

SARCOCYSTOSIS IN GOATS (*CAPRA HIRCUS*)

There are three reported species of *Sarcocystis* in domestic goats: *S. capracanis*, *S. hircicanis* (Figure 1), and *S. moulei*.

I. *SARCOCYSTIS CAPRACANIS* FISCHER, 1979

Distribution — Probably worldwide.
Definitive hosts — Dog (*Canis familiaris*), coyote (*Canis latrans*), red fox (*Vulpes vulpes*), and crab-eating fox (*Cerdocyon thous*).

A. Structure and Life Cycle
Sarcocysts are up to 1000 µm long and 100 µm wide. The sarcocyst wall is up to 3 µm thick with radial striations and finger-like villar protrusions (Figure 1A and C). Sarcocysts are found in virtually all skeletal muscles, in the CNS, and in the heart.[180]

Information on the life cycle is summarized in Table 1 and Figure 2. There are two generations of schizogony in blood vessels.

B. Pathogenicity
Sarcocystis capracanis is the most pathogenic species of *Sarcocystis* in goats.[98,158,366] It can cause fever, weakness, anorexia, weight loss, tremors, irritability, abortion, and death, depending on the number of sporocysts ingested. As few as 5000 sporocysts cause clinical disease, and 100,000 sporocysts are generally lethal.[158] Goats that recover from acute sarcocystosis remain unthrifty, have a dull, dry hair coat, and are predisposed to other infections.[169] One million sporocysts cause severe acute sarcocytosis and goats die between 18 and 21 d.

C. Protective Immunity
Goats immunized once with 100 or 1000 sporocysts develop subclinical infections and become refractive to lethal challenge with large numbers of *S. capracanis* sporocysts.[155a,169,170,179] This protective immunity persisted at least 274 days postinoculation (DPI). Immunization with 1000 sporocysts provided better protection than with 100, and ingestion of 10 sporocysts induced no protection.[155a,169]

II. *SARCOCYSTIS HIRCICANIS* HEYDORN AND UNTERHOLZNER, 1983

Distribution — India,[552] Turkey,[228] and the Federal Republic of Germany.[9,367]
Definitive host — Dog (*Canis familiaris*).

A. Structure and Life Cycle
Sarcocysts are up to 2.5 mm long. The sarcocyst wall is thin (<1 µm) and has hairy protrusions (Figure 1B and D). The villar protrusions are of the Type 7, long, filamentous, without tubules, and thus folded over the cyst wall. Little is known of the pathogenicity of this species. Of 2 goats fed 10 million sporocysts, 1 died 43 DPI and the other was euthanatized while moribund 35 DPI.

III. *SARCOCYSTIS MOULEI* NEVEU-LEMAIRE, 1912, AND OTHER MACROSCOPIC SARCOCYSTS IN DOMESTIC GOATS AND MOUNTAIN GOATS

Synonym — *S. caprifelis* El-Refaii et al., 1980.

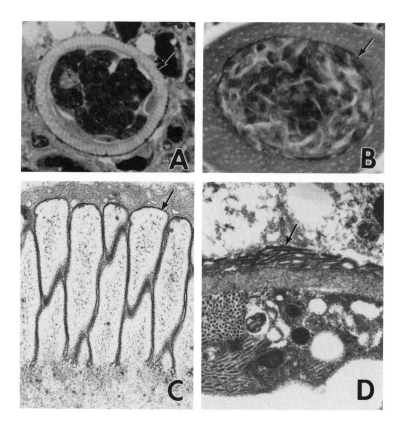

FIGURE 1. Sarcocyst walls of *S. capracanis* (A and C) and *S. hircicanis* (B and D) in skeletal muscles of goats. Arrows point to thick villi in *S. capracanis* and hair-like protrusions in *S. hircicanis*. ([A] Toluidine blue, [B] H. & E.; magnification [A and B] × 1000, [C] × 10,000, [D] × 16,000.)

Distribution — Unknown.
Definitive host — Cat (*Felis catus*).

There is considerable confusion concerning structure and life cycle of macroscopic sarcocysts in goats; therefore, a brief history is given.

Macroscopic sarcocysts were first reported in a goat from Germany in 1873.[527b] Moulé[504a] reported microscopic sarcocysts in goats from France. Neveu-Lemaire[527a] first named *S. moulei* from goats in France. His description of the sarcocysts is vague, but he said that the macroscopic sarcocysts (*S. moulei*) in goats were larger than those of sheep (*S. gigantea*).

Macroscopic sarcocysts from the mountain goat (*Capra siberica*) in the U.S.S.R. were named *S. orientalis*.[451] The sarcocysts were up to 7.5 mm long and up to 2.5 mm wide. Many of these sarcocysts were found in the esophagus. The authors found no macroscopic sarcocysts in domestic goats (*Capra hircus*) from the same region.[451]

The macroscopic and microscopic species of sarcocysts from goats were differentiated.[272a] The microscopic species was named *S. capracanis* and its structure and life cycle were described.

In India and the Federal Republic of Germany, investigators independently recognized some thin-walled sarcocysts with hair-like protrusions in muscles of the domestic goat.[9,552] The organism was named *S. hircicanis* and its life cycle was described.[367] Both *S. capracanis* and *S. hircicanis* are transmissible to canids, but not to felids.

A species of *Sarcocystis* from goats in Egypt transmissible to cats was named *S. caprifelis*.[198] Although the sarcocysts were not clearly described, two types were seen in histologic sections

Table 1
COMPARISON OF DEVELOPMENTAL STAGES OF *S. CAPRACANIS* AND *S. HIRCICANIS* IN GOATS

	S. capracanis[a]	*S. hircicanis*[b]
First-generation meronts		
Location	Gut-associated arteries	Arterioles, intestinal lymph nodes, liver
Duration (DPI)	15—17	17—18
Peak development (DPI)	10—12	
Size of meronts (μm)	26.1 × 17.4 (sections)	?[c]
No. of merozoites	Up to 80	Up to 100 or more
Size of merozoites (μm)	5.5—7.1 × 2.5 (sections)	6—7.5 × 2.5—3.0 (smear)
Second-generation meronts		
Location	Capillary endothelium of all organs	Capillary endothelium of several organs
Duration (DPI)	14—24	28—35
Peak development (DPI)	19	?
Size of meronts (μm)	18.8 × 10.1	?
No. of merozoites	4—36	?
Size of merozoites (μm)	5.5 × 1.5 (sections)	6—7.5 × 2.5—3.0 (smear)
Parasitemia		
Duration (DPI)	17—24	?
Intraleukocytic multiplication	Yes	?
Sarcocysts		
Size (μm)	Up to 1000	Up to 2500
Earliest seen (DPI)	30	43
Maturation time (DPI)	64	84
Wall (μm)	2—3 thick, striated	<1, thin, hairy projections
Location	Striated muscle, CNS	Striated muscles

[a] Data from Reference 180.
[b] Data from Reference 367.
[c] ? — unknown.

of goat diaphragm: thick-walled sarcocysts transmissible to dogs and considered to be *S. capracanis* and organisms with, "thin or smooth sarcocyst walls with many fibrils seen almost clear at the apex," transmissible to cats. Sarcocysts transmissible to cats were also found in a goat in Turkey.[543] Because no known species of *Sarcocystis* transmissible via cats is transmissible via dogs, it seems that a species of *Sarcocystis* in goats transmitted via cats is valid. Because the identity of *S. moulei* and *S. caprifelis* cannot be resolved retrospectively, we consider the cat-transmitted species to be *S. moulei*, and we synonymize *S. caprifelis* with it.

Three structurally distinct sarcocysts were found in other goats in Turkey:[228] *S. capracanis* with characteristic radially striated cyst wall, *S. hircicanis* with characteristic hairy villar protrusions and macroscopic sarcocysts with a thin (<1 μm) sarcocyst wall with knob- or button-like protrusions. The macroscopic sarcocysts were up to 4 mm long and 2 mm wide. However, this paper is confusing because it does not clearly distinguish the sarcocysts of goats from these sheep.

Another species of *Sarcocystis* is found in the American mountain goat (*Oreamnos americanus*) (see Chapter 12).

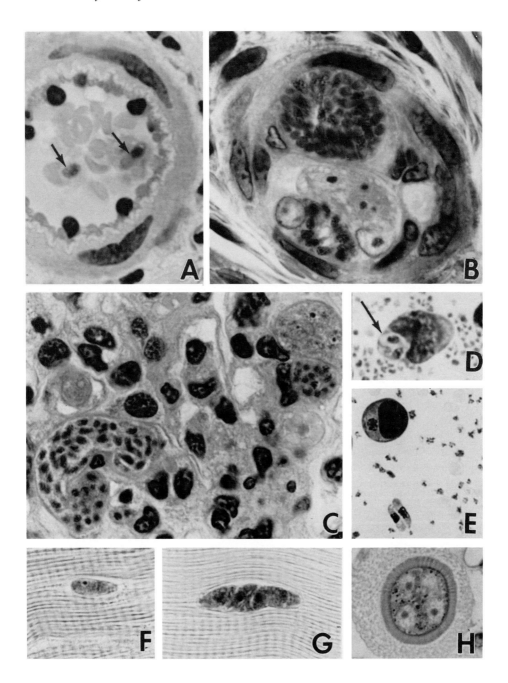

FIGURE 2. Developmental stages of *S. capracanis* in tissues of experimentally infected goats. (A) Two sporozoites (arrows) in lumen of an artery (5 DPI). (B) Three developing first-generation schizonts in an artery. The vascular lumen is occluded by hypertrophied endothelial cells and parasites (10 DPI). (C) Second-generation schizonts and merozoites in renal glomerulus (20 DPI) D. Dividing merozoite in a monocyte-like cell in peripheral blood (20 DPI). (E). An intracellular merozoite and two extracellular merozoites in peripheral blood (20 DPI). (F) Binucleate sarcocyst (35 DPI). (G) Sarcocyst with four nuclei (35 DPI). (H) Sarcocyst with well-developed wall (68 DPI). (Magnification × 1000.) (From Dubey, J. P., Speer, C. A., Epling, G. P., and Blixt, J. A., *Int. Goat Sheep Res.*, 2, 252, 1984. With permission.)

BIBLIOGRAPHY

Information pertinent to the subject matter of this chapter may be found in References 9, 17, 32, 86, 96, 98 to 100, 153, 155a, 156, 158, 168 to 170, 178, 179, 184, 198, 228, 272a, 280, 347a, 365 to 368, 382, 449, 451, 504a, 527a and b, 540, 543, 549, 552, 553, 565, 581a, 603, 620, 638, 644, and 672b.

Chapter 6

SARCOCYSTOSIS IN PIGS (*SUS SCROFA*)

There are three reported species in pigs: *S. miescheriana*, *S. suihominis* (Figure 1), and *S. porcifelis*.

I. *SARCOCYSTIS MIESCHERIANA* (KÜHN, 1865) LABBÉ, 1899

This is the type species of the genus.
Synonym — *S. suicanis* Erber, 1977.
Distribution — Probably worldwide.
Definitive hosts — Dog (*Canis familiaris*), raccoon (*Procyon lotor*), wolf (*Canis lupus*), red fox (*Vulpes vulpes*), and jackal (*Canis aureus*).[499]

A. Structure and Life Cycle

Sarcocysts are as large as 1500 µm long and 200 µm wide and are found in skeletal and cardiac muscles.[222] The sarcocyst wall is 3 to 6 µm thick and appears radially striated. The villar protrusions on the sarcocyst wall are up to 5 µm long and 1.3 µm wide (Figure 1A and C).[222,493] Unusually large bradyzoites, up to 20 µm long, are present within mature sarcocysts.

Information on endogenous development is summarized in Table 1. There are two known generations of schizogony which develop much more rapidly than those of cattle and sheep. First- and second-generation schizonts mature within 13 days postinoculation (DPI), and immature sarcocysts are found as early as 27 DPI.[20,364,375]

B. Pathogenicity

Sarcocystis miescheriana can cause weight loss and purpura of the skin, especially of the ears and buttocks, dyspnea, muscle tremors, abortion, and death, depending on the number of sporocysts ingested.[21,222,224,225,697] Ingestion of less than 1 million sporocysts generally results in subclinical infection. Stress of pregnancy may modify the severity of infection. Of 5 pregnant sows experimentally infected with 50,000 sporocysts each, all became ill: 2 aborted 12 to 14 DPI, 1 died, and 2 became moribund and were euthanatized.[224] A similar number of sporocysts given to finishing pigs caused no clinical signs, but resulted in weight gains that were 11 to 27% less than uninfected controls.[42] Administration of 25,000 or 15,000 sporocysts neither reduced weight gains nor produced clinical signs.

The effect of subclinical sarcocystosis on meat quality has been studied.[120-122] Pigs infected with 50,000 *S. miescheriana* sporocysts weighed 5 to 12 kg less than uninoculated pigs when examined 3 months postinoculation; however, the quality of meat as determined by water absorbing and water binding capacity, pH, rigor values, color brightness, and back fat ratios was not reduced compared with that of uninfected pigs.[120,121] Genetically determined stress sensitivity (measured by the halothane test) did not affect the response to *Sarcocystis* infection as measured by growth or meat quality.

C. Protective Immunity

Pigs immunized once with 1000 to 50,000 *S. miescheriana* sporocysts become refractive to lethal challenge with large numbers of *S. miescheriana* sporocysts.[593,606,608,679,697] This protection was found at 80 DPI, but not at 120 DPI.[679] Pigs given multiple low doses of sporocysts (trickle infections) also developed protection.[606] This protective effect resulting from administration of *S. miescheriana* sporocysts was not protective against a challenge infection with *S. suihominis* sporocysts.[225]

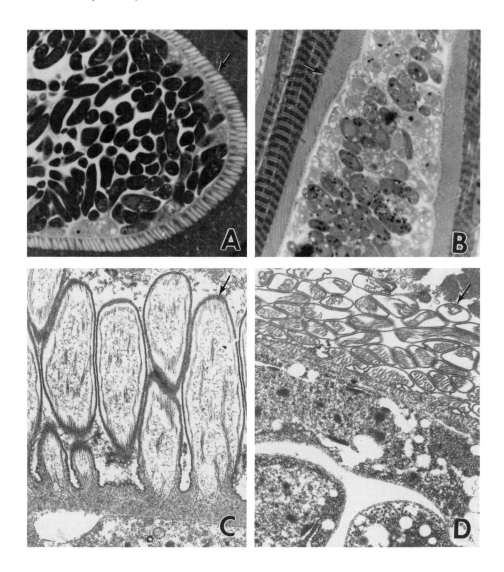

FIGURE 1. Sarcocysts of *S. miescheriana* (A and C) and *S. suihominis* (B and D) in skeletal muscles of pigs. The villar protrusions (arrow) of *S. suihominis* are thinner and longer than those of *S. miescheriana*. (Toluidine blue; magnification [A and B] × 1000, [C and D] × 16,000.)

II. *SARCOCYSTIS SUIHOMINIS* (TADROS AND LAARMAN, 1976) HEYDORN, 1977

Distribution — Europe.

Definitive hosts — Man (*Homo sapiens*) and nonhuman primates (*Macaca mulatta*, *Macaca irus*, *Pan troglodytes*, and *Papio cynocephalus*).

A. Structure and Life Cycle

Sarcocysts are up to 1500 μm long.[222] The sarcocyst wall is 4 to 9 μm thick[222] and appears hirsute, with villar protrusions up to 13 μm long (Figure 1B and D).[492] The bradyzoites are approximately 15 μm long.

Life cycle data are summarized in Table 1. The two generations of schizogony develop at

Table 1
COMPARISON OF DEVELOPMENTAL STAGES OF *SARCOCYSTIS* SPECIES IN PIGS

Intermediate host	*S. miescheriana*[a]	*S. suihominis*[b]
First-generation meronts		
Location	Liver endothelium	Liver endothelium
Duration (DPI)	6—7	5—6
Size of meronts (μm)	?[c]	?
No. of merozoites	?	?
Size of merozoites (μm)	?	?
Second-generation meronts		
Location	Many organs	Many organs
Duration (DPI)	7—13	7—20
Size of meronts (μm)	7—17 × 11—79	
No. of merozoites	20—94	50—90
Size of merozoites (μm)	?	? 6.5 × 3.2
Parasitemia		
Duration (DPI)	?	?
Blood cell multiplication	?	?
Sarcocysts		
Wall (μm)	Thick, striated, 2.7—6	Thick, striated, 4—9
First seen (DPI)	27	27
Maximum length (μm)	Up to 1500	Up to 1500
Infective (d)	58	56
Bradyzoites (μm)	17.1 × 3.8	15 × 4.5

[a] Data from References 20, 220, 325, and 364.
[b] Data from References 357, 362, and 492.
[c] ? — unknown.

about the same time as and are structurally similar to those of *S. miescheriana*.[357,360,362] Immature sarcocysts were first seen 27 DPI, and bradyzoites were seen 56 DPI.[492]

B. Pathogenicity

S. suihominis is pathogenic for pigs. Pigs fed 50,000 or more sporocysts became ill, and half of those fed 1 million sporocysts died.[357] Clinical signs in pigs were similar to those in pigs infected with *S. miescheriana*.[357]

III. *SARCOCYSTIS PORCIFELIS* DUBEY, 1976

The validity of this species is uncertain because it has been reported only once and therefore lacks confirmation. This species was named[146] for the parasite described from the U.S.S.R.[330] *Sarcocystis*-infected esophagi from pigs were fed to four cats. The cats passed sporulated oocysts 5 to 10 d after ingesting infected swine tissues. The sporocysts were 13.5 × 7.6 μm and each contained four sporozoites measuring 9.5 × 3.8 μm. The sporocysts from cat feces were fed to eight littermate pigs. All pigs became ill and one died 89 DPI. Sarcocysts were found in skeletal muscles and the heart. Other pigs were killed 3.5 months postinoculation, and sarcocysts were found in them.[330] The structure of the sarcocysts was not described.

IV. PREVALENCE OF *SARCOCYSTIS* AND ECONOMIC IMPACT

The overall prevalence of sarcocysts in pigs is relatively low (Table 2). The prevalence in

Table 2
PREVALENCE OF SARCOCYSTS IN PIGS

Country	No. exam.	Sows	Finishing pigs	No. infected	% infected	Method used	No. condemned	Ref.
Austria	712	348	364	130	18.3[a]	Digestion + squash	—[b]	374
Denmark	108,917	48,780	60,137	1,770	2.8, 0.67	Trichinoscope	18	333
Federal Republic of Germany	1,175	409	766	145	35.5[c]	Digestion	—	38
Japan	300	200	100	17	8.8, 0	Digestion + microscopic	—	602
Poland	3,953, 556	—	—	1,470	—	Trichinoscope	1,470	124
U.S.	192 (wild swine)	—	—	62	32	Digestion	—	19
U.S.	180	180	—	28	16.6	Digestion	—	567
U.S.	236	236	—	8	3.4	Digestion	—	152

[a] 13.6% *S. miescheriana* and 4.1% *S. suihominis*.
[b] — = no data.
[c] 48.3% had *S. miescheriana* and 60.7% had *S. suihominis* of those examined microscopically.

finishing- or market-age pigs is lower than in adult culled pigs. *S. suihominis* is more prevalent in the Federal Republic of Germany than in Austria, but information is not available from other countries.

Carcass condemnation for *Sarcocystis* has been reported,[124,334] but there is no documentation of any species of *Sarcocystis* in swine with a macroscopic sarcocyst. Therefore, the basis for documentation is uncertain. Natural clinical sarcocystosis in pigs has not been reported.

BIBLIOGRAPHY

Information pertinent to the subject matter of this chapter may be found in References 19 to 22, 28, 38, 42, 44, 45, 120 to 122, 124, 146, 147, 152, 221, 222, 224, 225, 235, 255, 325, 330, 333 to 335, 357, 360 to 362, 364, 365, 369, 375, 408b, 492, 494, 499, 508, 528, 530, 531, 548, 567, 583, 590, 593, 606, 608, 647a, 679, 679a, 681, 682, 690, 697, and 698.

Chapter 7

SARCOCYSTOSIS IN EQUIDS (*EQUUS* SP.)

I. INTRODUCTION

There is considerable confusion concerning the validity of different species of *Sarcocystis* in the horse and other equids. Four species of *Sarcocystis* have been named from equids: *S. bertrami*, *S. fayeri*, *S. equicanis*, and *S. asinus*, all with the dog as the definitive host. Perhaps there is only one species of *Sarcocystis* in horses, mules, zebras, and donkeys. An extensive review of the literature on this subject concluded that structurally *S. bertrami* is the only valid species.[376] However, this conclusion was based on examination of tissues from naturally infected horses in Austria. Until each species of *Sarcocystis* in equids has been extensively compared, it may be premature to lump all species with *S. bertrami*.

II. *SARCOCYSTIS BERTRAMI* DOFLEIN, 1901

This species was named after a brief description in a book.[134] Sarcocysts were 9 to 10 mm long and were found in the esophagus and leg muscles. No other information was given.

The dog (*Canis familiaris*) is the definitive host.[470] Dogs fed infected muscles shed sporocysts 12.2 to 13.8 × 9.2 to 9.9 μm.

Sarcocysts in naturally infected horses in the Federal Republic of Germany were up to 12 mm long. In experimentally infected horses the sarcocysts were 2 mm long at 378 days postinoculation (DPI), and up to 9 mm long at 1040 DPI.[470]

Sarcocysts in naturally[376] infected horses and donkeys in Austria were up to 15 mm long.[373] The sarcocyst wall had conical protrusions with their outward-pointing tips penetrating the sarcoplasm of the myocytes. The protrusions were 4.2 to 7.3 μm long and 0.8 to 2.1 μm wide, and the sarcocyst wall appeared hairy or striated. The bradyzoites were up to 22.4 μm long. The dog (but not the cat or man) was the definitive host. Sporocysts were 11.9 to 14.4 × 8.4 to 10.1 μm.

III. *SARCOCYSTIS EQUICANIS* ROMMEL AND GEISEL, 1975

Sarcocysts were microscopic, up to 350 μm long in histologic sections with smooth, thin (1 μm) wall (Figure 1B).[585] Ultrastructurally, the sarcocyst wall had hair- or villus-like protrusions up to 2 to 3 μm long and up to 0.5 μm thick embedded into myofibrils.[326] Bradyzoites were up to 10 μm.[326] Dogs fed infected muscles shed sporocysts 15 to 16.3 × 8.8 to 11.3 μm, with a prepatent period of 8 d.[585]

IV. *SARCOCYSTIS FAYERI* DUBEY, STREITEL, STROMBERG, AND TOUSSANT, 1977

Sarcocysts were found primarily in skeletal muscles and rarely in the heart. Sarcocysts in histologic sections are up to 990 μm long and 136 μm wide[150,667] with radially striated cyst walls 1 to 3 μm thick. The cyst wall has villar protrusions 2.2 to 3.1 μm long, with 34 to 55 loosely arranged microtubules (Figures 1A and 2). Bradyzoites are 12 to 16 μm long in sections and 15 to 20 μm in smears. The dog (but not the cat) shed sporocysts, 11 to 13 × 7 to 8.5 μm, with a prepatent period of 12 to 15 d.[150]

Two generations of schizonts were found in arteries or capillaries of the heart, brain, and kidney between 10 and 25 DPI. Immature sarcocysts were seen from 55 DPI, and some

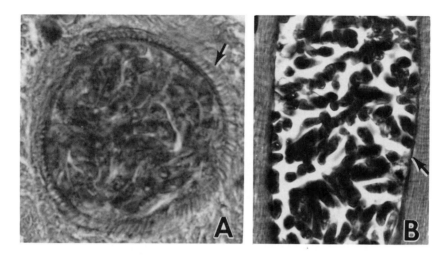

FIGURE 1. Sarcocysts of *S. fayeri* (A) and *S. equicanis* (B) in skeletal muscles of horses from the U.S. (A) and the Federal Republic of Germany (B). Note thick, striated wall in *S. fayeri* and thin wall in *S. equicanis*. (H. & E.; magnification × 1000.)

sarcocysts at 77 DPI had bradyzoites and were infectious for dogs.[261] At 156 DPI, the sarcocysts were up to 338 μm long in histologic sections.

V. PATHOGENICITY

Sarcocystis species of equids are only mildly pathogenic. In one study, ponies fed 1 million sporocysts of *S. fayeri* developed mild anemia and fever. A horse fed 10 million sporocysts developed stiff gait, but was otherwise clinically normal.[261] There are indications that pathogenicity may vary among different isolates of *Sarcocystis* or among horses. For example, a pony fed 2 million sporocysts of a Texas isolate of *Sarcocystis* became lethargic, stiff-legged, tired quickly, and had lost hair on its body, especially on its head and neck by 150 DPI (Figure 3).[262] This pony had a severe myositis associated with sarcocysts and developed an autoimmune anemia.[262,298] In another study, all 5 ponies fed 200,000 sporocysts of an isolate from the Federal Republic of Germany developed fever, apathy, anorexia, and muscle stiffness.[609]

VI. PREVALENCE OF NATURAL INFECTIONS AND CLINICAL DISEASE

Mature sarcocysts are common, and up to 90% of equids have been found infected.[150,227,376,411] Infections are more prevalent in older horses. The finding of mature sarcocysts in a 3-d-old foal from Canada[117] and in a 6-week-old foal in Britain[197] indicates transplacental infection.

Clinical myositis has been reported to be associated with the presence of sarcocysts in histologic sections.[292,604,650] However, in a controlled study of 91 horses with clinical myositis,[282a] sarcocysts were found in 12 muscle biopsies, but there was no association between the presence of sarcocysts and clinical myositis.

VII. REMARKS ON VALIDITY OF *SARCOCYSTIS* SPECIES IN EQUIDS

Whether there is more than one species of *Sarcocystis* in equids is uncertain at the present. Much of this confusion is due to inadequate descriptions, misinterpretations of original descriptions, the long period of growth of sarcocysts, techniques of measurement, and the stage of the sarcocyst studied. Are there thin- and thick-walled sarcocysts in horses? Both thin-walled

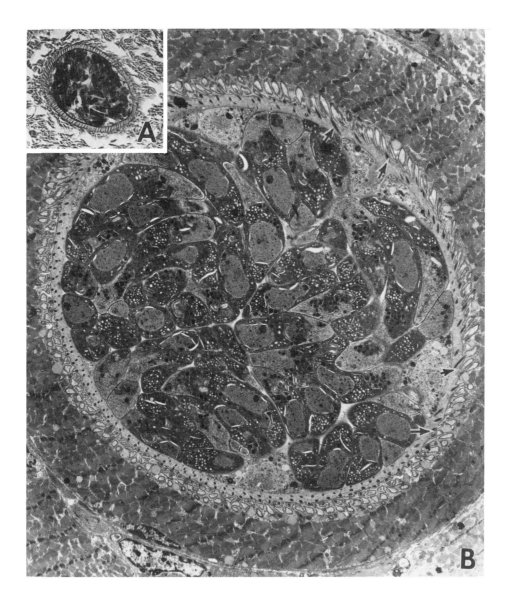

FIGURE 2. Sarcocysts of *S. fayeri* in the horse. ([A] H. & E.; magnification × 500.) (B) TEM of cross-section showing microtubules (arrows) of the villar protrusions extending into the interior of the sarcocyst. (Magnification ×9900.) (From Tinling, S. P., Cardinet, G. H., III, Blythe, L. L., and Vonderfecht, S. L., *J. Parasitol.*, 66, 458, 1980. With permission.)

sarcocysts (*S. equicanis*) and thick-walled sarcocysts (*S. fayeri*) have been found in horses in the Federal Republic of Germany,[227] and they are clearly demonstrable in Figure 1B, taken from the hepantotype slide supplied by Mehlhorn et al.[495]

The structural similarity of the sarcocysts in donkeys, zebras, and mules and the transmission of *Sarcocystis* from the donkey to a horse via dogs indicates that more than one of these equid species share the same *Sarcocystis* species.[119,294,376,462,467]

VIII. EQUINE PROTOZOAL MYELITIS (EPM)

A fatal protozoan disease in horses (EPM) was clinically recognized in early 1970 in New

FIGURE 3. A depressed pony with hair loss 6 months after inoculation with *S. fayeri*. (From Fayer, R., Hounsel, C., and Giles, R. C., *Vet. Rec.*, 113, 216, 1983. With permission.)

York,[595] Ohio,[144,145] Illinois,[118] and Pennsylvania.[26,27] Since that time EPM was reported in several areas of the U.S.,[88,136,185,451a,628] Canada,[88] Brazil,[18a] and in England in horses imported from the U.S.[478]

EPM affects the CNS and has been found in young and old horses of both sexes and many breeds. Common clinical signs include difficulty with gait, including falling, stumbling, recumbency, ataxia, spasticity, and paresis. Less common clinical signs are lameness, general weakness, unilateral hypermetia, unilateral facial paralysis, partial paralysis of the tongue, and muscle atrophy.[478]

Lesions are seen mainly in the CNS, particularly in the spinal cord. Focal hemorrhage and malacia are the characteristic lesions (Figure 4). Microscopically, there is necrosis and nonsuppurative myeloencephalitis characterized by infiltrations of mononuclear cells, giant cells, and eosinophils.[145]

The parasite has been seen only in the CNS. It is intracellular in neural cells,[145] vascular endothelial cells,[185] and leukocytes,[145,185] Only asexual stages are known; they consist of immature and mature schizonts and free merozoites; these resemble *Sarcocystis* in structure. EPM is located in the host cell cytoplasm without a parasitophorous vacuole. Schizonts are PAS-negative and have no well-defined cyst wall.

The life cycle is unknown. The parasite resembles *Sarcocystis*, but none of the *Sarcocystis* species from domestic animals have induced EPM.[265] The occurrence of isolated cases in herds of healthy horses indicates that *Sarcocystis* of an obscure animal may be involved.[265]

BIBLIOGRAPHY

Information pertinent to the subject matter of this chapter may be found in References 18a, 26, 27, 88, 117 to 119, 134, 136, 144, 145, 150, 185, 197, 227, 261, 262, 265, 282a, 292, 294, 298, 326, 365, 373, 376, 411, 451a, 462, 467, 470, 475 to 478, 585, 595, 604, 609, 628, 650, and 667.

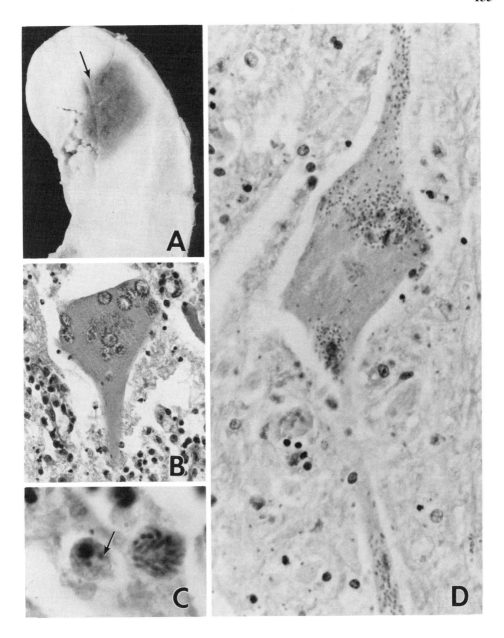

FIGURE 4. Equine protozoal myeloencephalitis (EPM) infection in naturally infected horses. (A) Area of necrosis (arrow) in medulla oblongata. (B) A heavily infected neuron in L_2 segment of spinal cord with numerous meronts. (C) Intracellular (arrow) and extracellular organisms in spinal cord. (D) Neuron with hundreds of organisms. ([B and D] H. & E.; magnification [B] × 125, [C] × 1250, [D] × 600.) (From Dubey, J. P., Davis, G. W., Koestner, A., and Kiryu, K., *J. Am. Vet. Med. Assoc.,* 165, 249, 1974. With permission.)

Chapter 8

SARCOCYSTOSIS IN WATER BUFFALO (*BUBALUS BUBALIS*)

There are two named species in buffalo: *S. levinei* and *S. fusiformis*.

I. *SARCOCYSTIS LEVINEI* DISSANAIKE AND KAN, 1978

Distribution — India,[85,127,307,387] the Philippines,[668-670] Brazil,[448] Romania, Turkey,[577] Egypt, and Malaysia.[133]

Definitive host — Dog (*Canis familiaris*).

A. Structure and Life Cycle

Sarcocysts are thin, spindle-shaped, from microscopic size up to 1150 μm long, and are found in skeletal muscles and heart (Figure 1A). The sarcocyst wall is thin (<1 μm) and has sloping villar protrusions with wavy outlines (Figure 2B).[133,399]

Little is known of its endogenous development and biology in the water buffalo. It appears to be mildly pathogenic. Of the 2 calves fed 2 million sporocysts, 1 became anemic, weak, and died 43 DPI; the other calf had no obvious clinical signs. Either 0.5 or 1 million sporocysts did not produce clinical signs in any of 6 other calves.[385]

II. *SARCOCYSTIS FUSIFORMIS* (RAILLIET, 1897) BERNARD AND BAUCHE, 1912

Distribution — India,[85,307] Egypt,[305,610] the Philippines,[6,668,670] Romania,[570] Brazil,[448] Turkey,[577] and probably other countries with the distribution of the host.

Definitive host — Cat (*Felis catus*).

A. Structure and Life Cycle

Sarcocysts are macroscopic, opaque, white, globular to spindle shaped, up to 32 mm long, and up to 8 mm wide. The sarcocyst wall is thin and has characteristic cauliflower-like villar protrusions (Figures 1B and 2A).[305,691] They are compartmentalized, but in older sarcocysts the centers are often empty. Most are found in the muscles of the esophagus.

Schizogonic stages are unknown. This species appears to be nonpathogenic. In one study, buffalo calves fed 500,000 or 5 million sporocysts did not develop clinical signs.[128] In another study, a calf fed 12,000 sporocysts became ill 38 DPI and died 77 DPI.[2] However, this report needs confirmation.

BIBLIOGRAPHY

Information pertinent to the subject matter of this chapter may be found in References 2, 6, 84, 85, 125 to 128, 132, 133, 307 to 307e, 310, 384 to 387a, 399, 448, 456a, 570, 577, 610, 644a, 645, 668 to 670, 691, and 695.

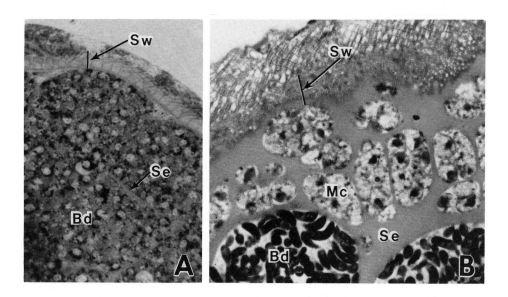

FIGURE 1. Sarcocysts of *S. levinei*-like organism (A) and *S. fusiformis* (B) from esophageal muscle from an Indian water buffalo. The sarcocyst wall (Sw) in A has wide-based villar protrusions, and in B the villar protrusions are highly branched. Bd, bradyzoites; Mc, metrocytes; Se, septa. (Toluidine blue; phase-contrast microscopy; magnification × 700.)

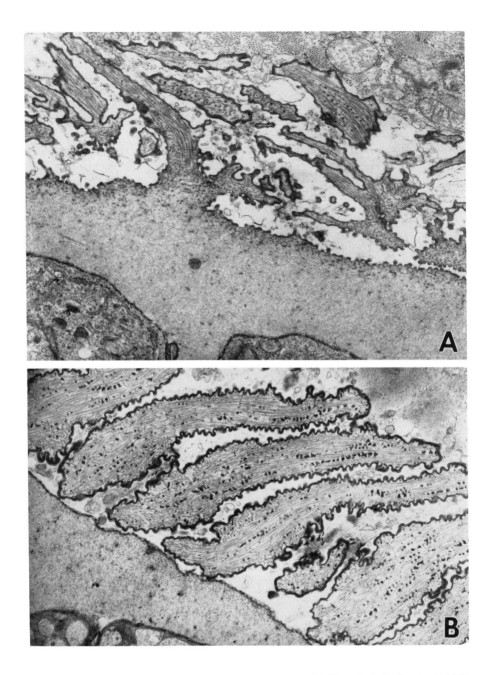

FIGURE 2. TEM of sarcocyst walls of the two *Sarcocystis* species of buffalo. (A) *S. fusiformis* with highly branched, cauliflower-like villar protrusions. (B) *S. levinei* with sloping, unbranched villi. (Magnification [A] × 11,200, [B] × 14,800.) (Courtesy of S. P. Kan and A. S. Dissanaike, University of Malaya, Kuala Lumpur, Malaysia.)

Chapter 9

SARCOCYSTOSIS IN CAMELS (*CAMELUS DROMEDARIUS* AND *C. BACTRIANUS*)

I. INTRODUCTION

There are two species of *Sarcocystis* in camels, but only one (*S. cameli*) is named. Mason[464] first reported sarcocysts in striated muscles, including the heart, of a camel from Cairo, Egypt. He saw two types of sarcocysts, one with a striated wall 1 to 2 µm thick, and the other with a smooth, thin wall (<1 µm). Mason[464] thought that they were both different stages of the same parasite and named them *S. cameli*. Therefore, in retrospect, it is not possible to decide which type of sarcocyst is *S. cameli*. Because the thick-walled striated sarcocyst has been found repeatedly, we have elected to call it *S. cameli*, and have left the thin-walled species unnamed (Figure 1).

II. *SARCOCYSTIS CAMELI* MASON, 1910

The species has been reported from Egypt,[206,306,464] Iran,[571] Sudan,[382] Afghanistan,[412] Morocco, and the U.S.S.R.[421] The dog (*Canis familiaris*) is the definitive host.

A. Structure and Life Cycle

The sarcocysts are found in striated muscles including the tongue, esophagus, and heart. They are up to 388 µm long. The sarcocyst wall is 1 to 2 µm thick and appears striated (Figure 1A). The villar protrusions are conical, sloping to straight, and 1.2 to 1.6 µm long and 0.5 µm wide at the base.[206,306] The bradyzoites are 8 to 12 µm long in histologic sections. Dogs shed sporocysts 12 × 9 µm 10 to 14 d after eating naturally infected camel meat.

BIBLIOGRAPHY

Information pertinent to the subject matter of this chapter may be found in References 206, 306, 371, 372, 382, 412, 421, 464, and 571.

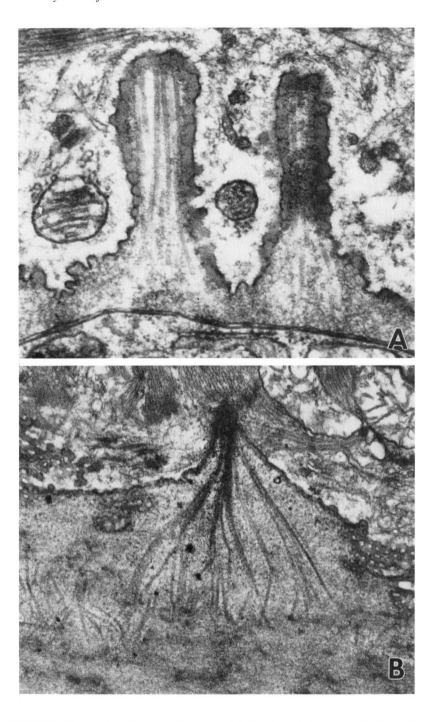

FIGURE 1. Ultrastructure of sarcocysts from naturally infected camels from Egypt. (A) *S. cameli* with characteristic branched villar protrusions. (B) *Sarcocystis* sp. with microtubules embedded in the granular layer. (Magnification [A] × 50,000, [B] × 30,000.) (From [A] Ghaffar, F. A., Entzeroth, R., Chobotar, B., and Scholtyseck, E., *Tropenmed. Parasitol.*, 30, 434, 1979; [B] Entzeroth, R., Ghaffar, F. A., Chobotar, B., and Scholtyseck, E., *Acta Vet. Acad. Sci. Hung.*, 29, 335, 1981. With permissions.)

Chapter 10

SARCOCYSTOSIS IN HUMANS AND OTHER PRIMATES

There are two known species of *Sarcocystis* for which humans serve as the definitive host, *S. hominis* and *S. suihominis*. Humans also serve as the accidental intermediate host for several unidentified species of *Sarcocystis*.

I. HUMANS (*HOMO SAPIENS*)

A. Intestinal Sarcocystosis

1. Sarcocystis hominis (Railliet and Lucet, 1891) Dubey, 1976

This species is acquired by ingesting uncooked beef containing *S. hominis* sarcocysts. The structure and life cycle are described in Chapter 3.

S. hominis is only mildly pathogenic for humans. A volunteer who ate raw beef from an experimentally infected calf developed nausea, stomachache, and diarrhea 3 to 6 h after ingesting beef; these symptoms lasted 24 to 36 h. The volunteer excreted *S. hominis* sporocysts between 14 and 18 d after ingesting the beef. During the period of patency he had diarrhea and stomachache.[358] Somewhat similar but milder symptoms were experienced by other volunteers who ate uncooked naturally infected beef.[8,369,583]

2. Sarcocystis suihominis (Tadros and Laarman, 1976) Heydorn, 1977

This species, acquired by eating undercooked pork, is more pathogenic than *S. hominis*. Its structure and life cycle are described in Chapter 6.

Human volunteers developed hypersensitivity-like symptoms: nausea, vomiting, stomachache, diarrhea, and dyspnea within 24 h of ingestion of uncooked pork from naturally or experimentally infected pigs. Sporocysts were shed 11 to 13 d after ingesting pork.[8,369,408b,556]

3. Natural Prevalence

Before the discovery of the life cycle of *Sarcocystis* and the recognition of cattle and pigs as sources of human infection, *Sarcocystis* sporocysts in human feces were referred to as *Isospora hominis*. Because of structural similarities between *S. hominis* and *S. suihominis* sporocysts, it is not possible to distinguish these two species by microscopic examination. Therefore, surveys do not distinguish between these two species.

Based on published reports, it appears that intestinal sarcocystosis is more common in Europe than on other continents.[60,123,273,300,308,335,660] *Sarcocystis* sporocysts were seen in 2% of 3500 fecal samples in France,[123] 1.6% of 1518 and 7.3% of 300 samples in the Federal Republic of Germany,[273,388a] and 10.4% of 125 fecal samples from 7- to 18-year-old children in Poland.[557] Enteritis was associated with *Sarcocystis* sporocysts in six cases in Thailand,[60] and two cases were reported from the People's Republic of China.[700]

B. Muscular Sarcocystosis

Sarcocysts have been found in striated muscles of human beings, mostly as incidental findings. Judging from the published reports, sarcocysts in humans are rare; only 40 documented cases were reported as of 1979.[23] The species of *Sarcocystis* and life cycle are unknown. Beaver et al.[23] reviewed case histories and histologic sections of reported cases of muscular sarcocystosis critically, concluding that

1. There were at least seven structurally distinct types of sarcocysts reported.
2. The parasite previously called *S. lindemani* is not a valid species of *Sarcocystis*.

3. Most reported cases were from Asia and Southeast Asia. Of the 40 histologically diagnosed reports they reviewed, 13 were from Southeast Asia, 8 from India, 5 from Central and South America, 4 from Europe, 4 from Africa, 3 from the U.S., 1 from China, and 2 were undetermined.

Six additional cases have been reported. Three cases were from India,[4,438] two from Malaysians of Indian origin,[402] and one from the U.S.[484a] Of the 46 confirmed cases, sarcocysts were found in the skeletal muscles of 35 and in the heart of 11.

The sources for these human infections are unknown, but the structure of sarcocysts suggests that other primates may be the true intermediate hosts for some.[23] Serologic surveys indicate that *Sarcocystis* infection may be more common than reported.[8,8a,660,665] For example, anti-*Sarcocystis* antibodies were found in 22.5 to 44.4% of 341 serum samples obtained from patients suspected of toxoplasmosis.[627] In one study, *Sarcocystis* bradyzoites were found in digests of 4 of 112 specimens of human muscle tissue.[335]

The clinical significance of sarcocysts in humans is unknown.

II. OTHER PRIMATES

There are numerous reports of muscular *Sarcocystis* infections in New World monkeys[403,409,569] and Old World monkeys.[193,348,401,403,456,461,491,664] Although only two species of *Sarcocystis* have been named, structural studies of sarcocysts indicate that there are more. In a review of histologic sections from 744 monkeys, including 374 captured wild monkeys, 4 structurally distinct sarcocysts were found in 21% of wild captive monkeys, but none were found in laboratory-raised monkeys.[403] The pathogenicity of *Sarcocystis* for monkeys is unknown, but myocarditis and myositis have been reported.[348,569,664]

Identification of sarcocysts of *Sarcocystis* species in primates is uncertain because their definitive hosts are not known. The two named species of *Sarocystis* in monkeys are *S. kortei* and *S. nesbiti*. Primates are definitive hosts for *S. hominis* and *S. suihominis*.

A. *Sarcocystis kortei* Castellani and Chalmers, 1909

Intermediate hosts include *Macaca mulatta*, *Erythrocebus patos*, *Cercopithecus mitis*, and probably others.

The sarcocysts in skeletal muscles are up to 800 μm with a 5- to 6-μm-thick sarcocyst wall. The bradyzoites are 9 to 14 × 3 to 4 μm in size.[456]

B. *Sarcocystis nesbiti* Mandour, 1969

Macaca mulatta is the intermediate host. The sarcocysts in skeletal muscles are up to 1100 μm with a 1.2- to 1.6-μm-thick, smooth sarcocyst wall. The bradyzoites are 7 to 9 × 2 to 3 μm.[456]

BIBLIOGRAPHY

Information pertinent to the subject matter of this chapter may be found in References 4, 8, 8a, 23, 60, 107, 123, 193, 273, 300, 308, 335, 348, 356, 357, 358, 369, 388a, 401 to 403, 408b, 409, 456, 461, 491, 547, 556, 557, 569, 583, 590, 627, 630, 657, 660, 664, 665, 690, and 700.

Chapter 11

SARCOCYSTOSIS IN DOGS, CATS, AND OTHER CARNIVORES

I. MUSCULAR SARCOCYSTOSIS

Sarcocysts have been found occasionally in muscles of dogs (*Canis familiaris*),[373a,601] domestic cats (*Felis catus*),[232a,373a,410] an Indian lion,[31] a skunk (*Mephitis mephitis*),[230] raccoons (*Procyon lotor*),[410a] foxes (*Vulpes corsac*),[546a] whales,[5] black bears (*Ursus americanus*),[114a] an Indian leopard (*Panthera pardus*),[632d] and other carnivores.[460,660]

Of these sarcocysts, the life cycle of only one *Sarcocystis* species in the fox, *S. corsaci*, is known, and it is unusual because *Vulpes corsac* acts both as the intermediate and the definitive host.[546a] Sarcocysts are up to 8.2 mm long, and the sarcocyst wall is 2.1 to 2.8 µm thick.

II. INTESTINAL SARCOCYSTOSIS

Dogs (Table 1) and cats (Table 2) are definitive hosts for numerous *Sarcocystis* species.

Coyotes are definitive hosts for *S. cruzi*,[151,244] *S. tenella*,[153] *S. capracanis*,[153] *S. odocoileocanis*,[202] *S. hemionilatrantis*,[380] and *S. alceslatrans*.[153]

Red foxes are definitive hosts for *S. cruzi*,[168] *S. tenella*,[10,168] *S. capracanis*,[168] and *S. capreoli*.[223]

Raccoons are definitive hosts for *S. miescheriana*,[567] possibly *S. cruzi*,[247] and *S. leporum*.[114] They are poor definitive hosts.[178,365]

Sporocysts of *Sarcocystis* are also found in the feces of naturally infected carnivores (Table 3).

Table 1
PREPATENT PERIODS AND SPOROCYST SIZE OF *SARCOCYSTIS* SPECIES IN FECES OF DOGS

Species	Intermediate host	Prepatent period (d)	Size of sporocysts (μm)	Ref.
S. arieticanis	Sheep	12 or more	15—16.5 × 9.8—10.5	367a
S. tenella	Sheep	8—9	14—15 × 9—10.5	164
S. cruzi	Cattle	8—33	14.5—17 × 9—11	160, 249
S. capracanis	Goat	10—11	12—16 × 9—11	180
S. hircicanis	Goat	12—15	15.0—17.3 × 10.5—11.3	367
S. bertrami	Horse	12—15	11—14.4 × 8—10.1	376
S. equicanis	Horse	8	15—16.3 × 8.8—11.3	585
S. fayeri	Horse	12—15	11—13 × 7.0—8.5	150
S. miescheriana	Pig	9—12	12.7 × 10.1 (12.2—13.2 × 9.8—10.4)	222
S. alceslatrans	Moose	11—15	14—17 × 8.5—11.5	105, 154
S. wapiti	Wapiti	10—12	15.7—17.4 × 9.6—11.9	637
S. sybillensis	Wapiti	14	15—17 × 10.5—12.0	176
S. cervicanis	Wapiti	?[a]	15.5—16.5 × 10.5—11.3	346
S. hemionilatrantis	Mule deer	9—12	14—16 × 9.5—11	177, 380
S. odocoileocanis	White-tailed deer	8—16	13.2—15.7 × 8.8—12.1	115
S. levinei	Water buffalo	12—34	15—16 × 10	133, 386
S. cameli	Camel	10—14	12 × 9	371
S. gracilis (?)	Roe deer	8—14	12—18 × 9—11.4	205, 216
S. sp.	Grant's gazelle	21	10.8—15.6 × 8.4—12.0	388
S. sp.	Chicken	9	11—12.2 × 8.5—10	512
S. sp.	Pheasant	11	13—15 × 9—11	680
S. sp.	Fallow deer	10—11	15.4 × 8.8	558

[a] ? — not determined.

Table 2
PREPATENT PERIODS AND SPOROCYST SIZE OF *SARCOCYSTIS* SPECIES IN FECES OF CATS

Species	Intermediate host	Prepatent period (d)	Size of sporocysts (μm)	Ref.
S. gigantea	Sheep	11—12	10.5—14.0 × 8.0—9.7	485b
S. medusiformis	Sheep	12—21	10.3—13.0 × 7.3—8.8	97
S. hirsuta	Cattle	8—10	11—14 × 7—9	159
S. fusiformis	Buffalo	8—14	11.5—14 × 9—10	133, 387
S. sp.	Grant's gazelle	11	12.6—18.0 × 8.4—12.0	388
S. muris	House mouse	5—11	8.7—11.7 × 7.5—9.0	597
S. cuniculi	Rabbit	12	11.6—14.5 × 8.7—10	518
S. cymruensis	Rat	4 or more	10.5 × 7.9	11
S. sp.	Chicken	6	11.2—12.2 × 8.5—9.3	680
S. leporum	Cottontail rabbit	10—25	13—16.7 × 9.3—11.1	251
S. odoi	Whitetailed deer	?[a]	11—15 × 9—11	178a
S. porcifelis	Pig	5—10	13.2—13.5 × 7.2—8.0	330

[a] ? — not determined.

Table 3
PREVALENCE OF *SARCOCYSTIS* SPOROCYSTS IN FECES OF PREDATORS

Host	Country	No. of feces examined	Percent positive	Ref.
Coyote (*Canis latrans*)	U.S.	82	20.7	7
	U.S.	150	14.0	244
	U.S.	17	88.0	106
	U.S.	169	52.7	151
Red fox (*Vulpes vulpes*)	Bulgaria	146	9.5	328
	U.K.	25	60.0	10
	U.K.	41	17.0	234
	U.S.	198	10.1	162
Bobcat (*Lynx rufus*)	U.S.	61	3.2	162
Raccoon (*Procyon lotor*)	U.S.	12	17.0	3
Cat (*Felis catus domesticus*)	Australia	71	1.4	103
	Brazil	100	9.0	538
	Federal Republic of Germany	694	4.3	41
	U.S.	16	6.6	187
	U.S.	1000	0.2	87

Table 3 (continued)
PREVALENCE OF *SARCOCYSTIS* SPOROCYSTS IN FECES OF PREDATORS

Host	Country	No. of feces examined	Percent positive	Ref.
	New Zealand	508	16.9	484
	New Zealand	66	13.6	482
Mountain lion (*Felis concolor*)	U.S.	12	16.6	162
Dog (*Canis familiaris*)	Australia	110	20.9	103
	Brazil	155	5.8	537
	New Zealand	481	58.8	483
	U.K.	66	39.3	437
	U.K.	20	25.0	437
	U.K.			
	Greyhound	33	24.0	234
	Sheepdog	123	36.5	
	U.S.	500	1.8	647
	Federal Republic of Germany	500	15.2	43
	Yugoslavia	322	10.5	626
Wolf (*Canis lupus*)	U.S.	72	3.0	203

Chapter 12

SARCOCYSTOSIS IN WILD RUMINANTS AND OTHER LARGE ANIMALS

I. INTRODUCTION

Sarcocysts are ubiquitous in cervids. Up to 100% of mule deer,[181,182,186,202,380,381,454a,560] reindeer,[130,311-324,340,689] white-tailed deer,[116,178a,202,203,209,210,405,447] fallow deer,[138,140,217,558] roe deer,[29,33,137,139,140,205,207,216,223,612,614] red deer or wapiti,[15,96,101,138,176,345-347,454a,458,527,560,689] moose,[105a,154,617] and other large ruminants[4a,395a,443,454a,660] have been found to have mature sarcocysts. Numerous species of Sarcocystis are named from cervids. Whether Sarcocystis species are host-specific for cervid species is not known. Until proven otherwise, species of Sarcocystis in each species of cervid are considered host specific.

II. ROE DEER (*CAPREOLUS CAPREOLUS*)

The number of Sarcocystis species in roe deer is not certain. There are at least two structurally distinct sarcocysts: thick-walled sarcocysts with villar protrusions containing microtubules, and thin-walled sarcocysts containing villar protrusions without microtubules.[207,216,223] There appears to be four species of Sarcocystis in roe deer[216] (Figure 1).

A. *Sarcocystis gracilis* Rátz, 1908[572]

Sarcocysts are up to 2 mm long and have 2.5- to 5.0-μm-long and 0.5- to 0.7-μm-wide villar protrusions of Type 10 (Figure 1A). Microtubules extend from the apex to the base of each villar protrusion.[216]

B. *Sarcocystis caproli* Levchenko, 1963 (Syn. *S. capreolicanis* Erber, Boch, and Barth, 1978)[223]

The sarcocysts have thin walls (<0.5 μm) with hair-like protrusions[223] (Figure 1B). Ultrastructurally, the villar protrusions are up to 8 μm long, lack microtubules, and are folded and are of Type 7. The dog and fox (but not the cat, marten, polecat, or raccoon) are the definitive hosts for Sarcocystis species in the roe deer.[205,216]

C. *Sarcocystis sibirica* Machuslkij, 1947

Sarcocysts are up to 1.6 mm long, and the sarcocyst wall is thin and has stubby projections 0.8×0.8 μm (Figure 1C). According to Entzeroth (personal communication) the dog is the definitive host of this species.

In addition to these three named species, an unnamed species has been reported[33,207] (Figure 1D). At least one species of Sarcocystis (which has not been closely identified or differentiated) is pathogenic for roe deer. All of 3 roe deer fed 100,000 sporocysts became ill; 1 died 49 days postinoculation (DPI), and 2 were euthanatized 33 and 45 DPI.[211] Individual merozoites were seen in the muscles of the roe deer 33 DPI, and schizonts were seen in roe deer 45 and 49 DPI.[211] In another study, a roe deer fawn fed 110,000 sporocysts died of acute sarcocystosis 11 DPI and a pregnant doe aborted after ingesting 50,000 sporocysts.[216]

III. FALLOW DEER (*CERVUS DAMA*)

The Sarcocystis species in the fallow deer have not been named. Only thin-walled sarcocysts are known from Europe.[217,558] They are found in the striated muscle, including the heart; they are up to 510 μm long. The sarcocyst wall is thin (<1.0 μm) and has folded protrusions that are

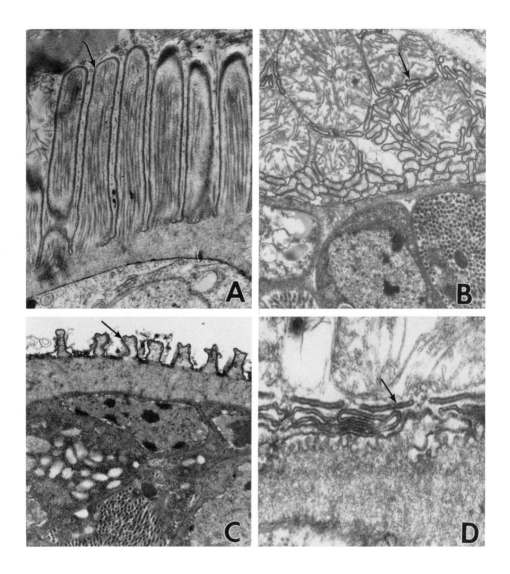

FIGURE 1. Ultrastructure of sarcocyst walls of four species of *Sarcocystis* in roe deer. Arrows point to distal end of villar protrusions. (A) *S. gracilis*, (B) *S. capreoli*, (C) *S. sibirica*, (D) *Sarcocystis* sp. (Magnification [A and C] × 13,000, [B] × 14,000, [D] × 42,000.) (Courtesy of Dr. R. Entzeroth, University of Bonn, Federal Republic of Germany.)

bifurcated at the tip.[217,558] The dog (not the cat) is the definitive host.[558] Nothing is known of the pathogenicity of this species.

IV. REINDEER (*RANGIFER TARANDUS TARANDUS*)

Sarcocystis infections were reported from the U.S.S.R.,[689] the U.S.,[130,340] and Norway.[311-324] Six species of *Sarcocystis* have been reported from the reindeer; their characteristics and biology are summarized in Table 1 and Figure 2. Nothing is known of their pathogenicity.

V. WHITE-TAILED DEER (*ODOCOILEUS VIRGINIANUS*)

There are three species in white-tailed deer; all are present in the U.S.

Table 1
SOME CHARACTERISTICS OF *SARCOCYSTIS* SPECIES IN REINDEER

Species	Definitive hosts	Sarcocyst shape and size (μm)	Villar protrusions shape and size[a] (μm)	Type
S. rangiferi Gjerde, 1984	Not known	Cigar shaped 400 × 21	Villiform 13.2 × 6.7	15
S. tarandi Gjerde, 1984	Not known	Spindle shaped 1000 × 80	Villiform 9.2 × 2.2	10
S. hardangeri Gjerde, 1984	Not known	Ovoid 1670 × 850	Inguiform 25 × 5 × 1	?[b]
S. tarandivulpes Gjerde, 1984	Foxes, raccoon, dog	Spindle shaped 870 × 60	Knob like 0.6—1.2 × 2.7—3.7	17
S. rangi Gjerde, 1984	Foxes	Thread like 8990 × 180	Hair like up to 12.6 × 0.3—0.6	7
S. grüneri Yakimoff and Sokoloff, 1934	Foxes, raccoon, dog	Sack like 580 × 140	Ribbon like up to 4.6 × 0.5 × 0.05	8

[a] Length × (large) diameter, or length × width × thickness.
[b] ? — not determined.

Courtesy of B. Gjerde, Norwegian College of Veterinary Medicine, Oslo, Norway.

A. *Sarcocystis odocoileocanis* Crum, Fayer, and Prestwood, 1981

The definitive hosts of this species include the dog (*Canis familiaris*),[115,202] red fox (*Vulpes vulpes*),[115] grey fox (*Urocyon cinereoargenteus*),[202,447] coyote (*Canis latrans*),[202] and possibly the wolf (*Canis lupus*).[203]

The sarcocysts in histologic sections of skeletal muscles are up to 620 μm long. The sarcocyst wall is 2 to 3 μm thick and appears striated. It has villar protrusions 1.9 to 4.0 × 1.0 to 3.2 μm (Type 15), with characteristic convoluted root-like invaginations at the base (Figure 41A, Chapter 1).[178a,209] The wall is distinctive, unlike any other seen in *Sarcocystis* sp. of livestock or cervids. The sporocysts are 13.2 to 15.7 × 8.8 to 12.1 μm. The bradyzoites are 9.5 to 14 × 2.8 to 4.6 μm.

S. odocoileocanis is mildly pathogenic.[115] All of 3 white-tailed deer fawns did not show clinical signs after ingesting 200,000 (2 deer) or 1 million (1 deer) sporocysts.

S. odocoileocanis is apparently infectious for cattle and sheep. Two sheep fed 1 million *S. odocoileocanis* sporocysts lost wool and body weight. *S. odocoileocanis*-like sarcocysts were found in histologic sections of cattle and sheep fed sporocysts, but the ultrastructure of the sarcocysts was not examined.[115] Because most *Sarcocystis* spp. of domestic herbivores are host-specific, this experiment needs confirmation and should include electron micrographs of the sarcocysts in cattle and sheep fed *S. odocoileocanis* sporocysts.

B. *Sarcocystis odoi* Dubey and Lozier, 1983

The definitive host is probably the cat (*Felis catus*).[115,178a] The sarcocysts are found in skeletal muscles. They are up to 1050 μm long in histologic sections and have a wall 5 to 10 μm thick. The villar protrusions on the sarcocyst wall are of Type 10 and measure 4.2 to 5.6 μm × 0.8 to 1.8 μm.[178] The bradyzoites are about 10 × 3 μm in sections.

C. *Sarcocystis* sp. Dubey and Lozier, 1983

Little is known of this organism. One sarcocyst with a wall 7 to 11 μm thick was found in the esophagus of one deer.[178a] The villar protrusions (Type 15) on the sarcocyst wall were thicker (5 to 8.4 × 3.2 to 4.6 μm) than those of *S. odoi* and had a wavy contour (Figure 40C, Chapter 1). The bradyzoites were 11 to 14.5 × 2.4 to 3.4 μm in sections.

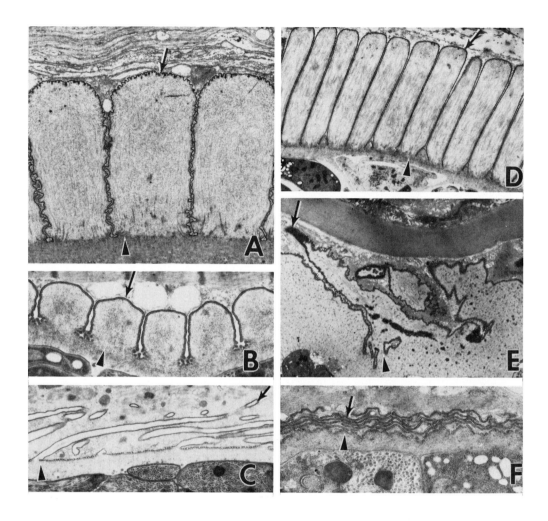

FIGURE 2. TEM of sarcocyst walls of the six *Sarcocystis* species in reindeer from Norway. Arrows point to tips of villi, and arrowheads point to bases of villi. (A) *S. rangiferi*, (B) *S. tarandivulpes*, (C) *S. rangi*, (D) *S. tarandi*, (E) *S. hardangeri*, (F) *S. grüneri*. (Magnification [A and D] × 2925, [B] × 9000, [C] × 8100, [E] × 1800, [F] × 11,250.) (Courtesy of B. Gjerde, Norwegian College of Veterinary Medicine, Oslo, Norway.)

VI. RED DEER OR WAPITI (*CERVUS ELAPHUS*)

There are three named species of *Sarcocystis* in the wapiti: *S. wapiti*, *S. sybillensis*, and *S. cervicanis*; all are transmissible via canids.[176,346,637]

A. *Sarcocystis wapiti* Speer and Dubey, 1982

This species has been reported from North America.[202,458,637] Its definitive hosts include the coyote (*Canis latrans*)[153] and the dog (*Canis familiaris*).[458] The sarcocysts are found in striated muscles and are up to 837 μm long. The sarcocyst wall is thin (<1 μm). The villar protrusions are stubby, finger-like, slender envaginations of Type 2 and are folded over the PVM (Figure 34C, Chapter 1). The bradyzoites are arranged in compartments and measure 14 to 18 × 1.5 to 4 μm. The sporocysts are 15.7 to 17.4 × 9.6 to 11.9 μm; the prepatent period is 10 to 12 d. Endogenous development is unknown.

B. *Sarcocystis sybillensis* Dubey, Jolley, and Thorne, 1983

This species has been reported from North America. The definitive host is the dog (*Canis familiaris*).[176] The sarcocysts are found in striated muscles and in the CNS. In histologic sections of striated muscles, they are up to 637 µm long. The sarcocyst wall is up to 8 µm thick and hairy. The villar protrusions are of Type 8, are up to 6.1 µm long, 0.3 µm wide at the base and 0.1 µm at the apex, have a central core, and the villar tips bifurcate (Figure 36C, Chapter 1). The bradyzoites are 9.5 to 12 × 2.5 to 4 µm. The sporocysts are 15 to 17 × 10.5 to 12 µm. The life cycle and pathogenicity of this species are unknown. Myositis and encephalitis have been found associated with sarcocysts.[176]

C. *Sarcocystis cervicanis* Hernandez-Rodriguez, Navarrete, and Martinez-Gomez, 1981

This species was reported from Europe.[345,346] Its definitive host is the dog (*Canis familiaris*).[346] The sarcocysts in the muscles are 150 to 200 µm long, and the sarcocyst wall is smooth and thin (0.5 µm). The villar protrusions are up to 1.4 µm long and 32 nm wide and bifurcate at the tip. Sporocysts are 16 × 10.9 µm.[348] This species appears similar to *S. wapiti*, but a critical comparison between the two species has not been made.

VII. MULE DEER (*ODOCOILEUS HEMIONUS*)

Four species have been named from mule deer from North America:[181,186,380] *S. hemionilatrantis* Hudkins and Kistner, 1977, *S. hemioni* Dubey and Speer, 1986, *S. youngi* Dubey and Speer, 1986, and *S. americana* Dubey and Speer, 1986. These species are distinguished by the structure of the sarcocyst wall (Figure 7E to H, Chapter 1).[182,186] In *S. hemioni*, the sarcocyst wall is 5 to 8.5 µm thick, and the villar protrusions on the sarcocyst wall are 3.6 to 7.1 µm long and 0.66 to 1.3 µm wide. Filaments and coarsely granular material extend from the base to the tip of villar protrusions (Figure 37B, Chapter 1). In *S. youngi*, the sarcocyst wall is 4 to 8 µm thick with stubby, finger-like envaginations. The villar protrusions are broad distally and narrow centrally or vice versa and measure 8.1 to 9.0 × 3.5 to 5.6 µm (Figure 40D, Chapter 1). In *S. americana*, the sarcocyst wall is 2 to 10 µm thick with hairy folded villar protrusions;[186] its ultrastructure is unknown. Of these four species, only the life cycle of *S. hemionilatrantis* is known.

A. *Sarcocystis hemionilatrantis* Hudkins and Kistner, 1977

The definitive hosts include the dog (*Canis familiaris*)[177,380,381] and coyote (*Canis latrans*).[380]

The sarcocysts are compartmentalized. They are found mostly in skeletal muscles and rarely in the heart. In histologic sections the sarcocysts are microscopic, up to 525 µm long. The sarcocyst wall is 1 to 2 µm thick, appears striated in thick sections, but has characteristic inverted T-shaped villar protrusions in thin sections (Figure 7F, Chapter 1). The villar protrusions are 1.8 to 2.4 µm long and 1.9 to 2.8 µm wide and are highly convoluted and branched at the base (Figure 41A, Chapter 1).[181] The bradyzoites are 13.3 to 16 × 2.6 to 3.2 µm in sections. The sporocysts are 14 to 16 × 9.5 to 11.0 µm.

There are at least three generations of schizogony.[177] First- and second-generation schizonts are in arteries and capillaries of the lung, heart, spleen, and several other organs. First-generation schizonts at 14 DPI were 14 to 39 × 14 to 25 µm and contained about 100 nuclei. The second-generation schizonts seen 24 to 39 DPI are 14 to 32 × 10 to 20 µm and contain 20 to 35 nuclei. Merozoites were seen in peripheral blood 24 DPI; one merozoite was binucleated, suggesting multiplication in the bloodstream.

A terminal generation of schizonts, characteristic of *S. hemionilatrantis*, was found in macrophages in the muscles (Figure 15, Chapter 1). Intramacrophageal schizonts are 10 to 28 × 7 to 14 µm and contain up to 40 nuclei.[177,414]

The sarcocysts at 63 DPI are up to 350 µm long and immature; the sarcocyst wall is thin (<1 µm) and not striated. Mature sarcocysts containing bradyzoites were seen 90 DPI and were up to 525 µm long and striated.[177]

S. hemionilatrantis is pathogenic for mule deer.[414] Of 12 mule deer fed 50,000, 250,000, or

1 million sporocysts (4 deer for each dose level), all became anorectic and uncoordinated. Nine died between 29 and 65 DPI, and three were euthanatized 41, 63, and 90 DPI.[414] The lesions and clinical signs were similar to those in cattle inoculated with 150,000 or more *S. cruzi* sporocysts, but the clinical disease was more protracted in mule deer than in cattle.

Data on natural epizootiology of *Sarcocystis* infections in mule deer indicated that *S. hemionilatrantis* affects the growth of fawns and, coupled with winter conditions, probably predisposes infected deer to predation.[182]

VIII. MOOSE (*ALCES ALCES*)

There are one named and two unnamed species of *Sarcocystis* in moose.

A. *Sarcocystis alceslatrans* Dubey, 1980

This species has been reported from North America[105a,154] and the German Democratic Republic.[617] The coyote (*Canis latrans*) and the dog (*Canis familiaris*) are the definitive hosts.[105,154]

The sarcocysts in skeletal muscles are up to 7 mm long and 0.7 mm wide.[105a] The sarcocyst wall is thin (<1 μm) and has villar protrusions up to 12.6 nm high. They have no microtubules.[105a,617] The bradyzoites are 10 to 13 μm long and 3 to 3.5 μm wide. The sporocysts are 14 to 17 × 8.5 to 10.5 μm. Colwell and Mahrt[105] described the gametogony of this species in the dog intestine.

B. *Sarcocystis* sp. Dubey, 1980

The sarcocysts in skeletal muscles were 85 μm in cross-section, with 3.5- to 10.5-μm-thick walls.[154] No other information is available.

C. *Sarcocystis* sp. Colwell and Mahrt, 1981

The sarcocysts in skeletal muscles were up to 2 mm long, the sarcocyst wall was thin (12 nm), and the primary sarcocyst wall branched out and folded with digitations up to 25 nm high.[105a] A secondary sarcocyst wall was present. The bradyzoites were 9 to 12 μm long.[105a]

IX. PRONGHORN (*ANTILOCAPRA AMERICANA*)

Sarcocystis infection is rare in the pronghorn. Sarcocysts have been reported in only 1 of 21 pronghorns.[154] They were up to 297 μm long and 95 μm wide. The sarcocyst wall was thin (<1 μm) and smooth.

X. UNNAMED DEER

Sarcocystis cervi Destombes, 1957,[129] is considered a *nomen nudum* because it is poorly described and the species of host deer is unknown.

XI. BIGHORN SHEEP (*OVIS CANADENSIS*)

Sarcocystis ferovis Dubey, 1983 is the only species reported. The definitive host of the species is the coyote (*Canis latrans*).[172] The sarcocysts in striated muscles are up to 780 μm long and the sarcocyst wall is thin (<1 μm), with characteristic mushroom-like Type 3 protrusions (Figure 35A, Chapter 1). The sporocysts in coyote feces are 13 to 15 × 9 to 11 μm. *S. ferovis* is not transmissible to ox, sheep, or goats.[172]

Table 2
SARCOCYSTIS INFECTION IN AFRICAN AND OTHER ANTELOPE

African antelope	Sarcocystis species	Maximum length	Wall	Ref.
Grant's gazelle (*Gazella granti*)	*S. woodhousei*, Dogiel, 1915	1.5 mm	?[a]	455
Korin or red-fronted gazelle (*Gazella rufifrons*)	*S. gazellae*, Balfour, 1913	4 mm	?	15a, 455
Coke's hartebeest (*Alcelaphus cokei*)	*S. bubalis*, Dogiel, 1915	2 mm	?	455
Defassa waterbuck (*Kobus defassa*)	*S. nelsoni*, Mandour and Keymer, 1970	5 mm in esophagus, microscopic in heart	4—5 µm, thin	455
Arkhar (*Ovis ammon polii*)	*S. gusevi*, Krylov and Saponzhnikov, 1965	?	?	442
Antelope (not named)	*S. ruandae*, Chiwy and Colback, 1926	?	?	442

[a] ? — Unknown.

XII. NORTH AMERICAN MOUNTAIN GOAT (*OREAMNOS AMERICANUS*)

Sarcocystis sp. infection appears to be common in this host. The sarcocysts were found in 11 of 15 (73%) goats in Alberta, Canada,[454] and in 24 of 56 (43%) goats in Washington, U.S. (Foreyt, personal communication). The sarcocyst wall is type 24 (Figure 43B, Chapter 1) and is distinctive.

XIII. BISON (*BISON BISON*)

S. cruzi is the only species reported.[153,154,163,258] Infection is common.[454a,560]

XIV. AFRICAN AND OTHER ANTELOPE

There are numerous reports of *Sarcocystis* infections in African antelope.[455] Information on named species is summarized in Table 2. Their life cycles are unknown. Some hosts have more than one type of sarcocyst. For example, macroscopic and microscopic species of *Sarcocystis* were found in Grant's gazelle, and dogs and cats were their definitive hosts.[388] Which species cycled through the cat and which species cycled through the dog was not determined.

XV. LLAMAS (*LAMA GLAMA*)

There are at least two species: *S. aucheniae* and *Sarcocystis* sp. (Figure 1, Chapter 1). *S. aucheniae* Brumpt, 1913, sarcocysts are macroscopic (Figure 1B, Chapter 1) with cauliflower-like villar protrusions.[607] Dogs (but not cats) fed macroscopic sarcocysts shed sporocysts 13.1 to 15.7 × 9.0 to 11.3 µm.

XVI. YAK (*POEPHAGUS GRUNNIENS*)

Two species have been named from the Chinese yak: *S. poephagicanis* and *S. poephagi*.[679a] The dog is the probable definitive host for *S. poephagicanis*. *S. poephagicanis* sarcocysts are microscopic, up to 490 μm long. The sarcocyst wall is thin (<1 μm) and smooth with Type 7 villar protrusions. *S. poephagi* sarcocysts are macroscopic and up to 4 cm long and up to 0.8 mm wide. The sarcocyst wall is up to 7 μm thick and striated with Type 10 villar protrusions.

XVII. SEA MAMMALS

S. leuti Blanchard, 1885, was reported from a sea lion (*Zalophus californianus*),[442] and *S. richardi* Hadwen, 1922, was reported from a seal (*Phocus richardi*);[340,501] little is known about these species of *Sarcocystis*.

S. balaenopteralis Akao, 1970, was described from a whale (*Balaenoptera borealis*) in Japan.[5] The sarcocysts were up to 20 cm long. The sarcocyst wall appears to be Type 1, but its description was based on frozen specimens. The bradyzoites were about 7.7 × 2.3 μm.

Chapter 13

SARCOCYSTOSIS IN BIRDS

I. INTRODUCTION

Numerous named (Table 10) and unnamed species of *Sarcocystis* have been reported from muscles of birds including chickens, ducks, geese, and turkeys (see References 46, 49, 61, 78, 81a, 110, 135, 142, 143, 237, 379, 394, 395, 446, 525, 627b, 646a, 655, 660, and 685), but it is not certain how many of these species are valid. Some species of *Sarcocystis* can use several species of birds as intermediate hosts,[54] and the life cycles of only two species of *Sarcocystis* that use birds as intermediate hosts are known.

II. CHICKENS (*GALLUS GALLUS*)

There are at least two species of *Sarcocystis* in chickens: *S. horvathi* (Figure 1) and *Sarcocystis* sp.[680] their structure and life cycles are poorly known.

Sarcocysts have been reported only from free-ranging fowl in Hungary,[573] the Federal Republic of Germany (FRG),[680] Papau New Guinae,[512] and Australia.[512] *S. horvathi* was named from the chicken in Hungary.[572,573] *S. horvathi* sarcocysts are up to 980 μm long.[680] The sarcocyst wall is 2.5 to 3.0 μm thick and appears radially striated. The villar protrusions are conical with microtubules (Figure 1). The bradyzoites are 9 to 12.5 × 2.5 to 3.0 μm. The definitive host is unknown. The dog, cat, polecat, marten, and goshawk were found not to be definitive hosts.[680]

Another species of *Sarcocystis* was reported, but was not named. Its bradyzoites were lanceolate and 14 to 17.5 × 2 to 2.5 μm. The villar protrusions were 1.5 to 2.0 × 0.8 μm. Both dogs and cats were found to be definitive hosts. Chickens fed 1000 or 10,000 sporocysts, either from dogs or cats, became infected. Immature sarcocysts were seen 23 days postinoculation (DPI), and sarcocysts were infectious for dogs and cats at 88 DPI.

An unnamed species of *Sarcocystis* was found in the muscles of a chicken from Papau New Guinae.[512] Its sarcocysts were up to 2 mm long and up to 45 μm wide. The sarcocyst wall was 1.5 μm wide and appeared striated. The villar protrusions were 1.5 μm long and 0.5 μm wide at the base and contained criss-crossing microtubules.[490] A dog shed sporocysts which measured 10 to 13 × 7.25 to 8.5 μm in its feces 7 d after ingesting naturally infected chicken tissues. These sporocysts were not infectious to laboratory-raised chickens.[512]

Both dogs and cats were reported to shed sporocysts after ingesting tissues of naturally infected chickens from the U.S.S.R.[329]

Sarcocysts were found in 45% of 78 free-ranging chickens in Papau New Guinae[512] and in 18.6% of 241 chickens in the FRG;[680] infection in battery-raised hens was not found. Granulomyositis due to degenerating sarcocysts was found in the naturally infected native chickens from Papau New Guinae and Australia.[512]

III. DUCKS

There are probably more than one species of *Sarcocystis* in ducks. Until recently, all sarcocysts in ducks and waterfowl were considered to be *S. riyeli*. *S. riyeli* (Stiles, 1893) Minchin, 1903, is widely distributed in North America. The striped skunk (*Mephitis mephitis*) is the definitive host.[63,683]

The domestic duck and wild mallard (*Anas* spp.) and possibly other ducks[379,646a,686] are intrmediate hosts. Sarcocysts up to 12mm long and 3mm wide are found in skeletal muscles, particularly of the breast.[646a] The sarcocyst wall has characteristic cauliflower-like, Type 23 villar projections[64] (Figure 43A, Chapter 1). Sporocysts are 10 to 14 × 5.5 to 9.5 μm.[64,69,683]

Table 1
AVIAN *SARCOCYSTIS* SPECIES WITH UNKNOWN LIFE CYCLES[a]

	Sarcocystis species	Intermediate host	Maximum length of sarcocysts	Wall thickness	Bradyzoite length (μm)	Ref.
1.	*S. ammodrami* (Splendore, 1907) Babudieri, 1932	Fringillid bird (*Pheucticus ludovicianus*)	?[b]	?	?	443
2.	*S. aramidis* Spendore, 1907	Rallid bird (*Aramides saracura*)	?	?	?	443
3.	*S. colii* Fantham, 1913	Red-faced African mouse bird (*Colius erthromelon*)	2.5 mm	?	5—7	232b
4.	*S. jacarinae* Barreto, 1940	Fringillid bird (*Volatinia jacarina*)	?	?	?	443
5.	*S. kirmsei* Garnham, Duggan, Sinden, 1979	Fire-baked pheasant (*Lophura diardi*)	1.1 mm (brain)	?	8—9	297
6.	*S. nontenella* (Eble, 1961) Levine and Tadros, 1980	Buzzard (*Buteo buteo*)	?	?	?	443
7.	*S. oliverioi* Pessoa, 1935	Psittacid bird (*Forpus passerinus*)	2.7 mm	?	?	551
8.	*S. platydactyli* Bertram, 1892	Gecko (*Tarentola mauritanica*)	?	?	?	443
9.	*S. turdi* Brumpt, 1913	European blackbird (*Turdus merula*)	?	?	?	443
10.	*S. setophagae* Crawley, 1914	Redstart (*Setophaga ruticilla*)	2.5 mm	?	4—5	112

[a] Validity of all these species is in doubt.
[b] ? — unknown.

Macroscopic sarcocysts have been found in up to 78& of adult ducks in North America.[110,142,686] *S. riyeli* is apparently only midly pathogenic; no clinical signs have been reported in experimentally infected ducks. Granulomatous myositis was found in 2 of 18 naturally infected ducks.[686]

Unnamed species are transmissible to ducks via the opossum[196] and cat,[329] but no other information is available on the type of *Sarcocystis* species transmitted.

IV. OTHER BIRDS

Sarcocysts are found in numerous species of birds including grackles (*Quiscalus quiscula*),[237,696] cowbirds (Corvidae),[49] geese (Anatidae),[142,685] and others.[48a,297] Of these, only the life cycle of *S. falcatula* is known.

S. falcatula Stiles, 1893, was first reported from North America. The opossum (*Didelphis virginiana*)[50,53] is the definitive host. Its intermediate hosts include cowbirds (*Molothrus ater*), sparrows (*Passer domesticus*), canaries (*Serinus canarius*), pigeons (*Columba livia*), and budgerigars (*Melopsittacus undulatus*).[54]

The sarcocysts are up to 3.2 mm long.[646a] The sarcocyst wall is thick and has 1- to 5-μm-long villar protrusions with microtubules that extend up to the zoites (Figure 38, Chapter 1).[56] The bradyzoites are about 7× 3 μm long. Sporocysts are 11.2 × 7.4 (9.6 to 12.0 × 6.0 to 8.4) μm.[56]

Schizogony occurs in blood vessels from day 2 to 5.5 months postinoculation.[631a,631b] Although two distinct phases of schizogony are found at 7 to 8 DPI and 28 DPI, all schizonts are structurally similar. There are probably numerous generations of schizonts. Furthermore, the site of schizogony shifts progressively from capillaries to venules to venous endothelial cells. The schizonts are about 15 to 18 μm in diameter, and the merozoites are 7.5 × 2.5 μm.

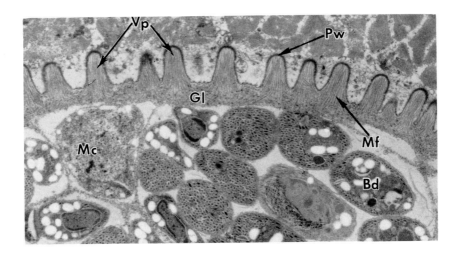

FIGURE 1. TEM of sarcocyst of *S. horvathi*. The primary sarcocyst wall (Pw) has stubby villar protrusions (Vp) with criss-crossing microtubules. (Magnification × 9000.) From Mehlhorn, H., Hartley, W. J., and Heydorn, A. O., *Protistologica*, 12, 451, 1976. With permission.)

Table 2
SPECIES OF *SARCOCYSTIS* WITH RAPTORIAL BIRDS AS DEFINITIVE HOSTS

	Sarcocystis species		
	S. rauschorum	**S. cernae**	**S. dispersa**
Intermediate host	Varying lemming (*Dicrostonyx richardsoni*)	Vole (*Microtus arvalis*)	House mouse (*Mus musculus*)
Schizonts	Hepatocytes, 5—6 d	Hepatocytes, 4—7 d	Hepatocytes, 4—8 d
Blood stages	Yes	Yes	Yes
Sarcocyst type			
Size	Macroscopic	Microscopic	Microscopic?
Sarcocyst wall	Smooth, <1 µm	Smooth, <1 µm	Smooth, <1 µm?
Maturation time (d)	28	?	20
Clinical disease	No	Yes	Yes
Definitive host	Snowy owl (*Nyctea scandiaca*)	Kestrel (*Falco tinnunculus*)	Barn owl, (*Tyto alba*), long-eared owl (*Asia otus*)
References	66	441	71, 73

Note: Other species that can cycle through birds are *S.* sp.[413] (*Mus musculus* and *Accipiter gentilis*), *S. scotti*[442] (*Mus musculus* and *Strix aluco*), *S.* sp.[511] (*Mus musculus* and *Tyto novaehollandiae*), and *S. sebeki*[659] (*Apodemus sylvaticus*, *Mus musculus*, and *Strix aluco*), but their developmental stages are unknown.

S. falcatula is pathogenic for its intermediate hosts. Budgerigars fed 40 sporocysts per gram of body weight died of acute sarcocystosis in 8 to 15 d, mainly due to pulmonary insufficiency.[631a,631b]

V. RAPTORIAL BIRDS

Sarcocysts have been found in the muscle of a bald eagle (*Haliaeetus leucocephalus*).[113] Raptorial birds are definitive hosts for certain species of *Sarcocystis* of small mammals (Table 2). Vultures (*Gyps* spp.) are definitive hosts for *Sarcocystis* of impala (*Aepyceros melampus*).[463]

Chapter 14

SARCOCYSTOSIS IN MISCELLANEOUS HOMIOTHERMIC ANIMALS

Information on sarcocysts of different species of *Sarcocystis* in rodents, marsupials, rabbits, and other small mammals is summarized in Table 1.

In addition, opossums are definitive hosts for a *Sarcocystis* sp. of ducks[196] and *S. falcatula*.[50]

Table 1
SARCOCYSTIS SPECIES IN SMALL MAMMALS

Intermediate host and *Sarcocystis* species	Sarcocysts Maximum length (μm, or stated)	Sarcocysts Wall thickness (μm) and type	Definitive host	Ref.
Mice				
House mouse (*Mus musculus*)				
S. muris (Railliet, 1886) Labbé, 1899	Several cm	Smooth, <1, Type 1 (Figure 34B, Chapter 1)	Cat (*Felis catus*), ferret (*Putorius putorius furo*)	589, 597
S. dispersa Černá, Kolářová, and Šulc, 1978	Microscopic	Smooth, <1, Type 18 (Figure 41B, Chapter 1)	Barn owl (*Tyto alba*), masked owl (*Tyto novaehollandiae*)	73, 619
S. crotali Entzeroth, Chobotar, and Scholtyseck, 1985	4 mm	Smooth, <1, Type 1	Mojave rattlesnake (*Crotalus scutylatus*)	218
S. scotti Levine and Tadros, 1980	?[a]	?	Tawny owl (*Strix aluco*)	442, 511, 660
S. sebeki Tadros and Laarman, 1976	Several cm	Smooth, <1, Type 1	Tawny owl (*Strix aluco*)	660, 655
S. dirumpens (Hoare, 1933) Häfner and Matuschka, 1984	25 mm	<1	Snake of genus *Bitis*	471
S. muriviperae Matuschka, Heydorn, Mehlhorn, Abd-Al-Aal, Diesing, and Biehler, 1987	8 mm	3.5, Type 18	Palestinian viper (*Vipera palaestinae*)	472
Deer mouse (*Peromyscus maniculatus*)				
S. idahoensis Bledsoe, 1980	4.8 mm	<1	Gopher snake (*Pituophis melanoleucus*)	35
S. peromysci Dubey, 1983	1.8 mm	2—5.5, hairy	?	174
Rats				
Rice rat (*Oryzomys capito*)				
S. azevedoi Shaw and Lainson, 1969	2.76 mm	1—1.8, striated	?	622
S. oryzomyos Shaw and Lainson, 1969	24.7 mm	3—3.2, striated	?	622
Moon rat (*Echinosorex gymnurus*)				
S. booliati Dissanaike and Poopalachelvam, 1975	5 mm	<0.5, Type 1	?	131, 397
Multimammate rat (*Mastomys natalensis*)				
S. dirumpens Häfner and Matuschka, 1984	25 mm	Thin, smooth	Snake (see Table 2, Chapter 15)	471

Table 1 (continued)
SARCOCYSTIS SPECIES IN SMALL MAMMALS

Intermediate host and *Sarcocystis* species	Sarcocysts		Definitive host	Ref.
	Maximum length (μm, or stated)	Wall thickness (μm) and type		

Rats

Intermediate host and *Sarcocystis* species	Maximum length	Wall thickness and type	Definitive host	Ref.
Malaysian rats (*Bunomys chrysocomus, Paruromys dominator*)				
S. *sulawesiensis* O'Donoghue, Watts, and Dixon, 1987	120	Thin, Type 5 (Figure 35C, Chapter 1)	?	533
Spiny rat (*Proechimyos guyannensis*)				
S. *proechimyos* Shaw and Lainson, 1969	3.9 mm	3.5—4, striated	?	622
Norway rat (*Rattus norvegicus*)				
S. *cymruensis* Ashford, 1978	50 mm	Thin, smooth, Type 1	Cat (*Felis catus*)	11
S. *murinotechis* Munday and Mason, 1980	300	Thick	Tiger snake (see Table 2, Chapter 15)	519
S. *singaporensis* Zaman and Colley (1975) 1976	1 mm	3—5, striated, Type 19 (Figure 41C, Chapter 1)	Python (*Python reticulatus*)	24, 693
S. *villivillosi* Beaver and Maleckar, 1981	1.1 mm	1.5, thick, Type 22, (Figure 42D, Chapter 1)	Python (*Python reticulatus*)	24
S. *zamani* Beaver and Maleckar, 1981	2 mm	1—3, Type 18	Python (*Python reticulatus*)	24

Voles

Intermediate host and *Sarcocystis* species	Maximum length	Wall thickness and type	Definitive host	Ref.
Vole (*Microtus arvalis*)				
S. *cernae* Levine, 1977	Microscopic	<1, smooth	Kestrel (*Falco tinnunculus*)	74, 441
Meadow vole (*M. pennsylvanicus*)				
S. *microti* Dubey, 1983	1224	Thick, Type 9	?	175
Long-tailed vole (*M. longicaudus*)				
S. *montanaensis* Dubey, 1983	648	<1, smooth, Type 1	?	175
(*M. savii*)				
S. *pitymysi* Splendore, 1918	?	?	?	441

Table 1 (continued)
SARCOCYSTIS SPECIES IN SMALL MAMMALS

Intermediate host and Sarcocystis species	Sarcocysts		Definitive host	Ref.
	Maximum length (μm, or stated)	Wall thickness (μm) and type		

Voles

Common European vole (*M. arvalis*), Short-tailed vole (*M. agrestis*)				
S. putorii (Railliet and Lucet, 1891) Tadros and Laarman, 1978	Macro-scopic	Thin, bristly, Type 9	Ferret (*Mustela putorious* var. *furo*) Common European weasel (*Mustela nivalis*) Stoat (*Mustela nivalis, erminea*) Mink (*Mustela lutreola*)	655, 660
European voles (*M. arvalis, M. oconomus, M. guentheri, Cletherionomys glareolus*)				
S. clethrionomyelaphis Matuschka, 1986	4.5 mm	3, Type 9	Aesculapian snake (*Elaphe longissima*)	469
S. dirumpens	See Table 2, Chapter 13, and house mouse			
S. muriviperae	See Table 2, Chapter 13, and house mouse			

Marsupials

Short-nosed rat kangaroo (*Bettongia lesueuri grayi*)				
S. bettongiae (Bourne, 1932) Bourne, 1934	10 mm	?	?	48a
Opossum (*Didelphis marsupialis*)				
S. didelphidis Scorza, Torrealba, and Dagert, 1957	935	5.2	?	614a
Four-eyed opossum (*Philander opossum*)				
S. garnhami Mandour, 1965	553	1.3—2.2, rose-thorn villi	?	622
Murine opossum (*Marmosa murinata*)				
S. marmosae Shaw and Lainson, 1969	2 mm	Spiny, villi 11.5—13.0 ¥ 2.6	?	622
Wallabies (*Macropus rufogriseus, Petrogale assimilis*)				
S. muscosa (Blanchard, 1885) Labbé, 1889	2 mm	Thin, Type 13 (Figure 39B, Chapter 1)	?	533a

Table 1 (continued)
SARCOCYSTIS SPECIES IN SMALL MAMMALS

Intermediate host and *Sarcocystis* species	Sarcocysts Maximum length (μm, or stated)	Wall thickness (μm) and type	Definitive host	Ref.
Other Small Mammals				
Nine-banded armadillo (*Dasypus novemcinctus*)				
S. dasypi Howells, Carvalho, Mello, and Rangel, 1975	1.53 mm	3.7—4.3, spinose	?	379a
S. diminuta Howells, Carvalho, Mello, and Rangel, 1975	224	2.6—3.5, hair-like projections	?	379a
European rabbit (*Oryctolagus cuniculus*)				
S. cuniculi Brumpt, 1913	Several mm	8—11, striated Type 10	Cat (*Felis catus*)	76, 112, 201, 653, 655
Cottontail rabbit (*Sylvilagus floridanus, S. nuttalli, S. pallistris*)				
S. leporum Crawley, 1914	2 mm	5—6, striated, Type 10	Cat (*Felis catus*)	112, 114, 201, 251, 518
Striped hamster (*Cricetulus griseus*)				
S. cricetuli Patton and Hindle, 1926	1.5 mm	?	?	547a
Guinea pig (*Cavia porcelluss*)				
S. caviae Almeida, 1928	?	?	?	442
Marmot (*Marmota baibacina*)				
S. baibacinacanis Umbetaliev, 1979	?	?	Dog (*Canis familiaris*) Fox (*Vulpes vulpes*) Wolf (*Canis lupus*)	
Collared anteater (*Tamandua tetradactyla*)				442
S. tamanduae Artigas and Uria, 1932	?	?	?	
Squirrels				
Richardson's ground squirrel (*Spermophilus richardsonii*)				
S. bozemanensis Dubey, 1983	300	Smooth, <1	?	173
S. campestris Cawthorn, Wobeser, and Gajadhar, 1983	4 mm	3.6 to 6.4, Type 9 (Figure 37A, Chapter 1)	American badger (*Taxidea taxus*)	64

Table 1 (continued)
SARCOCYSTIS SPECIES IN SMALL MAMMALS

Intermediate host and Sarcocystis species	Sarcocysts		Definitive host	Ref.
	Maximum length (μm, or stated)	Wall thickness (μm) and type		

Squirrels

Intermediate host and Sarcocystis species	Maximum length (μm, or stated)	Wall thickness (μm) and type	Definitive host	Ref.
Yellow suslik (*Spermophilus fulvus*)				
S. citellivulpes Pak, Perminora, and Yeshtokina, 1979	9 mm	1.5—3.5	Red fox (*Vulpes vulpes*)	546
Varying lemming (*Dicrostonyx richardsoni*)				
S. rauschorum Cawthorn, Gajadhar, and Brooks, 1984	122	<1, Type 1	Snowy owl (*Nyctea scandiaca*)	65—67
Red squirrel (*Tamiasciurus hudsonicus*)				
S. sp. Entzeroth, Chobotar, and Scholtyseck, 1983	Microscopic ?	<1, Type 1	?	212
Chipmunk (*Eutamias asiaticus*)				
S. eutamias Tanabe and Okinami, 1940	530	Not striated	?	661b
Eastern chipmunk (*Tamias striatus*)				
S. sp. Entzeroth, Scholtyseck, and Chobotar, 1983	75	Thin, Type 1	?	213

[a] ? — unknown.

Chapter 15

SARCOCYSTOSIS IN POIKILOTHERMIC ANIMALS

Sarcocysts of numerous species of *Sarcocystis* have been reported from snakes,[471,517,546b] lizards,[16,471] tortoises,[423] fishes,[233] and other species of poikilothermic animals.[423,471]

I. REPTILES

Information on 12 named species in reptiles is summarized in Table 1.

Of these 12 species, the definitive hosts for 3 species, *S. podarcicolubris*, *S. chalcidicolubris*, and *S. gallotiae* are known. Snakes of the genera *Coluber* and *Macroprotodon* shed *S. podarcicolubris* sporocysts 9.6 × 6 to 9 µm in their feces after eating infected lizards. *Coluber* snakes shed *S. chalcidicolubris* sporocysts, 9.4 to 11.7 × 8.2 to 9.9 µm after eating infected skinks, *Chalcides ocellatus*.[466,468a,471,473]

S. gallotiae has an unusual life cycle.[474] It can complete its life cycle in one host, *Gallatia galloti*. Its sarcocysts are found in the tails of the lizards. Sporocysts are produced in the intestine of lizards after eating the tails of other lizards; cannibalism is common in these lizards.[474]

Snakes serve as the definitive host for ten species of *Sarcocystis* which have mammals as intermediate hosts (Table 2).

None of the species of *Sarcocystis* are known to cause clinical illness in reptiles. Lainson and Shaw[423] reported granulomatous myositis in a naturally infected tortoise.

II. FISHES

Sarcocystis salvelini Fantham and Porter, 1943, was reported from the trout, *Salvelinus fontinalis*. Its sarcocysts were observed only once among several hundred trout examined in eastern Canada.[233] Sarcocysts up to 0.5 mm long were seen as whitish threads in the abdominal muscles. The bradyzoites were 5.2 to 8.8 × 1.5 to 2.5 µm. No other information is available.

Sarcocysts of an unnamed species were found in an eel (*Zoarces angularis*).[233] They were up to 1 mm long and their zoites were up to 15 µm long and up to 3.5 µm wide. No other information is available.

Table 1
REPTILES AS INTERMEDIATE HOSTS FOR *SARCOCYSTIS*

Species	Intermediate host	Size of bradyzoites (μm)	Size of sarcocyst (μm)	Sarcocyst wall thickness (μm)
S. platydactyli, Bertram, 1892	Gecko (*Tarentola mauritanica*)	5—6 × 1.5—2	400 × 2000	Striated, 7—10
S. gongyli Trinci, 1911	Ocellated skink (*Chalcides ocellatus*)	3—4 × 1.0	30-60 × 200—800	Striated, 2.5—3.5
S. pythonis, Tiegs, 1931	Python (*Morelia argus*)	4—7 × ?	1000	Thin (<1) ?
S. lacertae, Babudieri, 1932	Eurasian lizard (*Podarcis muralis*)	6.5—7.3 × 1.5—2	1000 × 1800—2000	Smooth, 2.5—3.2
S. scelopori, Ball, 1944	Western fence lizard (*Sceloporus occidentalis*)	5.2—6 × 1.5—2	180 × 600	Striated, 2—8
S. utae, Ball, 1944	Side-blotched lizard (*Uta stansburiana*)	5.5—7 × 1.5—2	120 × 950	Smooth, thin (<1)
S. chamaeleonis, Frank, 1966	Chameleon (*Chamaeleo fischeri*)	10—13 × 2	500—1000 × 15,000	Thin (<1), short protrusions
S. kinosterni, (Lainson & Shaw, 1971) 1972	Mud turtle (*Kinosternon scorpioides*)	18.4 × 1.7	170—230 × 8000	Smooth, <1
S. podarcicolubris, Matuschka, 1981	Eurasian lizards (*Podarcis* spp., *Lacerta* spp., *Algyroides nigropunctatus*	7.7—10 × 21	90—290 × 430—1300	Striated, 5
S. gallotiae, Matuschka & Mehlhorn, 1984	Eurasian lizard (*Gallotia galloti*)	16 × 12	43—140 × 110—370	Thick, 4
S. dugesii, Matuschka & Mehlhorn, 1984	Eurasian lizard (*Lacerta dugusii*)	12—16 × ?	40—75 × 220—450	Thick, 7
S. chalcidicolubris, Matuschka, 1987	Ocellated skink (*Chalcides ocellatus*)	10—12 × ?	124—275 × 27—53	Thin, irregular, 2

Modified from Matuschka, F. R., *Parasitol. Res.*, 73, 22, 1987.

Table 2
SPECIES OF SARCOCYSTIS UTILIZING REPTILES AS DEFINITIVE HOSTS

Sarcocystis species

	S. clethrionomyelaphis	*S. crotali*	*S. idahoensis*	*S. murinotechis*	*S. muriviperae*	*S. singaporensis*	*S. villivillosi*	*S. zamani*
Intermediate host	Voles (*Microtus arvalis*, *M. oeconomus*, *M. guentheri*, *Clethrionomys glareolus*)	House mouse (*Mus musculus*)	Deer mouse (*Peromyscus maniculatus*)	Norway rat (*Rattus norvegicus* and others)	House mouse (*Mus musculus*)	Norway rat (*Rattus norvegicus*)	Norway rat (*Rattus norvegicus*)	Norway rat (*Rattus norvegicus*)
Schizonts	Hepatocytes, 7—9 d	?	Hepatocytes	Endothelium, several organs	Hepatocytes, 9—10 d	Endothelium, several organs	Endothelium, several organs ?	Endothelium, several organs ?
Sarcocysts								
Length	Up to 4.5 mm	Up to 4 mm	Up to 4.8 mm	?	Up to 8 mm	Up to 1 mm	Up to 1.1 μm	Up to 2 mm
Cyst wall	3 μm, smooth ?	Thin, smooth ?	Thin, <1μm	Thin ?, 3.5 μm	3—5 μm, thick	Thick, striated	1.5 μm, thick, striated	1—3 μm, thick, smooth
Maturation time	3 months ?	2 months ?	50 d ?	?	?	2 months ?	2 months ?	2 months ?
Clinical disease	Yes	?	Yes	Yes	Yes	No	No	No
Definitive host	Aesculapian snake (*Elaphe longissima*)	Mojave's rattlesnake (*Crotalus scutylatus*)	Gopher snake (*Pituophis melanoleucus*)	Tiger snake (*Notechis ater*)	Palestinian viper (*Vipera palestiniae*)	Python (*Python reticulatus*)		
Reference	469, 496	218	35	519	472	24	24	24

Note: Other species that can cycle through snakes are *S. podarcicolubris* of lizards and *S. dirumpens* of house mouse, but their developmental stages are unknown.

Chapter 16

FRENKELIA AND RELATED GENERA

I. GENUS *FRENKELIA* BIOCCA, 1965

Definition — Asexual multiplication is in small mammals with tissue cysts in the CNS. Sexual stages have been found only in predatory birds.

The genus *Frenkelia* is discussed in detail here because it so closely resembles *Sarcocystis* that some authors have suggested synonymizing *Frenkelia* with *Sarcocystis*. In our opinion, the genus *Frenkelia* should be retained because it is unique. Its "cysts" are different from all sarcocysts; they have been found only in the CNS of small rodents.

There are two named species in this genus, *F. microti* and *F. glareoli*; the latter is the type species.

A. *Frenkelia microti* (Findlay and Middleton, 1934) Biocca, 1965

This species has been reported from Europe and North America. Its intermediate hosts include *Microtus agrestis*,[272] *M. arvalis*,[418] *Apodemus sylvaticus*,[418] *A. flavicollis*,[418] *A. agrarius*,[418] *Mesocricetus auratus*,[418] *Rattus norvegicus*,[588] *Mus musculus*,[418] *Mastomys natalensis*,[587] *Cricetus cricetus*,[587] *Chinchilla laniger*,[497,587] *Oryctolagus cuniculus*,[587] and probably *Lemmus lemmus*,[204] *Microtus modestus*,[287] *Ondatra zibethica*,[404] *Rattus* sp.,[343] and *Erethizon dorsatum*.[407]

The buzzard (*Buteo buteo*) is the definitive host. The definitive host for *F. microti* in North America is unknown, but hawks and owls are suspected.[344,447a] European workers[660] found that *F. microti* can cycle through the American red-tailed hawk, *Buteo borealis*.

Unlike *Sarcocystis*, *Frenkelia* is more specific for the definitive hosts than for the intermediate hosts. The sporocysts are 12 to 15 × 9 to 12 (mean 12 × 10) μm.[418] The schizonts are in the liver parenchymal cells 6 and 7 days postinoculation (DPI). The cysts are thin walled, lobulated, and measure up to 1 mm in diameter and are located in the brain (Figures 1 and 2). Cysts were first observed in the brain 23 DPI.[304]

B. *Frenkelia glareoli* (Erhardova, 1955) Biocca, 1965

This species has been found in Europe and in Japan.[293a] The European bank vole (*Clethrionomys glareolus*)[204,231,587] is its intermediate host. The buzzard (*Buteo buteo*) is the definitive host.[419]

The sporocysts are 11.3 to 13.8 × 7.8 to 10.0 (mean 12.5 × 8.8) μm.[587,588] The schizonts are in the liver parenchymal cells between 5 and 8 DPI; merozoites are 7.6 × 2.2 μm.[419] The cysts are found in brain, beginning 18 DPI. They are microscopic[408a,651] and up to 400 μm at 120 DPI.[419] Congenital transmission is suspected, but not proven.[419,656]

C. Pathogenicity of *Frenkelia*

F. microti and *F. glareoli* are only mildly pathogenic for their intermediate hosts. Although *Frenkelia* cysts may occupy as much as 3.6% of the brain, clinical signs are rarely observed except for diuresis.[204,287,343] Findlay and Middleton,[272] however, reported mortality in voles in England due to cerebral frenkeliosis; these observations have not been confirmed. The age of the voles at primary infection may be important in the pathogenesis of frenkeliosis. Kepka and Skofitsch[408a] reported mortality in voles older than 500 d, whereas younger voles survived natural *Frenkelia* infections. Experimentally infected voles (*M. agrestis*) developed inflammation in the liver, heart, lung, and skeletal muscles in association with schizogonic development of *F. microti*.

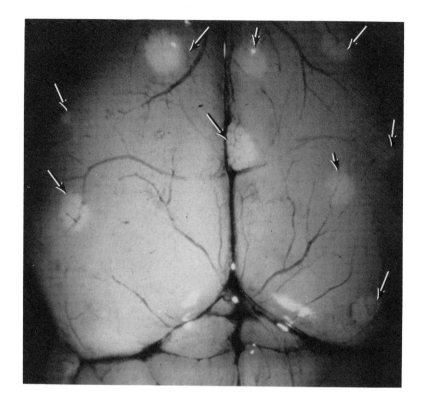

FIGURE 1. Macroscopic *Frenkelia microti* cysts in the brain of a vole (*Microtus agrestis*). (Courtesy of M. Rommel, School of Veterinary Medicine, Hannover, Federal Republic of Germany.)

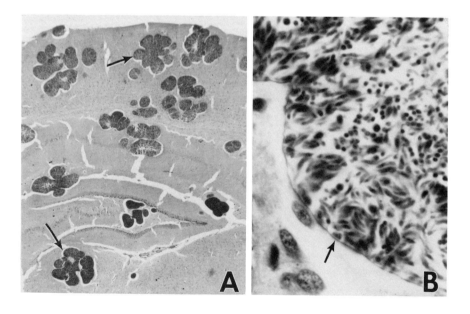

FIGURE 2. *Frenkelia microti* cysts (arrows) in the brain of *Microtus modestus*. (A) Numerous lobulated cysts. (B) Higher magnification to show smooth, thin cyst wall (arrow). (Magnification [A] × 32, [B] × 1000.) (Courtesy of J. K. Frenkel, University of Kansas Medical Center, Kansas City.)

Neither clinical illness nor lesions have been reported for infections with *Frenkelia* in the definitive host, *Buteo buteo*.

II. RELATED GENERA

Other cyst-forming coccidia with isosporan oocysts (containing two sporocysts, each with four sporozoites) are *Toxoplasma*, *Hammondia*, *Isospora* (or *Cystoisospora*), and *Besnoitia*. Their characteristics are summarized in Table 1 and Figure 3. All of these genera can be transmitted via cats or dogs. Also included in Table 1 is the genus *Caryospora*, the oocysts of which contain a single sporocyst with eight sporozoites. These are transmitted via reptiles and birds.

BIBLIOGRAPHY

Information relevant to *Frenkelia* may be found in References 30, 203a, 204, 231, 272, 287, 293a, 304, 343, 344, 404, 406, 407, 408a, 418, 419, 447a, 485c and d, 497, 586, 587, 588, 609a and b, 618, 651, and 656.

Table 1
CHARACTERISTICS OF SOME CYST-FORMING COCCIDIA

	Toxo-plasma	Hammondia	Isospora (Cystoiso-spora)	Besnoitia	Sarco-cystis	Frenkelia	Caryo-spora
Intermediate host	Wide[a] range	Probably wide range ?[b]	Probably wide range ?	Wide range	Wide range	Small mammals	Probably wide range ?
Tissue cyst							
Location	Many organs	Muscle	Many organs	Many organs	Muscle	Brain	Connective tissue
Type	Non-septate	Non-septate	Non-septate	Non-septate	Septate	Septate	Non-septate
Number of organisms in cyst	Many	Many	One	Many	Many	Many	One
Type of wall	Thin	Thin	Thick	Thick	Variable	Thin	Thick
Development in definitive host							
Intestine							
Schizogony	Yes	Yes	Yes	Yes	No	No	Yes
Gametogony	Yes	Yes	Yes	Yes	Yes	Yes	Yes
Oocyst sporulation	Outside of host	Outside of host	Outside of host	Outside of host	In host	In host	Outside of host
Extraintestinal cycle	Yes	No	Yes	No	No	No	?
Mode of transmission							
Carnivorism	+	+	+	+	+	+	+
Fecal	+	−	+	−	−	−	+
Congenital	+	−	−	?	−	−	?
Definitive hosts							

174 *Sarcocystosis of Animals and Man*

Intermediate hosts						
Carnivorism	+	−	−	+ or −	−	−	+
Fecal	+	+	+	+	+	+	+
Congenital	+	−	−	?	+	?	?

[a] Several orders or families.
[b] ? — unknown.

Modified from Dubey, J. P., in *Parasitic Protozoa*, Vol. 3, Kreier, J. P., Ed., Academic Press, New York, 1977. With permission.

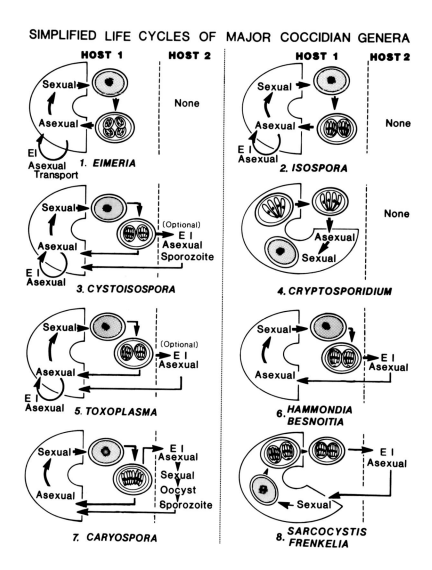

FIGURE 3. Simplified life cycles of major coccidian genera. For each genus, a diagrammatic representation of the intestinal tract appears on the left of the dotted line, under "Host 1" (the definitive host), where oocyst morphology is also shown. On the right, under "Host 2", the extraintestinal stages that develop in the intermediate host are listed in order of development. (From Fayer, R. and Dubey, J. P., *Int. J. Parasitol.*, 17, 615, 1987. With permission.)

REFERENCES

1. **Abbas, M. K. and Powell, E. C.,** Identification of surface antigens of *Sarcocystis muris* (coccidia), *J. Protozool.*, 30, 356, 1983.
2. **Achuthan, H. N.,** Sarcocystis and sarcocystosis in buffalo (*Bubalus bubalis*) calves, *Indian Vet. J.*, 60, 344, 1983.
3. **Adams, J. H., Levine, N. D., and Todd, K. S., Jr.,** *Eimeria* and *Sarcocystis* in raccoons in Illinois, *J. Protozool.*, 28, 221, 1981.
4. **Agarwal, P. K. and Srivastava, A. N.,** Sarcocystosis in man: a report of two cases, *Histopathology*, 7, 783, 1983.
4a. **Agrawal, R. D., Chauhan, P. P. S., and Ahluwalia, S. S.,** Occurrence of *Sarcocystis* sp. (Protozoa: Eimeriidae) in the oesophagus of a goral, *Naemorhedus goral, Indian J. Parasitol.*, 6, 115, 1982.
5. **Akao, S.,** A new species of *Sarcocystis* parasitic in the whale *Balaenoptera borealis*, *J. Protozool.*, 17, 290, 1970.
6. **Arambulo, P. V., III, Tongson, M. S., and Sarmiento, R. V.,** Sarcosporidiosis in Philippine carabaos, *Philipp. J. Vet. Med.*, 11, 53, 1972.
7. **Arther, R. G. and Post, G.,** Coccidia of coyotes in eastern Colorado, *J. Wildl. Dis.*, 13, 97, 1977.
8. **Aryeetey, M. E. and Piekarski, G.,** Serologische *Sarcocystis*-Studien an Menschen und Ratten, *Z. Parasitenkd.*, 50, 109, 1976.
8a. **Aryeetey, M. E. and Piekarski, G.,** Diaplazentarer Übergang von *Sarcocystis*-Antikörpern bei Mensch und Ratten, *Z. Parasitenkd.*, 56, 211, 1978.
9. **Aryeetey, M. E., Mehlhorn, H., and Heydorn, A. O.,** Electron microscopic studies on the development of *Sarcocystis capracanis* in experimentally infected goats, *Zentralbl. Bakteriol. Parasitenkd. Infektionskr. Hyg. I Abt. Orig. Reihe A*, 247, 543, 1980.
10. **Ashford, R. W.,** The fox, *Vulpes vulpes*, as a final host for *Sarcocystis* of sheep, *Ann. Trop. Med. Parasitol.*, 71, 29, 1977.
11. **Ashford, R. W.,** *Sarcocystis cymruensis* n. sp., a parasite of rats *Rattus norvegicus* and cats *Felis catus*, *Ann. Trop. Med. Parasitol.*, 72, 37, 1978.
11a. **Ashford, R. W.,** Who named *Sarcocystis muris*?, *Ann. Trop. Med. Parasitol.*, 72, 95, 1978.
12. **Atkinson, E. M. and Collins, G. H.,** Electrophoretic studies on three enzymes from *Sarcocystis* species in sheep, *Syst. Parasitol.*, 2, 213, 1981.
13. **Baetz, A. L., Barnett, D., Bryner, J. H., and Cysewski, S. J.,** Plasma progesterone concentration in the bovine before abortion or parturition in pregnant animals exposed to *Sarcocystis cruzi, Campylobacter fetus,* or *Aspergillus fumigatus*, *Am. J. Vet. Res.*, 41, 1767, 1980.
14. **Baetz, A. L., Crandell, S. E., Schmerr, M. J. F., Barnett, D., and Bryner, J. H.,** Plasma α-fetoprotein concentrations in pregnant cows exposed to *Sarcocystis cruzi, Campylobacter fetus,* or *Aspergillus fumigatus*, *Am. J. Vet. Res.*, 42, 2146, 1981.
15. **Balbo, T., Rossi, L., Lanfranchi, P., Meneguz, P. G., and Peirone, B.,** Sarcosporidiosis in red deer (*Cervus elaphus*) in regional park of "La Mandria", *Erkrankungen der Zootiere Verhandlungsbericht des 27*, Int. Symp. über die Erkrankungen der Zootiere, Akademie Verlag, Berlin, 1985, 159.
15a. **Balfour, A.,** A sarcocyst of a gazelle (*G. rufifrons*) showing differentiation of spores by vital staining. *Parasitology* 6, 52, 1913.
16. **Ball, G. H.,** Sarcosporidia in southern California lizards, *Trans. Am. Microsc. Soc.*, 63, 144, 1944.
17. **Barci, L. A. G., Amaral, do V., Santos, S. M., and Rebouças, M. M.,** Sarcocistose caprina: prevalência em animais provenientes do estado da Bahia-Brasil, com identificação do agente etiolóagico, *Biologico*, 49, 97, 1983.
18. **Barnett, D., Carter, J. K. Y., Hughes, D. E., Baetz, A. L., and Fayer, R.,** Practicable diagnosis of acute bovine sarcocystosis causally related to bovine abortion, *Annu. Proc. Am. Assoc. Vet. Lab. Diagn.*, 20, 131, 1977.
18a. **Barros de Lambardo, C. S., Barros de, S. S., and Santos dos, M. N.,** Equine protozoal myeloencephalitis in southern Brazil, *Vet. Rec.*, 119, 283, 1986.
19. **Barrows, P. L., Smith, H. M., Jr., Prestwood, A. K., and Brown, J.,** Prevalence and distribution of *Sarcocystis* sp among wild swine of southeastern United States, *J. Am. Vet. Med. Assoc.*, 179, 1117, 1981.
20. **Barrows, P. L., Prestwood, A. K., Adams, D. D., and Dykstra, M. J.,** Development of *Sarcocystis suicanis* Erber, 1977 in the pig, *J. Parasitol.*, 68, 674, 1982.
21. **Barrows, P. L., Prestwood, A. K., and Green, C. E.,** Experimental *Sarcocystis suicanis* infections: disease in growing pigs, *Am. J. Vet. Res.*, 43, 1409, 1982.
22. **Barutzki, D., Erber, M., and Boch, J.,** Möglichkeiten der Desinfektion bei Kokzidiose (Eimeria, Isospora, Toxoplasma, Sarcocystis), *Berl. Muench. Tieraerztl. Wochenschr.*, 94, 451, 1981.
23. **Beaver, P. C., Gadgil, R. K., and Morera, P.,** *Sarcocystis* in man: a review and report of five cases, *Am. J. Trop. Med. Hyg.*, 28, 819, 1979.

24. **Beaver, P. C. and Maleckar, J. R.**, *Sarcocystis singaporensis* Zaman and Colley, (1975) 1976, *Sarcocystis villivillosi* sp. n., and *Sarcocystis zamani* sp. n.: development, morphology, and persistence in the laboratory rat, *Rattus norvegicus*, *J. Parasitol.*, 67, 241, 1981.
25. **Becker, B., Mehlhorn, H., and Heydorn, A. O.**, Light and electron microscopic studies on gamogony and sporogony of 5 *Sarcocystis* species in vivo and in tissue cultures, *Zentralbl. Bakteriol. Parasitenkd. Infektionskr. Hyg., I Abt. Reihe Orig. A*, 244, 394, 1979.
26. **Beech, J.**, Equine protozoan encephalomyelitis, *Vet. Med. Small Anim. Clin.*, 69, 1562, 1974.
27. **Beech, J. and Dodd, D. C.**, Toxoplasma-like encephalomyelitis in the horse, *Vet. Pathol.*, 11, 87, 1974.
28. **Bergler, K. G., Erber, M., and Boch, J.**, Untersuchungen zur Überlebensfähigkeit von Sporozysten bzw. Oozysten von Sarcocystis, Toxoplasma, Hammondia und Eimeria unter Labor- und Freilandbedingungen, *Berl. Muench. Tieraerztl. Wochenschr.*, 93, 288, 1980.
28a. **Bergmann, V. and Kinder, E.**, Unterschiede in der Struktur der Zystenwand bei Sarkozysten des Schafes, *Monatsh., Veternaermed.*, 30, 772, 1975.
29. **Bergmann, V. and Kinder, E.**, Elektronenmikroskopische Untersuchungen zur Wandstruktur von Sarkozysten in der Skelettmuskulatur von Wildschwein und Reh, *Monatsh. Veternaermed.*, 31, 785, 1976.
30. **Bestetti, G. and Fankhauser, R.**, Doppelinfektion des Gehirns mit Frenkelia und Toxoplasma bei einem Chinchilla. Licht- und elektronenmikroskopische Untersuchungen, *Schweiz. Arch. Tierheilkd.*, 120, 591, 1978.
31. **Bhatavdekar, M. Y. and Purohit, B. L.**, A record of sarcosporidiosis in lion, *Indian Vet. J.*, 40, 44, 1963.
32. **Biocca, E., Balbo, T., Guarda, E., and Costantini, R.**, L'importanza della volpe (*Vulpes vulpes*) nella trasmissione della sarcosporidiosi dello stambecco (*Capra ibex*) nel Parco Nazionale del Gran Paradiso, *Parassitologia*, 17, 17, 1975.
32a. **Blanchard, R.**, Note sur les sarcosporidies et sur un essai de classification de ces sporozoaires, *Soc. Zool. Fr. Bull.*, 10, 244, 1885.
33. **Blažek, K., Schramlová, J., Ippen, R., and Kotrlý, A.**, Die Sarkosporidiose des Rehwildes (*Capreolus capreolus* L.), *Folia Parasitol. (Prague)*, 25, 99, 1978.
34. **Bledsoe, B.**, Sporogony of *Sarcocystis idahoensis* in the gopher snake, *Pituophis melanoleucus* (Daudin), *J. Parasitol.*, 65, 875, 1979.
35. **Bledsoe, B.**, *Sarcocystis idahoensis* sp. n. in deer mice *Peromyscus maniculatus* (Wagner) and gopher snakes *Pituophis melanoleucus* (Daudin), *J. Protozool.*, 27, 93, 1980.
36. **Bledsoe, B.**, Transmission studies with *Sarcocystis idahoensis* of deer mice (*Peromyscus maniculatus*) and gopher snakes (*Pituophis melanoleucus*), *J. Wildl. Dis.*, 16, 195, 1980.
37. **Boch, J., Laupheimer, K. E., and Erber, M.**, Drei Sarkosporidienarten bei Schlachtrindern in Süddeutschland, *Berl. Muench. Tieraerztl. Wochenschr.*, 91, 426, 1978.
38. **Boch, J., Mannewitz, U., and Erber, M.**, Sarkosporidien bei Schlachtschweinen in Süddeutschland, *Berl. Muench. Tieraerztl. Wochenschr.*, 91, 106, 1978.
39. **Boch, J., Bierschenck, A., Erber, M., and Weiland, G.**, Sarcocystis- und Toxoplasma-Infektionen bei Schlachtschafen in Bayern, *Berl. Muench. Tieraerztl. Wochenschr.*, 92, 137, 1979.
40. **Boch, J., Böhm, A., and Weiland, G.**, Die Kokzidien-Infektionen (Isospora, Sarcocystis, Hammondia, Toxoplasma) des Hundes, *Berl. Muench. Tieraerztl. Wochenschr.*, 92, 240, 1979.
41. **Boch, J. and Walter, D.**, Vier verschiedene Kokzidienarten bei Katzen in Süddcutschland, *Tieraerztl Umsch.*, 34, 749, 1979.
42. **Boch, J., Hennings, R., and Erber, M.**, Die wirtschaftliche Bedeutung der Sarkosporidiose (*Sarcocystis suicanis*) in der Schweinemast. Auswertung eines Feldversuches, *Berl. Muench. Tieraerztl. Wochenschr.*, 93, 420, 1980.
43. **Boch, J., Mannl, A., Weiland, G., and Erber, M.**, Die Sarkosporidiose des Hundes — Diagnose und Therapie, *Prakt. Tieraerzt*, 61, 636, 1980.
44. **Boch, J. and Erber, M.**, Vorkommen sowie wirtschaftliche und hygienische Bedeutung der Sarkosporidien von Rind, Schaf und Schwein, *Fleischwirtschaft*, 61, 1, 1981.
45. **Bogush, A. A. and Pyshko, I. I.**, Experimental infection of piglets with *Sarcocystis*. (In Russian), *Dostizh. Vet. Nauki Peredovogo Opyta-Zhivotnovodstvu*, 2, 46, 1976.
46. **Borst, G. H. A. and Zwart, P.**, Sarcosporidiosis in psittaciformes, *Z. Parasitenkd.*, 42, 293, 1973.
46a. **Botelho, G. G. and Lopes, C. W. G.**, Esporocistos de *Sarcocystis cruzi* (Apicomplexa: Sarcocystidae) em linfonodos mesentéricos de cães, *Arq. Univ. Fed. Rural Rio de Janeiro*, 7, 87, 1984.
47. **Böttner, A., Charleston, W. A. G., and Hopcroft, D.**, The structure and identity of macroscopically visible *Sarcocystis* cysts in cattle, *Vet. Parasitol.*, 24, 35, 1987.
48. **Böttner, A., Charleston, W. A. G., Pomroy, W. E., and Rommel, M.**, The prevalence and identity of *Sarcocystis* in beef cattle in New Zealand, *Vet. Parasitol.*, 24, 157, 1987.
48a. **Bourne, G.**, Sarcosporidia, *J. R. Soc. West. Aust.*, 19, 1, 1934.
49. **Box, E. D. and Duszynski, D. W.**, Survey for *Sarcocystis* in the brown-headed cowbird (*Molothrus ater*): a comparison of macroscopic, microscopic and digestion techniques, *J. Wildl. Dis.*, 13, 356, 1977.

50. **Box, E. D. and Duszynski, D. W.**, Experimental transmission of *Sarcocystis* from icterid birds to sparrows and canaries by sporocysts from the opossum, *J. Parasitol.*, 64, 682, 1978.
51. **Box, E. D. and McGuinness, T. B.**, Sarcocystis in beef from retail outlets demonstrated by digestion technique, *J. Parasitol.*, 64, 161, 1978.
52. **Box, E. D., Marchiondo, A. A., Duszynski, D. W., and Davis, C. P.**, Ultrastructure of *Sarcocystis* sporocysts from passerine birds and opossums: comments on classification of the genus *Isospora*, *J. Parasitol.*, 66, 68, 1980.
53. **Box, E. D. and Duszynski, D. W.**, *Sarcocystis* of passerine birds: sexual stages in the opossum (*Didelphis virginiana*), *J. Wildl. Dis.*, 16, 209, 1980.
54. **Box, E. D. and Smith, J. H.**, The intermediate host spectrum in a *Sarcocystis* species of birds, *J. Parasitol.*, 68, 668, 1982.
55. **Box, E. D.**, Recovery of *Sarcocystis* sporocysts from feces after oral administration, *Proc. Helminthol. Soc. Wash.*, 50, 348, 1983.
56. **Box, E. D., Meier, J. L., and Smith, J. H.**, Description of *Sarcocystis falcatula* Stiles, 1893, a parasite of birds and opossums, *J. Protozool.*, 31, 521, 1984.
57. **Bratberg, B. and Landsverk, T.**, Sarcocystis infection and myocardial pathological changes in cattle from south-eastern Norway, *Acta Vet. Scand.*, 21, 395, 1980.
58. **Bratberg, B., Helle, O., and Hilali, M.**, Sarcocystis infection in sheep from south-western Norway, *Acta Vet. Scand.*, 23, 221, 1982.
59. **Brehm, H. and Frank, W.**, Der Entwicklungskreislauf von *Sarcocystis singaporensis* Zaman und Colley, 1976 im End- und Zwischenwirt, *Z. Parasitenkd.*, 62, 15, 1980.
60. **Bunyaratvej, S., Bunyawongwiroj, P., and Nitiyanant, P.**, Human intestinal sarcosporidiosis: report of six cases, *Am. J. Trop. Med. Hyg.*, 31, 36, 1982.
60a. **Burgess, D. E.**, *Tritrichomonas foetus*: preparation of monoclonal antibodies with effector function, *Exp. Parasitol.*, 62, 266, 1986.
60b. **Burgess, D. E., Speer, C. A., and Reduker, D. W.**, Identification of antigens of *Sarcocystis cruzi* sporozoites, merozoites and bradyzoites with monoclonal antibodies, *J. Parasitol.*, 74, 1988, in press.
61. **Burtscher, H.**, Große Coccidien-Gewebszysten im Gehirn von *Gracula religiosa* (Aves: Sturnidae), *Zentrabl. Veterinaermed. Reihe B*, 30, 590, 1983.
62. **Carrigan, M. J.**, An outbreak of sarcocystosis in dairy cattle, *Aust. Vet. J.*, 63, 22, 1986.
63. **Cawthorn, R. J., Rainnie, D., and Wobeser, G.** Experimental transmission of *Sarcocystis* sp. (Protozoa: Sarcocystidae) between the shoveler (*Anas clypeata*) duck and the striped skunk (*Mephitis mephitis*), *J. Wildl. Dis.*, 17, 389, 1981.
64. **Cawthorn, R. J., Wobeser, G. A., and Gajadhar, A. A.**, Description of *Sarcocystis campestris* sp. n. (Protozoa: Sarcocystidae): a parasite of the badger *Taxidea taxus* with experimental transmission to the Richardson's ground squirrel, *Spermophilus richardsonii*, *Can. J. Zool.*, 61, 370, 1983.
65. **Cawthorn, R. J., Gajadhar, A. A., and Brooks, R. J.**, Description of *Sarcocystis rauschorum* sp. n. (Protozoa: Sarcocystidae) with experimental cyclic transmission between varying lemmings (*Dicrostonyx richardsoni*) and snowy owls (*Nyctea scandiaca*), *Can. J. Zool.*, 62, 217, 1984.
66. **Cawthorn, R. J. and Brooks, R. J.**, Light microscopical observations on sporogony of *Sarcocystis rauschorum* (Protozoa: Sarcocystidae) in snowy owls (*Nyctea scandiaca*), *Can. J. Zool.*, 63, 1455, 1985.
67. **Cawthorn, R. J. and Brooks, R. J.**, Histological observations on precystic merogony and metrocyte formation of *Sarcocystis rauschorum* (Apicomplexa: Sarcocystidae) in varying lemmings, *Dicrostonyx richardsoni*, *Can. J. Zool.*, 63, 2907, 1985.
68. **Cawthorn, R. J. Reduker, D. W., Speer, C. A., and Dubey, J. P.**, In vitro excystation of *Sarcocystis capracanis*, *Sarcocystis cruzi* and *Sarcocystis tenella* (Apicomplexa), *J. Parasitol.*, 72, 880, 1986.
69. **Cawthorn, R., Speer, C. A., and Blixt, J. A.**, Redescription of *Sarcocystis rileyi*, in preparation.
70. **Černá, Ž.**, Relationship of oocysts of "*Isospora buteonis*" from the barn-owl (*Tyto alba*) to muscle cysts of sarcosporidians from the house mouse (*Mus musculus*), *Folia Parasitol. (Prague)*, 23, 285, 1976.
71. **Černá, Ž.**, Cycle de développement sarcosporidien d'une coccidie, chez la souris, après infestation des animaux par des oocystes-sporocystes isolés de l'intestin de la chouette effraie (*Tyto alba*), *Protistologica*, 13, 401, 1977.
72. **Černá, Ž. and Sénaud, J.**, Sur un type nouveau de multiplication asexuée d'une sarcosporidie dans le foie de le souris, *C. R. Acad. Sci. Ser. D*, 285, 347, 1977.
72a. **Černá, Ž. and Loucková, M.**, *Microtus arvalis*, the intermediate host of a coccidian from the Kestrel (*Falco tinnunculus*), *Vestn. Cesk. Spol. Zool.*, 41, 1, 1977.
73. **Černá, Ž., Kolářová, I., and Šulc, P.**, Contribution to the problem of cyst-producing coccidians, *Folia Parasitol. (Prague)*, 25, 9, 1978.
74. **Černá, Ž., Kolářová, I., and Šulc, P.**, *Sarcocystis cernae* Levine, 1977, excystation, life-cycle and comparison with other heteroxenous coccidians from rodents and birds, *Folia Parasitol. (Prague)*, 25, 201, 1978.

75. Černá, Ž. and Kolárová, I., Contribution to the serological diagnosis of sarcocystosis, *Folia Parasitol.* (Prague), 25, 289, 1978.
76. Černá, Ž., Loučková, M., Nedvedova, H., and Vávra, J., Spontaneous and experimental infection of domestic rabbits by *Sarcocystis cuniculi* Brumpt, 1913, *Folia Parasitol.* (Prague), 28, 313, 1981.
77. Černá, Ž., Multiplication of merozoites of *Sarcocystis dispersa* Ccrná, Kolárová et Sulc, 1978 and *Sarcocystis cernae* Levine, 1977 in the blood stream of the intermediate host, *Folia Parasitol.* (Prague), 30, 5, 1983.
78. Černá, Ž. and Pecka, Z., Muscle sarcocystosis in pheasants and first records of the genus Sarcocystis in *Phasianus colchicus* Linné, 1758 in Czechoslovakia, *Folia Parasitol.* (Prague), 31, 85, 1984.
79. Červa, L., Mácha, J., Gut, J., and Procházková, Z., *Sarcocystis tenella*: A lectin in muscle cysts, *Z. Parasitenkd.*, 67, 349, 1982.
80. Červa, L. and Černá, Z., Indirect haemagglutination reaction with *Sarcocystis dispersa* antigen, *Folia Parasitol.* (Prague), 29, 219, 1982.
81. Céspedes, C. O. C., Mernio, M., and Alvarez, B. F., Primer reporte de *Sarcocystis* en fetos bovinos, *Acad. Cienc. Cuba Ser. Biol.*, No. 155, 2, 1981.
81a. Chabreck, R. H., Sarcosporidiosis in ducks in Louisiana, *Trans. North Am. Wildl. Conf.*, 30, 174, 1965.
82. Chalis, J. R. G. and Mitchell, B. F., Endocrinology of pregnancy and parturition, in *The Biological Basis of Reproductive and Developmental Medicine*, Washaw, J. B., Ed., Elsevier, New York, 1983, 105.
83. Chalis, J. R. G. and Lye, S. J., Parturition, in *Oxford Reviews of Reproductive Biology*, Vol. 8, Clark, J. R., Ed., Clarendon Press, Oxford, 1986, 61.
83a. Chaudhury, R. K., Kushwah, H. S., and Shah, H. L., Biochemistry of the sarcocyst of *Sarcocystis fusiformis* of buffalo *Bubalus bubalis*, *Vet. Parasitol.*, 17, 295, 1984—1985.
84. Chauhan, P. P. S., Bhatia, B. B., Agrawal, R. D., Katara, R. P., and Ahluwalia, S. S., On the gametogonic development of bubaline *Sarcocystis fusiformis* in pups — an experimental study, *Indian J. Exp. Biol.*, 15, 492, 1977.
85. Chauhan, P. P. S., Agrawal, R. D., and Arora, G. S., Incidence of *Sarcocystis fusiformis* in Indian buffaloes, *Indian J. Parasitol.*, 2, 123, 1978.
86. Chhabra, M. B. and Mahajan, R. C., *Sarcocystis* sp from the goat in India, *Vet. Rec.*, 103, 562, 1978.
87. Christie, E., Dubey, J. P., and Pappas, P. W., Prevalence of *Sarcocystis* infection and of other intestinal parasitisms in cats from a humane shelter in Ohio, *J. Am. Vet. Med. Assoc.*, 168, 421, 1976.
87a. Christie, E., Pappas, P. W., and Dubey, J. P., Ultrastructure of excystment of *Toxoplasma gondii* oocysts, *J. Protozool.*, 25, 438, 1978.
88. Clark, E. G., Townsend, H. G. G., and McKenzie, N. T., Equine protozoal myeloencephalitis: a report of two cases from western Canada, *Can. Vet. J.*, 22, 140, 1981.
89. Clegg, F. G., Beverley, J. K. A., and Markson, L. M., Clinical disease in cattle in England resembling Dalmeny disease associated with suspected Sarcocystis infection, *J. Comp. Pathol.*, 88, 105, 1978.
90. Cole, D. J. W., Attempted transmission of ovine *Sarcocystis* from cats to SPF lambs, *N.Z. J. Zool.*, 9, 48, 1982.
91. Collery, P. and Weavers, E., An outbreak of sarcocystosis in calves in Ireland, *Ir. Vet. J.*, 35, 159, 1981.
92. Collery, P., The pathogenesis of acute bovine sarcocystosis. I. Clinical signs and anaemia, *Ir. Vet. J.*, 41, 273, 1987.
93. Collery, P., The pathogenesis of acute bovine sarcocystosis. III. In-vitro studies on the interactions between erythrocytes and mononuclear phagocytes, *Ir. Vet. J.*, 1988, in press.
94. Collery, P., The pathogenesis of acute bovine sarcocystosis. IV. The effects of corticosteroid therapy on the course of the anaemia, *Ir. Vet. J.*, 1988, in press.
95. Collins, G. H. and Crawford, S. J. S., Sarcocystis in goats: prevalence and transmission, *N.Z. Vet. J.*, 26, 288, 1978.
96. Collins, G. H. and Charleston, W. A. G., Studies on *Sarcocystis* species. II. Infection in wild and feral animals — prevalence and transmission, *N.Z. Vet. J.*, 27, 134, 1979.
97. Collins, G. H., Atkinson, E., and Charleston, W. A. G., Studies on *Sarcocystis* species. III. The macrocystic species of sheep, *N.Z. Vet. J.*, 27, 204, 1979.
98. Collins, G. H. and Charleston, W. A. G., Studies on *Sarcocystis* species. IV. A species infecting dogs and goats; development in goats, *N.Z. Vet. J.*, 27, 260, 1979.
99. Collins, G. H., Host reaction to *Sarcocystis* in goats, *N.Z. Vet. J.*, 28, 244, 1980.
100. Collins, G. H., Sutton, R. H., and Charleston, W. A. G., Studies in *Sarcocystis* species. V. A species infecting dogs and goats; observations on the pathology and serology of experimental sarcocystosis in goats, *N.Z. Vet. J.*, 28, 156, 1980.
101. Collins, G. H., Charleston, W. A. G., and Wiens, B. G., Studies on *Sarcocystis* species. VI. A comparison of three methods for the detection of *Sarcocystis* species in muscle, *N.Z. Vet. J.*, 28, 173, 1980.

102. **Collins, G. H. and Charleston, W. A. G.,** Studies on *Sarcocystis* species. VII. The effect of temperature on the viability of macrocysts (*Sarcocystis gigantea*) of sheep, *N.Z. Vet. J.,* 28, 189, 1980.
103. **Collins, G. H., Emslie, D. R., Farrow, B. R. H., and Watson, A. D. J.,** Sporozoa in dogs and cats, *Aust. Vet. J.,* 60, 289, 1983.
104. **Collins, G. H.,** Sarcocystosis and meat industry. *Monogr. Massey Univ. N. Z.,* 1, 1980.
105. **Colwell, D. D. and Mahrt, J. L.,** Development of *Sarcocystis alceslatrans* Dubey, 1980, in the small intestine of dogs, *Am. J. Vet. Res.,* 44, 1813, 1983.
105a. **Colwell, D. D. and Mahrt, J. L.,** Ultrastructure of the cyst wall and merozoites of *Sarcocystis* from moose (*Alces alces*) in Alberta, Canada, *Z. Parasitenskd.,* 65, 317, 1981.
106. **Conder, G. A. and Loveless, R. M.,** Parasites of the coyote (*Canis latrans*) in central Utah, *J. Wildl. Dis.,* 14, 247, 1978.
107. **McLeod, R., Hirabayashi, R. N., Rothman, W., and Remington, J. S.,** Necrotizing vasculitis and *Sarcocystis*: a cause-and-effect relationship?, *South. Med. J.,* 73, 1380,1980.
108. **Cornelissen, A. W. C. A., Overdulve, J. P., and van der Ploeg, M.,** Determination of nuclear DNA of five eucoccidian parasites, *Isospora* (*Toxoplasma*) *gondii, Sarcocystis cruzi, Eimeria tenella, E. acervulina* and *Plasmodium berghei,* with special reference to gamontogenesis and meiosis in *I. (T.) gondii, Parasitology,* 88, 531, 1984.
109. **Corner, A. H., Mitchell, D., Meads, E. B., and Taylor, P. A.,** Dalmeny disease. An infection of cattle presumed to be caused by an unidentified protozoon, *Can. Vet. J.,* 4, 252, 1963.
110. **Cornwell, G.,** New waterfowl host records for *Sarcocystis rileyi* and a review of sarcosporidiosis in birds, *Avian Dis.,* 7, 212, 1963.
111. **Cosgrove, M., Wiggins, J. P., and Rothenbacher, H.,** *Sarcocystis* sp. in the eastern cottontail (*Sylvilagus floridanus*), *J. Wildl. Dis.,* 18, 37, 1982.
112. **Crawley, H.,** Two new Sarcosporidia, *Proc. Natl. Acad. Sci. (Philadelphia),* 66, 214, 1914.
113. **Crawley, R. R., Ernst, J. V., and Milton, J. L.,** *Sarcocystis* in a bald eagle (*Haliaeetus leucocephalus*), *J. Wildl. Dis.,* 18, 253, 1982.
114. **Crum, J. M. and Prestwood, A. K.,** Transmission of *Sarcocystis leporum* from a cottontail rabbit to domestic cats, *J. Wildl. Dis.,* 13, 174, 1977.
114a. **Crum, J. M., Nettles, V. F., and Davidson, W. R.,** Studies on endoparasites of the black bear (*Ursus americanus*) in the southeastern United States, *J. Wildl. Dis.,* 14, 178, 1978.
115. **Crum, J. M., Fayer, R., and Prestwood, A. K.,** *Sarcocystis* spp. in white-tailed deer. I. Definitive and intermediate host spectrum with a description of *Sarcocystis odocoileocanis* n. sp., *J. Wildl. Dis.,* 17, 567, 1981.
116. **Crum, J. M. and Prestwood, A. K.,** Prevalence and distribution of *Sarcocystis* spp. among white-tailed deer of the southeastern United States, *J. Wildl. Dis.,* 18, 195, 1982.
117. **Cunningham, C. C.,** Sarcocysts in the heart muscle of a foal, *Vet. Rec.,* 92, 684, 1973.
118. **Cusick, P. K., Sells, D. M., Hamilton, D. P., and Hardenbrook, H. J.,** Toxoplasmosis in two horses, *J. Am. Vet. Med. Assoc.,* 164, 77, 1974.
119. **Daly, T. J. M., Markus, M. B., and Biggs, H. C.,** *Sarcocystis* of domestic and wild equine hosts, *Proc. Elect. Microsc. Soc. S. Afr.,* 13, 71, 1983.
120. **Daugschies, A., Rommel, M., Schnieder, T., Hennig, M., and Kallweit, E.,** Effects of *Sarcocystis miescheriana* infection on carcass weight and meat quality of halothane-tested fattening pigs, *Vet. Parasitol.,* 25, 19, 1987.
121. **Daugschies, A., Rommel, M., Schnieder, T., Henning, M., and Kallweit, E.,** Effects of *Sarcocystis miescheriana* infection on carcass quality and on the water-binding capacity of the meat of halothane-tested fattening pigs, *Vet. Parasitol.,* 27, 231, 1988.
122. **Daugschies, A., Schnieder, T., Rommel, M., and Bickhardt, K.,** The effects of *Sarcocystis miescheriana* infections on blood enzymes and weight gain of stress-sensitive and stress insensitive pigs, *Vet. Parasitol.,* 27, 221, 1988.
123. **Deluol, A. M., Mechali, D., Cenac, J., Savel, J., and Coulaud, J. P.,** Incidence et aspects cliniques des coccidioses intestinales dans une consultation de médecine tropicale, *Bull. Soc. Pathol. Exot.,* 73, 259, 1980.
124. **Deryło, A. and Kinka, R.,** Skutki ekonomiczne występowania sarkosporydiozy u swin, *Med. Weter.,* 34, 729, 1978.
125. **Deshpande, A. V., Shastri, U. V., and Deshpande, M. S.,** Experimental infection of pups by feeding cysts of *Sarcocystis levinei* Dissanaike and Kan, 1978, *Indian J. Parasitol.,* 6, 331, 1982.
126. **Deshpandey, A. V., Shastri, U. V., and Deshpande, M. S.,** Ineffective treatment of *Sarcocystis levinei* infection in dogs, *Trop. Vet. Anim. Sci. Res.,* 1, 92, 1983.
127. **Deshpandey, A. V., Shastri, U. V., and Deshpande, M. S.,** Prevalence of sarcocysts in cattle and buffaloes in Marathwada (India), *Trop. Vet. Anim. Sci. Res.,* 1, 92, 1983.

128. **Dessouky, M. I., Mohamed, A. H., Nassar, A. M., and Hilali, M.,** Haematological and biochemical changes in buffalo calves inoculated with *Sarcocystis fusiformis* from cats, *Vet. Parasitol.*, 14, 1, 1984.
129. **Destombes, P.,** Les sarcosporidioses au Vietnam, *Bull. Soc. Pathol. Exot.*, 50, 221, 1957.
129a. **D'Haese, J., Mehlhorn, H., and Peters, W.,** Comparative electron microscope study of pellicular structures in coccidia (*Sarcocystis, Besnoitia* and *Eimeria*), *Int. J. Parasitol.* 7, 505, 1977.
130. **Dieterich, R. A.,** Current status of reindeer/caribou diseases in Alaska, in *Proc. 2nd Int. Reindeer/Caribou Symp., Roros, Norway, 1979,* Reimers, E., Gaare, E., and Skjenneberg, S., Eds., Direktoratet for vilt og Ferskvannsfisk, Trondheim, Norway, 1980, 438.
131. **Dissanaike, A. S. and Poopalachelvam, M.,** *Sarcocystis booliati* n. sp. and a parasite of undetermined taxonomic position, *Octoplasma garnhami* n. gen. n. sp., from the moonrat, *Echinosorex gymnurus, Southeast Asian J. Trop. Med. Public Health,* 6, 175, 1975.
132. **Dissanaike, A. S., Kan, S. P., Retnasabapathy, A., and Baskaran, G.,** Developmental stages of *Sarcocystis fusiformis* (Railliet, 1897) and *Sarcocystis* sp., of the water buffalo, in the small intestines of cats and dogs respectively, *Southeast Asian J. Trop. Med. Public Health,* 8, 417, 1977.
133. **Dissanaike, A. S. and Kan, S. P.,** Studies on *Sarcocystis* in Malaysia. I. *Sarcocystis levinei* n. sp. from the water buffalo *Bubalus bubalis, Z. Parasitenkd.,* 55, 127, 1978.
134. **Doflein, F. J. T.,** *Sarcocystis bertrami* n. sp., in *Die Protozoen als Parasiten und Krankheitserreger, nach biologischen Gesichtspunkten dargestellt,* Gustav Fischer., Jena, 1901, 219.
135. **Douglass, E. M. and Hansen, B.,** Sarcosporidiosis in blue and yellow tanagers, *Vet. Med. Small Anim. Clin.,* 74, 1534, 1979.
136. **Dorr, T. E., Higgins, R. J., Dangler, C. A., Madigan, J. E., and Witham, C. L.,** Protozoal myeloencephalitis in horses in California, *J. Am. Vet. Med. Assoc.,* 185, 801, 1984.
137. **Drost, S. and Graubmann, H. D.,** Der Sarkosporidienbefall beim Rehwild, *Monatsh. Veterinaermed.,* 29, 620, 1974.
138. **Drost, S. and Graubmann, H. D.,** Der Sarkosporidienbefall des Rot- und Damwildes, *Monatsh. Veterinaermed.,* 30, 587, 1975.
139. **Drost, S.,** Die Sarkosporidien des Schalenwildes. II. Sarkosporidien beim Rehwild, *Angew. Parasitol.,* 18, 121, 1977.
140. **Drost, S.,** Die Sarkosporidien des Schalenwildes. III. Sarkosporidien beim Rot- und Damwild, *Angew. Parasitol.,* 18, 219, 1977.
141. **Drost, S. and Brackmann, H.,** Zum Sarkosporidienvorkommen bei Rindern eines Sanitätsschlachtbetriebes, *Montsh. Veterinaermed.,* 33, 175, 1978.
142. **Drouin, T. E. and Mahrt, J. L.,** The prevalence of *Sarcocystis* Lankester, 1882, in some bird species in western Canada, with notes on its life cycle, *Can. J. Zool.,* 57, 1915, 1979.
143. **Drouin, T. E. and Mahrt, J. L.,** The morphology of cysts of *Sarcocystis* infecting birds in western Canada, *Can. J. Zool.,* 58, 1477, 1980.
144. **Dubey, J. P.,** Toxoplasmosis in horses, *J. Am. Vet. Med. Assoc.,* 165, 668, 1974.
145. **Dubey, J. P., Davis, G. W., Koestner, A., and Kiryu, K.,** Equine encephalomyelitis due to a protozoan parasite resembling *Toxoplasma gondii, J. Am. Vet. Med. Assoc.,* 165, 249, 1974.
146. **Dubey, J. P.,** A review of *Sarcocystis* of domestic animals and of other coccidia of cats and dogs, *J. Am. Vet. Med. Assoc.,* 169, 1061, 1976.
147. **Dubey, J. P. and Streitel, R. H.,** Shedding of *Sarcocystis* in feces of dogs and cats fed muscles of naturally infected food animals in the midwestern United States, *J. Parasitol.,* 62, 828, 1976.
148. **Dubey, J. P.,** *Toxoplasma, Hammondia, Besnoitia, Sarcocystis,* and other tissue cyst-forming coccidia of man and animals, in *Parasitic Protozoa,* Vol. 3, Kreier, J. P., Ed., Academic Press, New York, 1977, 101.
149. **Dubey, J. P.,** Taxonomy of Sarcocystis and of other coccidia of cats and dogs, *J. Am. Vet. Med. Assoc.,* 170, 778 and 782, 1977.
150. **Dubey, J. P., Streitel, R. H., Stromberg, P. C., and Toussant, M. J.,** *Sarcocystis fayeri* sp. n. from the horse, *J. Parasitol.,* 63, 443, 1977.
151. **Dubey, J. P., Fayer, R., and Seesee, F. M.,** *Sarcocystis* in feces of coyotes from Montana: prevalence and experimental transmission to sheep and cattle, *J. Am. Vet. Med. Assoc.,* 173, 1167, 1978.
152. **Dubey, J. P.,** Frequency of *Sarcocystis* in pigs in Ohio and attempted transmission to cats and dogs, *Am. J. Vet. Res.,* 40, 867, 1979.
153. **Dubey, J. P.,** Coyote as a final host of *Sarcocystis* species of goats, sheep, cattle, elk, bison, and moose in Montana, *Am. J. Vet. Res.,* 41, 1227, 1980.
154. **Dubey, J. P.,** *Sarcocystis* species in moose (*Alces alces*), bison (*Bison bison*), and pronghorn (*Antilocapra americana*) in Montana, *Am. J. Vet. Res.,* 41, 2063, 1980.
155. **Dubey, J. P., Speer, C. A., and Douglass, T. G.,** Development and ultrastructure of first-generation meronts of *Sarcocystis cruzi* in calves fed sporocysts from coyote feces, *J. Protozool.,* 27, 380, 1980.
155a. **Dubey, J. P.,** Development of immunity to sarcocystosis in dairy goats, *Am. J. Vet. Res.,* 42, 800, 1981.
156. **Dubey, J. P.,** Abortion and death in goats included with *Sarcocystis* sporocysts from coyote feces, *J. Am. Vet. Med. Assoc.,* 178, 700, 1981.

157. **Dubey, J. P.,** Early developmental stages of *Sarcocystis cruzi* in calf fed sporocysts from coyote feces, *J. Protozool.,* 28, 431, 1981.
158. **Dubey, J. P., Weisbrode, S. E., Speer, C. A., and Sharma, S. P.,** Sarcocystosis in goats: clinical signs and pathologic and hematologic findings, *J. Am. Vet. Med. Assoc.,* 178, 683, 1981.
159. **Dubey, J. P.,** Development of ox-cat cycle of *Sarcocystis hirsuta, Proc. Helminthol. Soc. Wash.,* 49, 295, 1982.
160. **Dubey, J. P.,** Development of ox-coyote cycle of *Sarcocystis cruzi, J. Protozool.,* 29, 591, 1982.
161. **Dubey, J. P.,** Quantitative parasitemia in calves fed *Sarcocystis cruzi* sporocysts from coyotes, *Am. J. Vet. Res.,* 43, 1085, 1982.
162. **Dubey, J. P.,** *Sarcocystis* and other coccidia in foxes and other wild carnivores from Montana, *J. Am. Vet. Med. Assoc.,* 181, 1270, 1982.
163. **Dubey, J. P.,** Sarcocystosis in neonatal bison fed *Sarcocystis cruzi* sporocysts derived from cattle, *J. Am. Vet. Med. Assoc.,* 181, 1272, 1982.
164. **Dubey, J. P., Speer, C. A., Callis, G., and Blixt, J. A.,** Development of sheep-canid cycle of *Sarcocystis tenella, Can. J. Zool.,* 60, 2464, 1982.
165. **Dubey, J. P., Speer, C. A., and Epling, G. P.,** Sarcocystosis in newborn calves fed *Sarcocystis cruzi* sporocysts from coyotes, *Am. J. Vet. Res.,* 43, 2147, 1982.
166. **Dubey, J. P. and Bergeron, J. A.,** *Sarcocystis* as a cause of placentitis and abortion in cattle, *Vet. Pathol.,* 19, 315, 1982.
167. **Dubey, J. P.,** Clinical sarcocystosis in calves fed *Sarcocystis hirsuta* sporocysts from cats, *Vet. Pathol.,* 20, 90, 1983.
168. **Dubey, J. P.,** Experimental infections of *Sarcocystis cruzi, Sarcocystis tenella, Sarcocystis capracanis* and *Toxoplasma gondii* in red foxes (*Vulpes vulpes*), *J. Wildl. Dis.,* 19, 200, 1983.
169. **Dubey, J. P.,** Immunity to sarcocystosis: modifications of intestinal coccidiosis, and disappearance of sarcocysts in dairy goats, *Vet. Parasitol.,* 12, 23, 1983.
170. **Dubey, J. P.,** Impaired protective immunity to sarcocystosis in pregnant dairy goats, *Am. J. Vet. Res.,* 44, 132, 1983.
171. **Dubey, J. P.,** Microgametogony of *Sarcocystis hirsuta* in the intestine of the cat, *Parasitology,* 86, 7, 1983.
172. **Dubey, J. P.,** *Sarcocystis ferovis* sp. n. from the bighorn sheep (*Ovis canadensis*) and coyote (*Canis latrans*), *Proc. Helminthol. Soc. Wash.,* 50, 153, 1983.
173. **Dubey, J. P.,** *Sarcocystis bozemanensis* sp. nov. (Protozoa: Sarcocystidae) and *S. campestris* from the Richardson's ground squirrel (*Spermophilus richardsonii*), in Montana, U.S.A., *Can. J. Zool.,* 61, 942, 1983.
174. **Dubey, J. P.,** *Sarcocystis peromysci* n. sp. and *S. idahoensis* in deer mice (*Peromyscus maniculatus*) in Montana, *Can. J. Zool.,* 61, 1180, 1983.
175. **Dubey, J. P.,** *Sarcocystis montanaensis* and *S. microti* sp. n. from the meadow vole (*Microtus pennsylvanicus*), *Proc. Helminthol. Soc. Wash.,* 50, 318, 1983.
176. **Dubey, J. P., Jolley, W. R., and Thorne, E. T.,** *Sarcocystis sybillensis* sp. nov. from the North American elk (*Cervus elaphus*), *Can. J. Zool.,* 61, 737, 1983.
177. **Dubey, J. P., Kistner, T. P., and Callis, G.,** Development of *Sarcocystis* in mule deer transmitted through dogs and coyotes, *Can. J. Zool.,* 61, 2904, 1983.
178. **Dubey, J. P. and Blagburn, B. L.,** Failure to transmit *Sarcocystis* species of ox, sheep, goats, moose, elk, and deer to raccoons, *Am. J. Vet. Res.,* 44, 1079, 1983.
178a. **Dubey, J. P. and Lozier, S. M.,** *Sarcocystis* infection in the white-tailed deer (*Odocoileus virgininaus*) in Montana: intensity and description of *Sarcocystis odoi* n. sp., *Am. J. Vet. Res.,* 44, 1738, 1983.
179. **Dubey, J. P.,** Protective immunity to *Sarcocystis capracanis*-induced abortion in dairy goats, *J. Protozool.,* 31, 553, 1984.
180. **Dubey, J. P., Speer, C. A., Epling, G. P., and Blixt, J. A.,** *Sarcocystis capracanis*: development in goats, dogs and coyotes, *Int. Goat Sheep Res.,* 2, 252, 1984.
181. **Dubey, J. P. and Speer, C. A.,** Prevalence and ultrastructure of three types of *Sarcocystis* in mule deer, *Odocoileus hemionus* (Rafinesque), in Montana, *J. Wildl. Dis.,* 21, 219, 1985.
182. **Dubey, J. P. and Kistner, T. P.,** Epizootiology of *Sarcocystis* infections in mule deer fawns in Oregon, *J. Am. Vet. Med. Assoc.,* 188, 1181, 1985.
183. **Dubey, J. P., Leek, R. G., and Fayer, R.,** Prevalence, transmission, and pathogenicity of *Sarcocystis gigantea* of sheep, *J. Am. Vet. Med. Assoc.,* 188, 151, 1986.
184. **Dubey, J. P. and Livingston, C. W., Jr.,** *Sarcocystis capracanis* and *Toxoplasma gondii* infections in range goats from Texas, *Am. J. Vet. Res.,* 47, 523, 1986.
185. **Dubey, J. P. and Miller, S.,** Equine protozoal myeloencephalitis in a pony, *J. Am. Vet. Med. Assoc.,* 188, 1311, 1986.
186. **Dubey, J. P. and Speer, C. A.,** *Sarcocystis* infections in mule deer (*Odocoileus hemionus*) in Montana and the descriptions of three new species, *Am. J. Vet. Res.,* 47, 1052, 1986.
187. **Dubey, J. P., Miller, S., Powell, E. C., and Anderson, W. R.,** Epizootiologic investigations on a sheep farm with *Toxoplasma gondii*-induced abortions, *J. Am. Vet. Med. Assoc.,* 188, 155, 1986.

188. **Dubey, J. P. and Towle, A.,** Toxoplasmosis in sheep: a review and annotated bibliography, Misc. Publ. No. 10, Commonwealth Institute of Parasitology, London, 1986, 1.
189. **Dubey, J. P., Perry, A., and Kennedy, M. J.,** Encephalitis caused by a *Sarcocystis*-like organism in a steer, *J. Am. Vet. Med. Assoc.,* 191, 231, 1987.
190. **Dubey, J. P.,** Lesions in sheep inoculated with *Sarcocystis tenella* sporocysts from canine feces *Vet. Parasitol.,* 26, 237, 1988.
191. **Dubey, J. P., Lindsay, D. S., Speer, C. A., Fayer, R. and Livingston, C. W., Jr.,** *Sarcocystis arieticanis* and other *Sarcocystis* species in sheep in the United States, *J. Parasitol.,* 1988, in press.
192. **Dubey, J. P., Fayer, R., and Speer, C. A.,** Experimental *Sarcocystis hominis* infection in cattle: lesions and ultrastructure, *J. Parasitol.,* 1988, in press.
192a. **Dubey, J. P. and Sheffield, H. G.,** *Sarcocystis sigmodontis* n. sp. from the cotton rats (*Sigmodon hispidus*), *J. Parasitol.,* in press.
193. **Dubin, I. N. and Wilcox, A.,** Sarcocystis in *Macaca mulatta, J. Parasitol.,* 33, 151, 1947.
193a. **Dubremetz, J. F. and Torpier, G.,** Freeze fracture study of the pellicle of an eimerian sporozoite (protozoa, coccidia), *J. Ultrastruct. Res.,* 62, 94, 1978.
194. **Dubremetz, J. F. and Dissous, C.,** Characterisitc proteins of micronemes and dense granules from *Sarcocystis tenella* zoites (Protozoa, Coccidia), *Mol. Biochem. Parasitol.,* 1, 279, 1980.
195. **Dubremetz, M. J. F., Porchet-Henneré, E., and Parenty, M. D.,** Croissance de *Sarcocystis tenella* en culture cellulaire., *C. R. Acad. Sci. Ser. D.,* 280, 1793, 1975.
196. **Duszynski, D. W. and Box, E. D.,** The opossum (*Didelphis virginiana*) as a host for *Sarcocystis debonei* from cowbirds (*Molothrus ater*) and grackles (*Cassidix mexicanus, Quiscalus quiscula*), *J. Parasitol.,* 64, 326, 1978.
197. **Edwards, G. T.,** Prevalence of equine sarcocystis in British horses and a comparison of two detection methods, *Vet. Rec.,* 115, 265, 1984.
198. **El-Refaii, A. H., Abdel-Baki, G., and Selim, M. K.,** Sarcosporidia in goats of Egypt, *J. Egypt. Soc. Parasitol.,* 10, 471, 1980.
199. **Elsasser, T. H., Hammond, A. C., Rumsey, T. S., and Fayer, R.,** Perturbed metabolism and hormonal profiles in calves infected with *Sarcocystis cruzi, Domest. Anim. Endocrinol.,* 3, 277, 1986.
200. **Elsasser, T. H., Rumsey, T. S., Hammond, A. C., and Fayer, R.,** Influence of parasitism on plasma concentrations of growth hormone, somatomedin-C and somatomedin-binding proteins in calves, *J. Endocrinol.,* 116, 191, 1988.
201. **Elwasila, M., Entzeroth, R., Chobotar, B., and Scholtyseck, E.,** Comparison of the structure of *Sarcocystis cuniculi* of the European rabbit (*Oryctolagus cuniculus*) and *Sarcocystis leporum* of the cottontail rabbit (*Sylvilagus floridanus*) by light and electron microscopy, *Acta Vet. Hung.,* 32, 71, 1984.
202. **Emnett, C. W. and Hugghins, E. J.,** *Sarcocystis* of deer in South Dakota, *J. Wildl. Dis.,* 18, 187, 1982.
203. **Emnett, C. W.,** Prevalence of *Sarcocystis* in wolves and white-tailed deer in northeastern Minnesota, *J. Wildl. Dis.,* 22, 193, 1986.
203a. **Enmar, A.,** Studies on a parasite resembling Toxoplasma in the brain of the bank-vole, *Clethrionomys glareolus, Ark. Zool.,* 15, 381, 1963.
204. **Enemar, A.,** M-organisms in the brain of the Norway lemming, *Lemmus lemmus, Ark. Zool.,* 18, 9, 1965.
205. **Entzeroth, R.,** Untersuchungen an Sarkosporidien (Mieschersche Schläuche) des einheimischen Rehwildres (*Capreolus capreolus* L.), *Z. Jagdwiss.,* 27, 247, 1981.
206. **Entzeroth, R., Abdel Ghaffar, F., Chobotar, B., and Scholtyseck, E.,** Fine structural study of *Sarcocystis* sp. from Egyptian camels (*Camelus dromedarius*), *Acta Vet. Acad. Sci. Hung.,* 29, 335, 1981.
207. **Entzeroth, R.,** A comparative light and electron microscope study of the cysts of *Sarcocystis* species of roe deer (*Capreolus capreolus*), *Z. Parasitenkd.,* 66, 281, 1982.
208. **Entzeroth, R.,** Ultrastructure of gamonts and gametes and fertilization of *Sarcocystis* sp. from the roe deer (*Capreolus capreolus*) in dogs, *Z. Parasitenkd.,* 67, 147, 1982.
209. **Entzeroth, R., Chobotar, B., and Scholtyseck, E.,** Ultrastructure of *Sarcocystis* sp. from the muscle of a white-tailed deer (*Odocoileus virginianus*), *Z. Parasitenkd.,* 68, 33, 1982.
210. **Entzeroth, R., Stuht, N., Chobotar, B., and Scholtyseck, E.,** *Sarcocystis* of white-tailed deer (*Odocoileus virginianus*) and its transmission to the dog (*Canis familiaris*), *Tropenmed. Parasitol.,* 33, 111, 1982.
211. **Entzeroth, R.,** Electron microscope study of merogony preceding cyst formation of *Sarcocystis* sp. in roe deer (*Capreolus capreolus*), *Z. Parasitenkd.,* 69, 447, 1983.
212. **Entzeroth, R., Chobotar, B., and Scholtyseck, E.,** Ultrastructure of a *Sarcocystis* species from the red squirrel (*Tamiasciurus hudsonicus*) in Michigan, *Protistologica,* 19, 91, 1983.
213. **Entzeroth, R., Scholtyseck, E., and Chobotar, B.,** Ultrastructure of *Sarcocystis* sp. from the eastern chipmunk (*Tamias striatus*), *Z. Parasitenkd.,* 69, 823, 1983.
214. **Entzeroth, R.,** Electron microscope study of host-parasite interactions of *Sarcocystis muris* (Protozoa, Coccidia) in tissue culture and in vivo, *Z. Parasitenkd.,* 70, 131, 1984.
215. **Entzeroth, R.,** Invasion and early development of *Sarcocystis muris* (Apicomplexa, Sarcocystidae) in tissue cultures, *J. Protozool.,* 32, 446, 1985.

216. **Entzeroth, R.,** Light-, scanning-, and transmission electron microscope study of the cyst wall of *Sarcocystis gracilis* Ràtz, 1909 (Sporoza, Coccidia), from the roe deer (*Capreolus capreolus* L.), *Arch. Protistenkd.*, 129, 183, 1985.
217. **Entzeroth, R., Chobotar, B., Scholtyseck, E., and Neméseri, L.,** Light and electron microscope study of *Sarcocystis sp.* from the fallow deer (*Cervus dama*), *Z. Parasitenkd.*, 71, 33, 1985.
218. **Entzeroth, R., Chobotar, B., and Scholtyseck, E.,** *Sarcocystis crotali* sp. n. with the Mojave rattlesnake (*Crotalus scutulatus scutulatus*)-mouse (*Mus musculus*) cycle, *Arch. Protistenkd.*, 129, 19, 1985.
219. **Entzeroth, R., Chobotar, B., and Scholtyseck, E.,** Electron microscope study of gamogony of *Sarcocystis muris* (Protozoa, Apicomplexa) in the small intestine of cats (*Felis catus*), *Protistologica,* 21, 399, 1985.
220. **Entzeroth, R., Dubremetz, J. F., Hodick, D., and Ferreira, E.,** Immunoelectron microscopic demonstration of the exocytosis of dense granule contents into the secondary parasitophorous vacuole of *Sarcocystis muris* (Protozoa, Apicomplexa), *Eur. J. Cell. Biol.*, 41, 182, 1986.
220a. **Entzeroth, R. and Goerlich, R.,** Monoclonal antibodies against cystozoites of *Sarcocystis muris* (Protozoa, Apicomplexa), *Parasitol. Res.*, 73, 568, 1987.
221. **Erber, M. and Boch, J.,** Untersuchungen über Sarkosporidien des Schwarzwildes. Sporozystenausscheidung durch Hund, Fuchs und Wolf, *Berl. Muench. Tieraerztl. Wochenschr.*, 89, 449, 1976.
222. **Erber, M.,** Möglichkeiten des Nachweises und der Diffenzierung von zwei *Sarcocystis*-Arten des Schweines, *Berl. Muench. Tieraerztl. Wochenschr.*, 90, 480, 1977.
223. **Erber, M., Boch, J., and Barth, D.,** Drei Sarkosporidienarten des Rehwildes, *Berl. Muench. Tieraerztl. Wochenschr.*, 91, 482, 1978.
224. **Erber, M., Meyer, J., and Boch, J.,** Aborte beim Schwein durch Sarkosporidien (*Sarcocystis suicanis*), *Berl. Muench. Tieraerztl. Wochenschr.*, 91, 393, 1978.
225. **Erber, M. and Geisel, O.,** Untersuchungen zur Klinik und Pathologie der *Sarcocystis-suicanis*-Infektion beim Schwein, *Berl. Muench. Tieraerztl. Wochenschr.*, 92, 197, 1979.
226. **Erber, M. and Burgkart, M.,** Wirtschaftliche Verluste durch Sarkosporidiose (*Sarcocystis ovicanis* und *S.* spec.) bei der Mast von Schaflämmern, *Prakt. Tieraerzt,* 62, 422, 1981.
227. **Erber, M. and Geisel, O.,** Vorkommen und Entwicklung von 2 Sarkosporidienarten des Pferdes, *Z. Parasitenkd.*, 65, 283, 1981.
228. **Erber, M. and Göksu, K.,** Sarcosporidia in goats in Turkey and the differentiation of species, *Recent German Research on Problems of Parasitology, Animal Health and Animal Breeding in the Tropics and Subtropics,* Selected Reports on DFG-Supported Research, Markel, H. and Bittner, A., Eds., Deutsche Forschungssgemeinschaft, Bonn, W. Germany, 1984, 21.
229. **Erber, M.,** Life cycle of *Sarcocystis tenella* in sheep and dog, *Z. Parasitenkd.*, 68, 171, 1982.
230. **Erdman, L. F.,** Sarcocystis in striped skunks, *Iowa State Univ. Vet.*, 40, 112, 1978.
231. **Erhardová, B.,** Nález cizopasníku podobných toxoplasmé v mozku norníka rudého — *Clethrionomys glareolus* Schr., *Biologie,* 4, 251, 1955.
232. **Euzéby, J., Lestra, T., and Gauthey, M.,** Note de recherche: sur les affinités taxonomiques des sarcosporidies. *Bull. Soc. Sci. Vet. Med. Comp. Lyon,* 74, 207, 1972.
232a. **Everitt, J. I., Basgall, E. J., Hooser, S. B., and Todd, K. S., Jr.,** *Sarcocystis* sp. in the striated muscle of domestic cats, *Felis catus, Proc. Helminthol. Soc. Wash.*, 54, 279, 1987.
232b. **Fantham, H. B.,** *Sarcocystis colii,* n. sp., a sarcosporidian occurring in the red-faced African mouse bird, *Colius erythromelon, Proc. Philos. Soc. Cambridge,* 17, 221, 1913.
233. **Fantham, H. B. and Porter, A.,** On ostrich Plasmodium and Sarcocystis of Canadian fishes. III. *Plasmodium struthionis,* sp. n. from Sudanese ostriches and *Sarcocystis salvelini,* sp. n., from Canadian speckled trout (*Salvelinus fontinalis*), together with a record of a *Sarcocystis* in the eel pout (*Zoarces angularis*), *Proc. Zool. Soc. London Ser. B,* 113, 25, 1943.
234. **Farmer, J. N., Herbert, I. V., Partridge, M., and Edwards, G. T.,** The prevalence of *Sarcocystis* spp. in dogs and red foxes, *Vet. Rec.*, 102, 78, 1978.
235. **Farooqui, A. A., Adams, D. D., Hanson, W. L., and Prestwood, A. K.,** Studies on the enzymes of *Sarcocystis suicanis*: purification and characterization of an acid phosphatase, *J. Parasitol.*, 73, 681, 1987.
236. **Fayer, R.,** Sarcocystis: development in cultured avian and mammalian cells, *Science,* 168, 1104, 1970.
237. **Fayer, R. and Kocan, R. M.,** Prevalence of *Sarcocystis* in grackles in Maryland, *J. Protozool.*, 18, 547, 1971.
238. **Fayer, R.,** Gametogony of Sarcocystis sp. in cell culture, *Science,* 175, 65, 1972.
239. **Fayer, R. and Johnson, A. J.,** Development of *Sarcocystis fusiformis* in calves infected with sporocysts from dogs, *J. Parasitol.*, 59, 1135, 1973.
240. **Fayer, R. and Leek, R. G.,** Excystation of *Sarcocystis fusiformis* sporocysts from dogs, *Proc. Helminthol. Soc. Wash.*, 40, 294, 1973.
241. **Fayer, R.,** Development of *Sarcocystis fusiformis* in the small intestine of the dog, *J. Parasitol.*, 60, 660, 1974.
242. **Fayer, R. and Johnson, A. J.,** *Sarcocystis fusiformis*: development of cysts in calves infected with sporocysts from dogs, *Proc. Helminthol. Soc. Wash.*, 41, 105, 1974.

243. **Fayer, R.,** Effects of refrigeration, cooking, and freezing on *Sarcocystis* in beef from retail food stores, *Proc. Helminthol. Soc. Wash.,* 42, 138, 1975.
244. **Fayer, R., and Johnson, A. J.,** *Sarcocystis fusiformis* infection in the coyote (*Canis latrans*), *J. Infect. Dis.,* 131, 189, 1975.
245. **Fayer, R. and Johnson, A. J.,** Effect of amprolium on acute sarcocystosis in experimentally infected calves, *J. Parasitol.,* 61, 932, 1975.
246. **Fayer, R. and Thompson, D. E.,** Cytochemical and cytological observations on *Sarcocystis* sp. propagated in cell culture, *J. Parasitol.,* 61, 466, 1975.
247. **Fayer, R., Johnson, A. J., and Hildebrandt, P. K.,** Oral infection of mammals with *Sarcocystis fusiformis* bradyzoites from cattle and sporocysts from dogs and coyotes, *J. Parasitol.,* 62, 10, 1976.
248. **Fayer, R., Johnson, A. J., and Lunde, M.,** Abortion and other signs of disease in cows experimentally infected with *Sarcocystis fusiformis* from dogs, *J. Infect. Dis.,* 134, 624, 1976.
249. **Fayer, R.,** Production of *Sarcocystis cruzi* sporocysts by dogs fed experimentally infected and naturally infected beef, *J. Parasitol.,* 63, 1072, 1977.
250. **Fayer, R.,** The first asexual generation in the life cycle of *Sarcocystis bovicanis, Proc. Helminthol. Soc. Wash.,* 44, 206, 1977.
251. **Fayer, R. and Kradel, D.,** *Sarcocystis leporum* in cottontail rabbits and its transmission to carnivores, *J. Wildl. Dis.,* 13, 170, 1977.
252. **Fayer, R. and Lunde, M. N.,** Changes in serum and plasma proteins and in IgG and IgM antibodies in calves experimentally infected with *Sarcocystis* from dogs, *J. Parasitol.,* 63, 438, 1977.
253. **Fayer, R. and Lynch, G. P.,** Pathophysiological changes in urine and blood from calves experimentally infected with *Sarcocystis cruzi, Parasitology,* 79, 325, 1979.
254. **Fayer, R.,** Multiplication of *Sarcocystis bovicanis* in the bovine bloodstream, *J. Parasitol.,* 65, 980, 1979.
255. **Fayer, R., Heydorn, A. O., Johnson, A. J., and Leek, R. G.,** Transmission of *Sarcocystis suihominis* from humans to swine to nonhuman primates (*Pan troglodytes, Macaca mulatta, Macaca irus*), *Z. Parasitenkd.,* 59, 15, 1979.
256. **Fayer, R. and Leek, R. G.,** *Sarcocystis* transmitted by blood transfusion, *J. Parasitol.,* 65, 890, 1979.
257. **Fayer, R. and Prasse, K. W.,** Hematology of experimental acute *Sarcocystis bovicanis* infection in calves. I. Cellular and serologic changes, *Vet. Pathol.,* 18, 351, 1981.
258. **Fayer, R., Dubey, J. P., and Leek, R. G.,** Infectivity of *Sarcocystis* spp. from bison, elk, moose, and cattle for cattle via sporocysts from coyotes, *J. Parasitol.,* 68, 681, 1982.
259. **Fayer, R., Leek, R. G., and Lynch, G. P.,** Attempted transmission of *Sarcocystis bovicanis* from cows to calves via colostrum, *J. Parasitol.,* 68, 1127, 1982.
260. **Fayer, R., Lynch, G. P., Leek, R. G., and Gasbarre, L. C.,** Effects of sarcocystosis on milk production of dairy cows, *J. Dairy Sci.,* 66, 904, 1982.
261. **Fayer, R. and Dubey, J. P.,** Development of *Sarcocystis fayeri* in the equine, *J. Parasitol.,* 68, 856, 1982.
262. **Fayer, R., Hounsel, C., and Giles, R. C.,** Chronic illness in a sarcocystis infected pony, *Vet. Rec.,* 113, 216, 1983.
263. **Fayer, R. and Dubey, J. P.,** Protective immunity against clinical sarcocystosis in cattle, *Vet. Parasitol.,* 15, 187, 1984.
264. **Fayer, R. and Dubey, J. P.,** Bovine sarcocystosis, *Comp. Contin. Educ. Pract. Vet.,* 8, 130, 1986.
265. **Fayer, R. and Dubey, J. P.,** Comparative epidemiology of coccidia: clues to the etiology of equine protozoal myeloencephalitis, *Int. J. Parasitol.,* 17, 615, 1987.
266. **Fayer, R., Andrews, C., and Dubey, J. P.,** Lysates of *Sarcocystis cruzi* bradyzoites stimulate RAW 264.7 macrophages to produce tumor necrosis factor (cachectin), *J. Parasitol.,* 74, 660, 1988.
267. **Fayer, R.,** Influence of parasitism on growth of cattle possibly mediated through tumor necrosis factor, *Beltsville Symp. in Agricultural Research, Biomechanisms Regulating Growth and Development,* Kluwer Academic, Boston, 1988, 437.
268. **Fayer, R. and Dubey, J. P.,** *Sarcocystis* induced abortion and fetal death, in *Progress in Clinical and Biological Research,* Vol. 281, Proc. NIH Conf. Transplacental Effects on Fetal Health, Scarpelli, D. G. and Magaki, G., Eds., Alan R. Liss, New York, 1988, 153.
269. **Ferguson, H. W. and Ellis, W. A.,** Toxoplasmosis in a calf, *Vet. Rec.,* 104, 392, 1979.
270. **Ferguson, H. W.,** Toxoplasmosis in a calf, *Vet. Rec.,* 105, 135, 1980.
271. **Filho, M. T. de J. and Miraglia, T.,** Histochemical observations on the *Sarcocystis fusiformis* cysts in ox hearts, *Acta Histochem.,* 59, 160, 1977.
272. **Findlay, G. M. and Middleton, A. D.,** Epidemic disease among voles (*Microtus*) with special reference to *Toxoplasma, J. Anim. Ecol.,* 3, 150, 1934.
272a. **Fischer, G.,** Die Entwicklung von *Sarcocystis capracanis* n. spec. in der Ziege, Ph.D., thesis, Freie Universität Berlin, 1979, 1.
273. **Flentje, B., Jungmann, R., and Hiepe, T.,** Vorkommen von Isospora-hominis-Sporozysten beim Menschen, *Dtsch. Gesundheitswes.,* 30, 523, 1975.

273a. **Foggin, C. M.,** Sarcocystis infection and granulomatous myositis in cattle in Zimbabwe, *Zimbabw. Vet. J.*, 11, 8, 1980.
274. **Ford, G. E.,** Prey-predator transmission in the epizootiology of ovine sarcosporidiosis, *Aust. Vet. J.*, 50, 38, 1974.
275. **Ford, G. E.,** Transmission of sarcosporidiosis from dogs to sheep maintained specific pathogen free, *Aust. Vet. J.*, 51, 408, 1975.
276. **Ford, G. E.,** Immunity of sheep to homologous challenge with dog-borne *Sarcocystis* species following varying levels of prior exposure, *Int. J. Parasitol.*, 15, 629, 1985.
277. **Ford, G. E.,** Biochemical characterization for identification of ovine sarcosporidia, *Aust. J. Biol. Sci.*, 39, 31, 1986.
278. **Ford, G. E.,** Completion of the cycle for transmission of sarcosporidiosis between cats and sheep reared specific pathogen free, *Aust. Vet. J.*, 63, 42, 1986.
279. **Ford, G. E.,** Role of the dog, fox, cat and human as carnivore vectors in the transmission of the sarcosporidia that affect sheep meat production, *Aust. J. Agric. Res.*, 37, 79, 1986.
280. **Ford, G. E., Fayer, R., Adams, M., O'Donoghue, P. J., Dubey, J. P., and Baverstock, P. R.,** Genetic characterization by isoenzyme markers of North American and Australasian isolates of species of *Sarcocystis* (Protozoa: Apicomplexa) from mice, sheep, goats and cattle, *Syst. Parasitol.*, 9, 163, 1987.
280a. **Ford, G. E.,** Hosts of two canid genera, the red fox and the dog, as alternate vectors in the transmission of *Sarcocystis tenella* from sheep, *Vet. Parasitol.*, 26, 13, 1987.
281. **Foreyt, W. J.,** Evaluation of decoquinate, lasalocid, and monensin against experimentally induced sarcocystosis in calves, *Am. J. Vet. Res.*, 47, 1674, 1986.
282. **Foreyt, W. J., Parish, S. M., and Leathers, C. W.,** Bovine sarcocystosis: how would you handle an outbreak?, *Vet. Med.*, 81, 275, 1986.
282a. **Fransen, J. L. A., Degryse, A. D. A. Y., Van Mol, K. A. C., and Ooms, L. A. A.,** Sarcocystis und chronische Myopathien bei Pferden., *Berl. Muench. Tieraerztl. Wochenschr.*, 100, 229, 1987.
283. **Frelier, P., Mayhew, I. G., Fayer, R., and Lunde, M. N.,** Sarcocystosis: a clinical outbreak in dairy calves, *Science*, 195, 1341, 1977.
284. **Frelier, P. F., Mayhew, I. G., and Pollock, R.,** Bovine sarcocystosis: pathologic features of naturally occurring infection with *Sarcocystis cruzi*, *Am. J. Vet. Res.*, 40, 651, 1979.
285. **Frelier, P. F.,** Experimentally induced bovine sarcocystosis: correlation of in vitro lymphocyte function with structural changes in lymphoid tissue, *Am. J. Vet. Res.*, 41, 1201, 1980.
286. **Frelier, P. F. and Lewis, R. M.,** Hematologic and coagulation abnormalities in acute bovine sarcocystosis, *Am. J. Vet. Res.*, 45, 40, 1984.
287. **Frenkel, J. K.,** Infections with organisms resembling Toxoplasma, together with the description of a new organism: *Besnoitia jellisoni*, *Atti Congr. Int. Microbiol. Roma*, 5, 426, 1953.
288. **Frenkel, J. K., Heydorn, A. O., Mehlhorn, H., and Rommel, M.,** Sarcocystinae: *Nomina dubia* and available names, *Z. Parasitenkd.*, 58, 115, 1979.
289. **Frenkel, J. K., Heydorn, A. O., Mehlhorn, H., and Rommel, M.,** Clear communication or arbitrary ambiguity, *Z. Parasitenkd.*, 62, 199, 1980.
290. **Frenkel, J. K., Mehlhorn, H., and Heydorn, A. O.,** Protozoan *Nomina dubia:* to arbitrarily restrict or replace. The case of *Sarcocystis* spp., *J. Parasitol.*, 70, 813, 1984.
291. **Frenkel, J. K., Mehlhorn, H., and Heydorn, A. O.,** Beyond the oocyst: over the molehills and mountains of coccidialand, *Parasitol. Today*, 3, 250, 1987.
292. **Freudenberg, H.,** Zur Klinik der Myositis sarcosporidica des Pferdes, *Tieraerztl. Umsch.*, 11, 91, 1956.
293. **Fujino, T., Koga, S., and Ishii, Y.,** Stereoscopical observations on the sarcosporidian cyst wall of *Sarcocystis bovicanis*, *Z. Parasitenkd.*, 68, 109, 1982.
293a. **Fujita, O., Oku, Y., and Ohbayashi, M.,** *Frenkelia* sp. from the red-backed vole, *Clethrionomys rufocanus bedfordiae*, in Hokkaido, Japan, *Jpn. J. Vet. Res.*, 36, 69, 1988.
294. **Gadaev, A.,** On sarcocysts of ass (*Equus asinus*) (In Russian), *Uzb. Biol. Zh.*, 1, 47, 1978.
295. **Gajadhar, A. A., Yates, W. D. G., and Allen, J. R.,** Association of eosinophilic myositis with an unusual species of *Sarcocystis* in a beef cow, *Can. J. Vet. Res.*, 51, 373, 1987.
295a. **Galfre, G., Howe, S. C., Milstein, C., Butcher, G. W., and Howard, J. C.,** Antibodies to major histocompatibility antigens produced by hybrid cell lines, *Nature (London)*, 266, 550, 1977.
296. **Gardner, C. H., Fayer, R., and Dubey, J. P.,** An Atlas of Protozoan Parasites in Animal Tissues, Agricultural Handbook No. 651, U.S. Department of Agriculture, Washington, D. C., 1988, in press.
297. **Garnham, P. C. C., Duggan, A. J., and Sinden, R. E.,** A new species of *Sarcocystis* in the brain of two exotic birds, *Ann. Parasitol. (Paris)*, 54, 393, 1979.
298. **Gasbarre, L. C.,** Demonstration of anti-horse red blood cell antibodies in a sarcocystis infected pony, *Vet. Rec.*, 111, 15, 1982.
299. **Gasbarre, L. C., Suter, P., and Fayer, R.,** Humoral and cellular immune responses in cattle and sheep inoculated with *Sarcocystis*, *Am. J. Vet. Res.*, 45, 1592, 1984.

300. **Gauert, B., Jungmann, R., and Hiepe, T.**, Beziehungen zwischen Intestinalstörungen und parasitären Darminfektionen des Menschen unter besonderer Berücksichtigung von Sarcocystis-Befall, *Dtsch. Gesundheitswes.*, 38, 62, 1983.

301. **Gestrich, R., Schmitt, M., and Heydorn, A.O.**, Pathogenität von *Sarcocystis tenella*-Sporozysten aus den Fäzes von Hunden für Lämmer, *Berl. Muench. Tieraerztl. Wochenschr.*, 87, 362, 1974.

302. **Gestrich, R. and Heydorn, A. O.**, Untersuchungen zur Überlebensdauer von Sarkosporidienzysten im Fleisch von Schlachttieren, *Berl. Muench. Tieraerztl. Wochenschr.*, 87, 475, 1974.

303. **Gestrich, R., Heydorn, A. O., and Baysu, N.**, Beiträge zum Lebenszyklus der Sarkosporidien. VI. Untersuchungen zur Artendifferenzierung bei *Sarcocystis fusiformis* und *Sarcocystis tenella*, *Berl. Muench. Tieraerztl. Wochenschr.*, 88, 191 und 201, 1975.

303a. **Gestrich, R., Mehlhorn, H., and Heydorn, A. O.**, Licht- und elektronenmikroskopische Untersuchungen an Cysten von *Sarcocystis fusiformis* in der Muskulatur von Kälbern nach experimenteller Infektion mit Oocysten und Sporocysten der großen Form von *Isospora bigemina* der Katze, *Zentralbl. Bakteriol. Parasitenkd. Infektionskr. Hyg. I Abt. Orig. Reihe A*, 233, 261, 1975.

304. **Geisel, O., Kaiser, E., Vogel, O., Krampitz, H. E., and Rommel, M.**, Pathomorphologic findings in short-tailed voles (*Microtus agrestis*) experimentally-infected with *Frenkelia microti, J. Wildl. Dis.*, 15, 267, 1979.

305. **Ghaffar, F. A., Hilali, M., and Scholtyseck, E.**, Ultrastructural study of *Sarcocystis fusiformis* (Railliet, 1897) infecting the Indian water buffalo (*Bubalus bubalis*) of Egypt, *Tropenmed. Parasitol.*, 29, 289, 1978.

306. **Ghaffar, F. A., Entzeroth, R., Chobotar, B., and Scholtyseck, E.**, Ultrastructural studies of Sarcocystis sp. from the camel (*Camelus dromedarius*) in Egypt, *Tropenmed. Parasitol.*, 30, 434, 1979.

307. **Ghoshal, S. B., Joshi, S. C., and Shah, H. L.**, A note on the natural occurrence of *Sarcocystis* in buffaloes (*Bubalus bubalis*) in Jabalpur region, M.P., *Indian Vet. J.*, 63, 165, 1986.

307a. **Ghosal, S. B., Joshi, S. C., and Shah, H. L.**, Development of sarcocysts of *Sarcocystis levinei* in water buffalo infected with sporocysts from dogs, *Vet. Parasitol.*, 26, 165, 1987.

307b. **Ghosal, S. B., Joshi, S. C., and Shah, H. L.**, Development of sarcocysts of *Sarcocystis fustiformis* in the water buffalo (*Bubalus bubalis*) experimentally infected with sporocysts from cats, *Indian J. Anim. Sci.*, 57, 413, 1987.

307c. **Ghosal, S. B., Joshi, S. C., and Shah, H. L.**, Sporocyst output in cats fed sarcocysts of *Sarcocystis fusiformis* of the buffalo, *Indian J. Anim. Sci.*, 57, 1100, 1987.

307d. **Ghosal, S. B., Joshi, S. C., and Shah, H. L.**, Morphological studies of the sarcocyst of *Sarcocystis levinei* of the naturally infected water buffaloes (*Bubalus bubalis*), *Indian Vet. J.*, 64, 915, 1987.

307e. **Ghosal, S. B., Joshi, S. C., and Shah, H. L.**, Sporocyst output in dogs fed sarcocysts of *Sarcocysti levinei* of the buffalo (*Bubalus bubalis*), *Vet. Parasitol.*, 28, 173, 1988.

308. **Giboda, M. and Rakár, J.**, First record of "*Isospora hominis*" in Czechoslovakia, *Folia Parasitol. (Prague)*, 25, 16, 1978.

309. **Giles, R. C., Tramontin, R., Kadel, W. L., Whitaker, K., Miksch, D., Bryant, D. W., and Fayer, R.**, Sarcocystosis in cattle in Kentucky, *J. Am. Vet. Med. Assoc.*, 176, 543, 1980.

310. **Gill, H. S., Singh, A., Vadehra, D. V., and Sethi, S. K.**, Shedding of unsporulated isosporan oocysts in feces by dogs fed diaphragm muscles from water buffalo (*Bubalus bubalis*) naturally infected with *Sarcocystis, J. Parasitol.*, 64, 549, 1978.

311. **Gjerde, B.**, A light microscopic comparison of the cysts of four species of *Sarcocystis* infecting the domestic reindeer (*Rangifer tarandus*) in northern Norway, *Acta Vet. Scand.*, 25, 195, 1984.

312. **Gjerde, B.**, Sarcocystis infection in wild reindeer (*Rangifer tarandus*) from Hardangervidda in southern Norway: with a description of the cysts of *Sarcocystis hardangeri* n. sp., *Acta Vet. Scand.*, 25, 205, 1984.

313. **Gjerde, B.**, *Sarcocystis hardangeri* and *Sarcocystis rangi* n. sp. from the domestic reindeer (*Rangifer tarandus*) in northern Norway, *Acta Vet. Scand.*, 25, 411, 1984.

314. **Gjerde, B.**, The fox as definitive host for *Sarcocystis* sp. Gjerde, 1984 from skeletal muscle of reindeer (*Rangifer tarandus*) with a proposal for *Sarcocystis tarandivulpes* n. sp. as replacement name, *Acta Vet. Scand.*, 25, 403, 1984.

315. **Gjerde, B.**, The raccoon dog (*Nyctereutes procyonoides*) as definitive host for *Sarcocystis* spp. of reindeer (*Rangifer tarandus*), *Acta Vet. Scand.*, 25, 419, 1984.

316. **Gjerde, B. and Bratberg, B.**, The domestic reindeer (*Rangifer tarandus*) from northern Norway as intermediate host for three species of *Sarcocystis, Acta Vet. Scand.*, 25, 187, 1984.

317. **Gjerde, B.**, The fox as definitive host for *Sarcocystis rangi* from reindeer (*Rangifer tarandus tarandus*), *Acta Vet. Scand.*, 26, 140, 1985.

318. **Gjerde, B.**, Ultrastructure of the cysts of *Sarcocystis rangi* from skeletal muscle of reindeer (*Rangifer tarandus tarandus*), *Rangifer*, 5, 43, 1985.

319. **Gjerde, B.**, Ultrastructure of the cysts of *Sarcocystis grueneri* from cardiac muscle of reindeer (*Rangifer tarandus tarandus*), *Z. Parasitenkd.*, 71, 189, 1985.

320. **Gjerde, B.**, Ultrastructure of the cysts of *Sarcocystis hardangeri* from skeletal muscle of reindeer (*Rangifer tarandus tarandus*), *Can. J. Zool.*, 63, 2676, 1985.

321. **Gjerde, B.**, Ultrastructure of the cysts of *Sarcocystis tarandivulpes* from skeletal muscle of reindeer (*Rangifer tarandus tarandus*), *Acta Vet. Scand.*, 26, 91, 1985.

322. **Gjerde, B.,** Ultrastructure of the cysts of *Sarcocystis rangiferi* from skeletal muscle of reindeer (*Rangifer tarandus tarandus*), *Can. J. Zool.*, 63, 2669, 1985.
323. **Gjerde, B.,** Ultrastructure of the cysts of *Sarcocystis tarandi* from skeletal muscle of reindeer (*Rangifer tarandus tarandus*), *Can. J. Zool.*, 63, 2913, 1985.
324. **Gjerde, B.,** Scanning electron microscopy of the sarcocysts of six species of *Sarcocystis* from reindeer (*Rangifer tarandus tarandus*), *Acta Pathol. Microbiol. Scand. Sect. B*, 94, 309, 1986.
325. **Göbel, E., Katz, M., and Erber, M.,** Licht- und elektronenmikroskopische Untersuchungen zur Entwicklung von Muskelzysten von *Sarcocystis suicanis* in Hausschweinen nach experimenteller Infektion, *Zentralbl. Bakteriol. Parasitenkd. Infektionskr. Hyg. I Abt. Orig. Reihe A*, 241, 368, 1978.
326. **Göbel, E. and Rommel, M.,** Licht- und elektronenmikroskopische Untersuchungen an Zysten von *Sarcocystis equicanis* in der ösophagusmuskulatur von Pferden, *Berl. Muench. Tieraerztl. Wochenschr.*, 93, 41, 1980.
327. **Godoy, G. A., Volcán, G. S., and Medrano, P. C. E.,** Informacion adicional sobre *Sarcocystis bovicanis*, Heydorn, Gestrich, Mehlhorn y Rommel, 1975 (*S. fusiformis*, Railliet, 1897), en el Estado Bolivar, Venezuela, *Rev. Inst. Med. Trop. Sao Paulo*, 21, 207, 1979.
327a. **Göksu, K.,** Koyunlarda sarcosporidiosis' in yayilisi üzerine arastirmalar, *Istanbul Univ. Vet. Fak. Dergisi*, 1, 110, 1975.
328. **Golemansky, V.,** Observations des oocystes et des spores libres de *Sarcocystis* sp. (*Protozoa: Coccidia*) dans le gros intestin du renard commun (*Vulpes vulpes* L.) en Bulgarie, *Acta Protozool.*, 14, 291, 1975.
329. **Golubkov, V. I.,** Infection of dogs and kittens with *Sarcocystis* from chickens and ducks (in Russian), *Veterinariya*, 77, 55, 1979.
330. **Golubkov, V. I., Rybaltovskii, D. V., and Kislyakova, Z. I.,** The source of infection for swine *Sarcocystis* (in Russian), *Veterinariya*, 11, 85, 1974.
331. **Gomes, A. G. and Lima, J. D.,** *Sarcocystis* (Lankester, 1882) em bovinos de Minas Gerais; ocorrência e métodos de diagnóstico, *Arq. Esc. Vet. Univ. Fed. Minas Gerais*, 34, 83, 1982.
331a. **Gorman, T. R., Alcaíno, H. A., Munoz, H., and Cunazza, C.,** *Sarcocystis* sp. in guanaco (*Lama guanicoe*) and effect of temperature on its viability, *Vet. Parasitol.*, 15, 95, 1984.
332. **Granstrom, D. E., Ridley, R. K., Baoan, Y., Gershwin, L. J., Nesbitt, P. M., and Wempe, L. A.,** Type I hypersensitivity as a component of eosinophilic myositis (muscular sarcocystosis) in cattle, *Am. J. Vet. Res.*, in press.
333. **Greve, E.,** Forekomsten af *Sarcocystis miescheriana* hos svin. En undersøgelse af infektionsfrekvens og geografisk spredning i et sjaellandsk område, *Nord. Veterinaermed.*, 25, 545, 1973.
334. **Greve, E.,** Zur Diagnostik der Sarkosporidiose des Schweines bei der routinemäßigen Fleischuntersuchung, *Monatsh. Veterinaermed.*, 35, 150, 1980.
335. **Greve, E.,** Sarcosporidiosis — an overlooked zoonosis. Man as intermediate and final host, *Dan. Med. Bull.*, 32, 228, 1985.
336. **Groulade, P. and Vallee, A.,** Encéphalite chez un agneau avec presence de toxoplasmes, *Bull. Acad. Vet. Fr.*, 32, 135, 1959.
337. **Gupta, S. L. and Gautam, O. P.,** *Sarcocystis* infection in goats of Hissar and its transmission to dogs, *Indian J. Parasitol.*, 6, 73, 1982.
338. **Gut, J.,** Effectiveness of methods used for the detection of sarcosporidiosis in farm animals, *Folia Parasitol. (Prague)*, 29, 289, 1982.
339. **Gut, J.,** Infection of mice immunized with formolized cystozoites of *Sarcocystis dispersa* Cerná, Kolárová et Sulc, 1978, *Folia Parasitol. (Prague)*, 29, 285, 1982.
340. **Hadwen, S.,** Cyst-forming protozoa in reindeer and caribou, and a sarcosporidian parasite of the seal (*Phoca richardi*), *J. Am. Vet. Med. Assoc.*, 61, 374, 1922.
341. **Hartley, W. J. and Blakemore, W. F.,** An unidentified sporozoan encephalomyelitis in sheep, *Vet. Pathol.*, 11, 1, 1974.
342. **Hartley, W. J.,** Foetal protozoan infection in cattle, Proc. 13th World Congr. Dis. Cattle, Durban, S. Afr., 1, 512, 1984.
342a. **Haskard, D., Cavender, D., Beatty, P., Springer, T. and Ziff, M.,** T lymphocyte adhesion to endothelial cells: mechanisms demonstrated by anti-LFA-1 monoclonal antibodies, *J. Immunol.*, 137, 2901, 1986.
342b. **Hasselmann, G. E.,** Alteracoes pathologicas do myocardio na sarcosporideose, *Bull. Inst. Bras. Scienz.*, 2, 319, 1926.
343. **Hayden, D. W., King, N. W., and Murthy, A. S. K.,** Spontaneous Frenkelia infection in a laboratory-reared rat, *Vet. Pathol.*, 13, 337, 1976.
344. **Henry, D. P.,** *Isospora buteonis* sp. nov. from the hawk and owl, and notes on *Isospora lacazii* (Labbé) in birds, *Univ. Calif. Publ. Zool.*, 37, 291, 1932.
345. **Hernandez, S., Martinez, F., Calero, R., Moreno, T., and Navarrete, I.,** Parasitos del ciervo (*Cervus elaphus*) en Cordoba. I. Primera relacion, *Rev. Iber. Parasitol.*, 40, 93, 1980.
346. **Hernandez Rodriguez, S., Navarrete, I., and Martínez Gómez, F.,** *Sarcocystis cervicanis*, nueva especie parasita del ciervo (*Cervus elaphus*), *Rev. Iberi. Parasitol.*, 41, 43, 1981.

347. **Hernandez-Rodriguez, S., Martínez-Gómez, F., Navarrete, I., and Acosta-García, I.,** Estudio al microscopio optico y electronico del quiste de *Sarcocystis cervicanis, Rev. Iber. Parasitol.,* 41, 351, 1981.

347a. **Hernandez-Rodriguez, S., Martinez-Gomez, F., Lopez-Rodriguez, R., and Navarrete, I.,** Morfologiá y biologia de *Sarcocystis capracanis* Fischer, 1979, primera cita en España, *Rev. Iber. Parasitol.,* 46, 7, 1986.

348. **Hernandez-Jauregui, P., Silva-Lemoine, E., and Giron-Rojas, H.,** Miocarditis por sarcosporidios en un macaco Rhesus, Estudio di microscopia electrónica y de luz, *Arch. Invest. Med.,* 14, 139, 1983.

349. **Heydorn, A. O. and Rommel, M.,** Beiträge zum Lebenszyklus der Sarkosporidien. II. Hund und Katze als Überträger der Sarkosporidien des Rindes, *Berl. Muench. Tieraerztl. Wochenschr.,* 85, 121, 1972.

350. **Heydorn, A. O. and Rommel, M.,** Beiträge zum Lebenszyklus der Sarkosporidien. IV. Entwicklungsstadien von *S. fusiformis* in der Dünndarmschleimhaut der Katze, *Berl. Muench. Tieraerztl. Wochenschr.,* 85, 333, 1972.

351. **Heydorn, A. O., Ipczynski, V., Muhs, E. O., and Gestrich, R.,** Zystenstadien aus der Muskulatur von *Isospora hominis*-infizierten Kälbern, *Berl. Muench. Tieraerztl. Wochenschr.,* 87, 278, 1974.

352. **Heydorn, A. O., Gestrich, R., and Ipczynski, V.,** Zum Lebenszyklus der kleinen Form von *Isospora bigemina* des Hundes. II. Entwicklungsstadien im Darm des Hundes, *Berl. Muench. Tieraerztl. Wochenschr.,* 88, 449, 1975.

353. **Heydorn, A. O., Gestrich, R., Mehlhorn, H., and Rommel, M.,** Proposal for a new nomenclature of the Sarcosporidia, *Z. Parasitenkd.,* 48, 73, 1975.

354. **Heydorn, A. O., Mehlhorn, H., and Gestrich, R.,** Licht- und elektronenmikroskopische Untersuchungen an Cysten von *Sarcocystis fusiformis* in der Muskulatur von Kälbern nach experimenteller Infektion mit Oocysten und Sporocysten der großen Form von *Isospora bigemina* des Hundes. II. Die Feinstruktur der Cystenstadien, *Zentralbl. Bakteriol. Parasikenkd. Infektionskr. Hyg. I Abt. Orig. Reihe A,* 232, 123, 1975.

354a. **Heydorn, A. O., Mehlhorn, H., and Gestrich. R.,** Licht- und elektronenmikroskopische Untersuchungen an Cysten von *Sarcocystis fusiformis* in der Muskulatur von Kälbern nach experimenteller Infektion mit Oocysten und Sporocysten von *Isospora hominis* Railliet et Lucet, 1891. II. Die Feinstruktur der Metrocysten und Metozoiten, *Zentralbl. Bakteriol. Parasitenkd. Infektionskr. Hyg. I Abt. Orig. Reihe A,* 232, 373, 1975.

355. **Heydorn, A. O. and Gestrich, R.,** Beiträge zum Lebenszyklus der Sarkosporidien. VII. Entwicklungsstadien von *Sarcocystis ovicanis* im Schaf, *Berl. Muench. Tieraerztl. Wochenschr.,* 89, 1, 1976.

356. **Heydorn, A. O., Gestrich, R., and Janitschke, K.,** Beiträge zum Lebenszyklus der Sarkosporidien VIII. Sporozysten von *Sarcocystis bovihominis* in den Fäzes von Rhesusaffen (*Macaca rhesus*) und Pavianen (*Papio cynocephalus*), *Berl. Muench. Tieraerztl. Wochenschr.,* 89, 116, 1976.

357. **Heydorn, A. O.,** Beiträge zum Lebenszyklus der Sarkosporidien. IX. Entwicklungszyklus von *Sarcocystis suihominis* n. spec., *Berl. Muench. Tieraerztl. Wochenschr.,* 90, 218, 1977.

358. **Heydorn, A. O.,** Sarkosporidieninfiziertes Fleisch als mögliche Krankheitsursache für den Menschen, *Arch. Lebensmittelhyg.,* 28, 27, 1977.

359. **Heydorn, A. O. and Mehlhorn, H.,** Licht- und elektronenmikroskopische Untersuchungen an zwei Typen von Riesenschizonten aus dem Dünndarm des Schafes, *Zentralbl. Bakteriol. Parasitenkd. Infektionskr. Hyg. I. Abt. Orig. Reihe A,* 237, 124, 1977.

360. **Heydorn, A. O. and Mehlhorn, H.,** Light and electron microscopic studies on *Sarcocystis suihominis.* II. The schizogony preceding cyst formation, *Zentralbl. Bakteriol. Parasitenkd. Infektionskr. Hyg. I Abt. Orig. Reihe A,* 240, 123, 1978.

361. **Heydorn, A. O., Döhmen, H., Funk, G., Pähr, H., and Zientz, H.,** Zur Verbreitung der Sarkosporidieninfektion beim Hausschwein, *Arch. Lebensmittelhyg.,* 29, 184, 1978.

362. **Heydorn, A. O. and Ipczynski, V.,** Zur Schizogonie von *Sarcocystis suihominis* im Schwein, *Berl. Muench. Tieraerztl. Wochenschr.,* 91, 154, 1978.

363. **Heydorn, A. O.,** Zur Widerstandsfähigkeit von *Sarcocystis bovicanis*-Sporozysten, *Berl. Muench. Tieraerztl. Wochenschr.,* 93, 267, 1980.

364. **Heydorn, A. O., Matuschka, F. R., and Ipczynski, V.,** Zur Schizogonie von *Sarcocystis suicanis* im Schwein, *Berl. Muench. Tieraerztl. Wochenschr.,* 94, 49, 1981.

365. **Heydorn, A. O. and Matuschka, F. R.,** Zur Endwirtspezifität der vom Hund übertragenen Sarkosporidienarten, *Z. Parasitenkd.,* 66, 231, 1981.

365a. **Heydorn, A. O., Haralambidis, S., and Matuschka, F. R.,** Zur Chemoprophylaxe und Therapie der aukten Sarkosporidiose, *Berl. Muench. Tieraerztl. Wochenschr.,* 94, 229, 1981.

366. **Heydorn, A. O. and Haralambidis, S.,** Zur Entwicklung von *Sarcocystis capracanis* Fischer, 1979, *Berl. Muench. Tieraerztl. Wochenschr.,* 95, 265, 1982.

367. **Heydorn, A. O., and Unterholzner, J.,** Zur Entwicklung von *Sarcocystis hircicanis* n. sp., *Berl. Muench. Tieraerztl. Wochenschr.,* 96, 275, 1983.

367a. **Heydorn, A. O.,** Zur Entwicklung von *Sarcocystis arieticanis* n. sp., *Berl. Muench. Tieraerztl. Wochenschr.,* 98, 231, 1985.

367b. **Heydorn, A. O. and Karaer, Z.**, Zur Schizogonie von *Sarcocystis ovicanis*, *Berl. Muench. Tieraerztl. Wochenschr.*, 99, 185, 1986.

368. **Heydorn, A. O. and Mehlhorn, H.**, Fine structure of *Sarcocystis arieticanis* Heydorn, 1985 in its intermediate and final hosts (sheep and dog), *Zentralbl. Bakteriol. Parasitenkd. Infektionskr. Hyg. I Abt. Orig. Reihe A*, 264, 353, 1987.

369. **Hiepe, F., Hiepe, T., Hlinak, P., Jungmann, R., Horsch, R., and Weidauer, B.**, Experimentelle Infektion des Menschen und von Tieraffen (*Cercopithecus callitrichus*) mit Sarkosporidien-Zysten von Rind und Schwein, *Arch. Exp. Veterinaermed.*, 33, 819, 1979.

370. **Hiepe, F., Litzke, L. F., Scheibner, G., Jungmann, R., Hiepe, T., and Montag, T.**, Untersuchungen zur toxischen Wirkung von Extrakten aus *Sarcocystis-ovifelis*-Makrozysten auf Kaninchen, *Monatsh. Veterinaermed.*, 36, 908, 1981.

371. **Hilali, M. and Mohamed, A.**, The dog (*Canis familiaris*) as the final host of *Sarcocystis cameli* (Mason, 1910), *Tropenmed. Parasitol.*, 31, 213, 1980.

372. **Hilali, M., Imam, El S., and Hassan, A.**, The endogenous stages of *Sarcocystis cameli* (Mason, 1910), *Vet. Parasitol.*, 11, 127, 1982.

373. **Hilali, M. and Nasser, A. M.**, Ultrastructure of *Sarcocystis* spp. from donkeys *(Equus asinus)* in Egypt, *Vet. Parasitol.*, 23, 179, 1987.

373a. **Hill, J. E., Chapman, W. L., Jr., and Prestwood, A. K.**, Intramuscular *Sarcocystis* sp. in two cats and a dog, *J. Parasitol.*, 74, 724, 1988.

374. **Hinaidy, H. K., Burgu, A., Supperer, R., and Kallab, K.**, Sarkosporidienbefall des Rindes in Österreich, *Wien. Tieraerztl. Monatsschr.*, 66, 181, 1979.

375. **Hinaidy, H. K. and Supperer, R.**, Sarkosporidienbefall des Schweines in Österreich, *Wien. Tieraerztl. Monatsschr.*, 66, 281, 1979.

376. **Hinaidy, H. K. and Loupal, G.**, *Sarcocystis bertrami* Doflein, 1901, ein Sarkosporid des Pferdes, *Equus caballus, Zentralbl. Veterinaermed. Reihe B*, 29, 681, 1982.

377. **Hofer, J., Boch, J., and Erber, M.**, Zelluläre und humorale Reaktionen bei Mäusen nach experimenteller *Sarcocystis muris-* und *S. dispersa*-Infektion, *Berl. Muench. Tieraerztl. Wochenschr.*, 95, 169, 1982.

378. **Hong, C. B., Giles, R. C., Newman, L. E., and Fayer, R.**, Sarcocystosis in an aborted bovine fetus, *J. Am. Vet. Med. Assoc.*, 181, 585, 1982.

379. **Hoppe, D. M.**, Prevalence of macroscopically detectable *Sarcocystis* in North Dakota ducks, *J. Wildl. Dis.*, 12, 27, 1976.

379a. **Howells, R. E., Carvalho, A. D. V., Mello, M. N. and Rangel, N. M.**, Morphological and histochemical observations on *Sarcocystis* from the nine-banded armadillo, *Dasypus novemcinctus*, *Ann. Trop. Med. Parasitol.*, 69, 463, 1975.

380. **Hudkins, G. and Kistner, T. P.**, *Sarcocystis hemionilatrantis* (sp. n.) life cycle in mule deer and coyotes, *J. Wildl. Dis.*, 13, 80, 1977.

381. **Hudkins-Vivion, G., Kistner, T. P., and Fayer, R.**, Possible species differences between Sarcocystis from mule deer and cattle, *J. Wildl. Dis.*, 12, 86, 1976.

382. **Hussein, H. S. and Warrag, M.**, Prevalence of *Sarcocystis* in food animals in the Sudan, *Trop. Anim. Health Prod.*, 17, 100, 1985.

383. **Imes, G. D., Jr. and Migaki, G.**, Eosinophilic myositis in cattle — pathology and incidence, *Proc. 71st Annu. Meet. U.S. Livestock Sanit. Assoc.*, 1967, 111.

383a. **Jacobs, L., Remington, J. S., and Melton, M.L.**, A survey of meat samples from swine, cattle, and sheep for the presence of encysted *Toxoplasma, J. Parasitol.*, 46, 23, 1960.

384. **Jain, A. K., Gupta, S. L., Singh, R. P., and Mahajan, S. K.**, Experimental *Sarcocystis levinei* infection in buffalo calves, *Vet. Parasitol.*, 21, 51, 1986.

385. **Jain, P. C. and Shah, H. L.**, Cross-transmission studies of *Sarcocystis cruzi* of the cattle to the buffalo-calves, *Indian J. Anim. Sci.*, 55, 27, 1985.

385a. **Jain, P. C. and Shah, H. L.**, Prevalence and seasonal variations of *Sarcocystis* of cattle in Madhya Pradesh, *Indian J. Anim. Sci.*, 55, 29, 1985.

386. **Jain, P. C. and Shah, H. L.**, Experimental study on gametogonic development of *Sarcocystis levinei* in the small intestine of dogs, *Indian J. Anim. Sci.*, 56, 314, 1986.

387. **Jain, P. C. and Shah, H. L.**, Gametogony and sporogony of *Sarcocystis fusiformis* of buffaloes in the small intestine of experimentally infected cats, *Vet. Parasitol.*, 21, 205, 1986.

387a. **Jain, P. C. and Shah, H. L.**, Comparative morphology of oocysts and sporocysts of bovine and bubaline *Sarcocystis* in Madhya Pradesh, *Indian J. Anim. Sci.*, 57, 849, 1987.

387b. **Jain, P. C. and Shah, H. L.**, *Sarcocystis hominis* in cattle in Madhya Pradesh and its public health importance, *Indian Vet. J.*, 64, 650, 1987.

388. **Janitschke, K., Protz, D., and Werner, H.,** Beitrag zum Entwicklungszyklus von Sarkosporidien der Grantgazelle (*Gazella granti*), *Z. Parasitenkd.*, 48, 215, 1976.
388a. **Janitschke, K.,** Neue Erkenntnisse über die Kokzidien-Insktionen des Menschen. II. Isospora-Infektion, *Bundesgesundheitsblatt*, 18, 419, 1975.
388b. **Jantzen, B. and Entzeroth, R.,** Exocystosis of dense granules of cyst merozoites (cystozoites) of *Sarcocystis cuniculi* (Protozoa, Apicomplexa) in cultures, *Parasitol. Res.*, 73, 472, 1987.
388c. **Jeffrey, M., O'Toole, D., Smith, T., and Bridges, A. W.,** Immunocytochemistry of ovine sporozoan encephalitis and encephalomyelitis, *J. Comp. Pathol.*, 98, 213, 1988.
389. **Jensen, R., Alexander, A. F., Dahlgren, R. R., Jolley, W. R., Marquardt, W. C., Flack, D. E., Bennett, B. W., Cox, M. F., Harris, C. W., Hoffmann, G. A., Troutman, R. S., Hoff, R. L., Jones, R. L., Collins, J. K., Hamar, D. W., and Cravans, R. L.,** Eosinophilic myositis and muscular sarcocystosis in the carcasses of slaughtered cattle and lambs, *Am. J. Vet. Res.*, 47, 587, 1986.
390. **Jerrett, I. V., McOrist, S., Waddington, J., Browning, J. W., Malecki, J. C., and McCausland, I. P.,** Diagnostic studies of the fetus, placenta and maternal blood from 265 bovine abortions, *Cornell Vet.*, 74, 8, 1984.
391. **Johnson, A. J., Hildebrandt, P. K., and Fayer, R.,** Experimentally induced Sarcocystis infection in calves: pathology, *Am. J. Vet. Res.*, 36, 995, 1975.
392. **Jolley, W. R., Jensen, R., Hancock, H. A., and Swift, B. L.,** Encephalitic sarcocystosis in a newborn calf, *Am. J. Vet. Res.*, 44, 1908, 1983.
393. **Jungmann, R., Bergmann, V., Hiepe, T., and Nedjari, T.,** Untersuchungen zur septikämisch verlaufenden experimentellen *Sarcocystis-bovicanis*-Infektion des Rindes, *Monatsh. Veterinaermed.*, 32, 885, 1977.
394. **Kaiser, I. A. and Markus, M. B.,** *Sarcocystis* in the avian intermediate host, *Proc. Electron Microsc. Soc. S. Afr.*, 11, 115, 1981.
395. **Kaiser, I. A. and Markus, M. B.,** Species of *Sarcocystis* in wild South African birds, *Proc. Electron Microsc. Soc. S. Afr.*, 13, 103, 1983.
395a. **Kaliner, G., Grootenhuis, J. G., and Protz, D.,** A survey for sarcosporidial cysts in east African game animals, *J. Wildl. Dis.*, 10, 237, 1974.
396. **Kalyakin, V. N. and Zasukhin, D. N.,** Distribution of Sarcocystis (Protozoa: Sporozoa) in vertebrates, *Folia Parasitol. (Prague)*, 22, 289, 1975.
397. **Kan, S. P. and Dissanaike, A. S.,** Ultrastructure of *Sarcocystis booliati* Dissanaike and Poopalachelvam, 1975 from the moonrat, *Echinosorex gymnurus*, in Malaysia, *Int. J. Parasitol.*, 6, 321, 1976.
398. **Kan, S. P. and Dissanaike, A. S.,** Ultrastructure of *Sarcocystis* sp. from the Malaysian house rat, *Rattus rattus diardii*, *Z. Parasitenkd.*, 52, 219, 1977.
399. **Kan, S. P. and Dissanaike, A. S.,** Studies on *Sarcocystis* in Malaysia. II. Comparative ultrastructure of the cyst wall and zoites of *Sarcocystis levinei* and *Sarcocystis fusiformis* from the water buffalo, *Bubalus bubalis*, *Z. Parasitenkd.*, 57, 107, 1978.
400. **Kan, S. P.,** Ultrastructure of the cyst wall of *Sarcocystis* spp. from some rodents in Malaysia, *Int. J. Parasitol.*, 9, 475, 1979.
401. **Kan, S. P., Prathap, K., and Dissanaike, A. S.,** Light and electron microstructure of a *Sarcocystis* sp. from the Malaysian long-tailed monkey, *Macaca fascicularis*, *Am. J. Trop. Med. Hyg.*, 28, 634, 1979.
402. **Kan, S. P.,** A review of sarcocystosis with special reference to human infection in Malaysia, *Trop. Biomed.*, 2, 167, 1985.
403. **Karr, S. L., Jr. and Wong, M. M.,** A survey of *Sarcocystis* in nonhuman primates, *Lab. Anim. Sci.*, 25, 641, 1975.
404. **Karstad, L.,** *Toxoplasma microti* (the M-organism) in the muskrat (*Ondatra zibethica*), *Can. Vet. J.*, 4, 249, 1963.
405. **Karstad, L. and Trainer, D. O.,** *Sarcocystis* in white-tailed deer, *Bull. Wildl. Dis. Assoc.*, 5, 25, 1969.
406. **Kennedy, M. J. and Frelier, P. F.,** *Frenkelia* sp. from the brain of a porcupine (*Erethizon dorsatum*) from Alberta, Canada, *J. Wildl. Dis.*, 22, 112, 1986.
407. **Kepka, O. and Scholtyseck, E.,** Weitere Untersuchungen der Feinstruktur von *Frenkelia* spec. (=M-Organismus, Sporozoa), *Protistologica*, 6, 249, 1970.
408. **Kepka, O. and Österreicher, H. D.,** Zur Häufigkeit von Sarkosporidien in Rindern der Steiermark, *Wien. Tieraerztl. Monatsschr.*, 66, 184, 1979.
408a. **Kepka, O. and Skofitsch, G.,** Zur Epidemiologie von Frenkelia (Apicomplexa, Protozoa) der mitteleuropäischen Waldrötelmaus (*Clethrionomys glareolus*), *Mitt. Naturwiss. Ver. Steiermark.*, 109, 283, 1979.
408b. **Kimmig, P., Piekarski, G., and Heydorn, A. O.,** Zur Sarkosporidiose (*Sarcocystis suihominis*) des Menschen (II), *Immun. Infekt.*, 7, 170, 1979.
409. **Kimura, T., Ito, J., Suzuki, M., and Inokuchi, S.,** *Sarcocystis* found in the skeletel muscle of common squirrel monkeys, *Primates*, 28, 247, 1987.
410. **Kirkpatrick, C. E., Dubey, J. P., Goldschmidt, M. H., Saik, J. E., and Schmitz, J. A.,** *Sarcocystis* sp. in muscles of domestic cats, *Vet. Pathol.*, 23, 88, 1986.

410a. **Kirkpatrick, C. E., Hamir, A. N., Dubey, J. P., and Rupprecht, C. E.,** *Sarcocystis* in muscles of raccoons (*Procyon lotor* L.), *J. Protozool.,* 34, 445, 1987.
411. **Kirmse, P.,** Sarcosporidiosis in equines of Morocco, *Br. Vet. J.,* 142, 70, 1986.
412. **Kirmse, P. and Mohanbabu, B.,** *Sarcocystis* sp. in the one-humped camel (*Camelus dromedarius*) from Afghanistan, *Br. Vet. J.,* 142, 73, 1986.
413. **Kolárová, L.,** Mouse (*Mus musculus*) as intermediate host of *Sarcocystis* sp. from the goshawk (*Accipiter gentilis*), *Folia Parasitol. (Prague),* 33, 15, 1986.
414. **Koller, L. D., Kistner, T. P., and Hudkins, G. G.,** Histopathologic study of experimental *Sarcocystis hemionilatrantis* infection in fawns, *Am. J. Vet. Res.,* 38, 1205, 1977.
415. **Koudela, B.,** Purification of cystozoites of *Sarcocystis* sp. by the chromatographic gel spheron, *Folia Parasitol (Prague),* 32, 295, 1985.
416. **Koudela, B. and Steinhauser, L.,** Evaluation of vitality of sarcocysts in beef by the DAPI fluorescence test, *Acta Vet. (Brno.),* 53, 193, 1984.
417. **Kühn, J.,** Untersuchungen über die Trichinenkrankheit der Schweine, *Mittheilungen des Landwirthschaftlichen Institutes der Universität du Halle,* 1865, 1.
418. **Krampitz, H. E. and Rommel, M.,** Experimentelle Untersuchungen über das Wirtsspektrum der Frenkelien der Erdmaus, *Berl. Muench. Tieraerztl. Wochenschr.,* 90, 17, 1977.
419. **Krampitz, H. E., Rommel, M., Geisel, O., and Kaiser, E.,** Beiträge zum Lebenszyklus der Frenkelien. II. Die ungeschlechtliche Entwicklung von *Frenkelia clethrionomyobuteonis* in der Rötelmaus, *Z. Parasitenkd.,* 51, 7, 1976.
420. **Kunde, J. M., Jones, L. P., and Craig, T. M.,** Protozoal encephalitis in a bovine fetus, *Southwest. Vet.,* 33, 231, 1980.
421. **Kuraev, G. T.,** Sarcocystis infection in dromedaries and bactrian camels in Kazakhstan (In Russian), *Veterinariya,* 7, 41, 1981.
422. **Labbé, A.,** Sporozoa, *Das Tierreich,* 5, 116, 1899.
422a. **Laemmli, U. K.,** Cleavage of structural proteins during the assembly of the head of bacteriophage T4, *Nature (London),* 227, 680, 1970.
423. **Lainson, R. and Shaw, J. J.,** *Sarcocystis gracilis* n. sp. from the Brazilian tortoise *Kinosternon scorpioides,* *J. Protozool.,* 18, 365, 1971.
424. **Lainson, R.,** A note on sporozoa of undetermined taxonomic position in an armadillo and a heifer calf, *J. Protozool.,* 19, 582, 1972.
425. **Lainson, R. and Shaw, J. J.,** *Sarcocystis* in tortoises: a replacement name, *Sarcocystis kinosterni,* for the homonym *Sarcocystis gracilis,* Lainson and Shaw, 1971, *J. Protozool.,* 19, 212, 1972.
426. **Landsverk, T., Gamlem, H., and Svenkerud, R.,** A *Sarcocystis*-like protozoon in a sheep with lymphadenopathy and myocarditis, *Vet. Pathol.,* 15, 186, 1978.
427. **Landsverk, T.,** An outbreak of sarcocystosis in a cattle herd, *Acta Vet. Scand.,* 20, 238, 1979.
428. **Lankester, E. R.,** On *Drepanidium ranarum* the cell-parasite of the frog's blood and spleen (Gaule's Würmschen), *Q. J. Microsc. Sci.,* 22, 53, 1882.
428a. **Larsen, R. A., Kyle, J. E., Whitmire, W. M., and Speer, C. A.,** Effect of nylon wool purification on infectivity and antigenicity of *Eimeria falciformis* sporozoites and merozoites, *J. Parasitol.,* 70, 597, 1984.
429. **Last, M. J. and Powell, E. C.,** Separation of *Sarcocystis muris* and *Isospora felis* in mice used by Powell and McCarley (1975) in studies on the life cycle of *S. muris, J. Parasitol.,* 64, 162, 1978.
430. **Leek, R. G., Fayer, R., and Johnson, A. J.,** Sheep experimentally infected with *Sarcocystis* from dogs. I. Disease in young lambs, *J. Parasitol.,* 63, 642, 1977.
431. **Leek, R. G. and Fayer, R.,** Sheep experimentally infected with *Sarcocystis* from dogs. II. Abortion and disease in ewes, *Cornell Vet.,* 68, 108, 1978.
432. **Leek, R. G. and Fayer, R.,** Infectivity of *Sarcocystis* in beef and beef products from a retail food store, *Proc. Helminthol. Soc. Wash.,* 45, 135, 1978.
433. **Leek, R. G. and Fayer, R.,** Survival of sporocysts of *Sarcocystis* in various media, *Proc. Helminthol. Soc. Wash.,* 46, 151, 1979.
434. **Leek, R. G. and Fayer, R.,** Amprolium for prophylaxis of ovine *Sarcocystis, J. Parasitol.,* 66, 100, 1980.
435. **Leek, R. G. and Fayer, R.,** Experimental *Sarcocystis ovicanis* infection in lambs: salinomycin chemoprophylaxis and protective immunity, *J. Parasitol.,* 69, 271, 1983.
436. **Leek, R. G.,** Infection of sheep with frozen sporocysts of *Sarcocystis ovicanis, Proc. Helminthol. Soc. Wash.,* 53, 297, 1986.
437. **Leguia, G. and Herbert, I. V.,** The prevalence of *Sarcocystis* spp in dogs, foxes and sheep and *Toxoplasma gondii* in sheep and the use of the indirect haemagglutination reaction in serodiagnosis, *Res. Vet. Sci.,* 27, 390, 1979.
437a. **Leier, H., Boch, J., and Erber, M.,** Möglichkeiten der Immunisierung von Mäusen gegen sarkosporidien (*Sarcocystis muris*) mit abgeschwächten Sporozysten, *Berl. Muench. Tieraerztl. Wochenschr.,* 95, 231, 1982.

438. **Lele, V. R., Dhopavkar, P. V., and Kher, A.,** Sarcocystis infection in man (a case report), *Indian J. Pathol. Microbiol.*, 29, 87, 1986.
439. **Lerche, M. and Brochwitz, H.,** Sarkosporidienbefall des Rindes und Perimyositis eosinophilica, *Dtsch. Tieraerztl. Wochenschr.*, 64, 251, 1957.
440. **Levine, N. D.,** Nomenclature of *Sarcocystis* in the ox and sheep and of fecal coccidia of the dog and cat, *J. Parasitol.*, 63, 36, 1977.
441. **Levine, N. D.,** *Sarcocystis cernae* n. sp., replacement name for *Sarcocystis* sp. Cerná and Loucková, 1976, *Folia Parasitol. (Prague)*, 24, 316, 1977.
442. **Levine, N. D. and Tadros, W.,** Named species and hosts of *Sarcocystis* (Protozoa: Apicomplexa: Sarcocystidae), *Syst. Parasitol.*, 2, 41, 1980.
443. **Levine, N. D.,** The taxonomy of *Sarcocystis* (Protozoa, Apicomplexa) species, *J. Parasitol.*, 72, 372, 1986.
444. **Levine, N. D. and Ivens, V.,** *The Coccidian Parasites (Protozoa, Apicomplexa) of Artiodactyla*, Illinois Biological Monograph No. 55, University of Illinois Press, Urbana, IL, 1986.
445. **Levine, N. D. and Baker, J. R.,** The Isospora-Toxoplasma-Sarcocystis confusion, *Parasitol. Today*, 3, 101, 1987.
446. **Levine, N. D.,** *The Protozoan Phylum Apicomplexa*, CRC Press, Boca Raton, FL, 1988.
447. **Lindsay, D. S., Blagburn, B. L., Mason, W. H., and Frandsen, J. C.,** Prevalence of *Sarcocystis odocoileocanis* from white-tailed deer in Alabama and its attempted transmission to goats, *J. Wildl. Dis.*, 24, 154, 1988.
447a. **Lindsay, D. S., Ambrus, S. I., and Blagburn, B. L.,** *Frenkelia* sp.-like infection in the small intestine of a red-tailed hawk, *J. Wildl. Dis.*, 23, 677, 1987.
448. **Lopes, C. W. G., Araújo, J. L. de B., and Pereira, M. J. S.,** *Sarcocystis levinei* (Apicomplexa: Sarcocystidae) in the water buffalo (*Bubalus bubalis*) in Brazil, *Arq. Univ. Fed. Rural Rio de Janeiro*, 5, 21, 1982.
449. **Lopez-Rodriguez, R., Hernandez, S., Navarrette, I., and Martinez-Gomez, F.,** Sarcocistosis experimental en la cabra (*Capra hircus*). II. Signos clínicos e índices de eritrocitos, *Rev. Iber. Parasitol.*, 46, 115, 1986.
450. **Lunde, M. N. and Fayer, R.,** Serologic tests for antibody to Sarcocystis in cattle, *J. Parasitol.*, 63, 222, 1977.
450a. **Mácha, J., Procházková, Z., Cerva, L., and Gut, J.,** Isolation and characterization of a lectin from *Sarcocystis gigantea*, *Mol. Biochem. Parasitol.*, 16, 243, 1985.
451. **Machul'skii, S. N. and Miskaryan, N. D.,** Sarcosporidiosis of wild artiodactyla (in Russian), *Tr. Buryat-Mong. Zoovet. Inst.*, 13, 297, 1958.
451a. **Madigan, J. E. and Higgins, R. J.,** Equine protozoal myeloencephalitis, *Vet. Clin. North Am. Equine Pract.*, 3, 397, 1987.
452. **Madsen, S. C. and Hugghins, E. J.,** Studies on *Sarcocystis* of wild ungulates in South Dakota, *Proc. S. Dakota Acad. Sci.*, 58, 169, 1979.
453. **Mahaffey, E. A., George, J. W., Duncan, J. R., Prasse, K. W., and Fayer, R.,** Hematologic values in calves infected with *Sarcocystis cruzi*, *Vet. Parasitol.*, 19, 275, 1986.
454. **Mahrt, J. L. and Fayer, R.,** Hematologic and serologic changes in calves experimentally infected with *Sarcocystis fusiformis*, *J. Parasitol.*, 61, 967, 1975.
454a. **Mahrt, J. L. and Colwell, D. D.,** *Sarcocystis* in wild ungulates in Alberta, *J. Wildl. Dis.*, 16, 571, 1980.
455. **Mandour, A. M. and Keymer, I. F.,** *Sarcocystis* infection in African antelopes, *Ann. Trop. Med. Parasitol.*, 64, 513, 1970.
456. **Mandour, A. M.,** *Sarcocystis nesbitti* n. sp. from the rhesus monkey, *J. Protozool.*, 16, 353, 1969.
456a. **Manuel, M. F., Misa, G. A., and Yoda, T.,** Histomorphological studies of bubaline Sarcocystis in the Philippines, *Philipp. J. Vet. Med.*, 22, 24, 1983.
457. **Marcela Mercado Pezzat, M. V. Z.,** Frecuencia de *Sarcocystis* spp. en corazones de bovinos, *Veterinaria (Mexico City)*, 11, 6, 1971.
458. **Margolin, J. H. and Jolley, W. R.,** Experimental infection of dogs with *Sarcocystis* from wapiti, *J. Wildl. Dis.*, 15, 259, 1979.
459. **Markus, M. B.,** Flies as natural transport hosts of *Sarcocystis* and other coccidia, *J. Parasitol.*, 66, 361, 1980.
460. **Markus, M. B. and Daly, T. J. M.,** Specificity of *Sarcocystis* for the intermediate host, *Proc. 3rd European Multicolloquium Parasitology* (Cambridge, England), 1980, 141.
461. **Markus, M. B., Kaiser, I. A., and Daly, T. J. M.,** *Sarcocystis* of the vervet monkey *Cercopithecus pygerythrus*, *Proc. Electron Microsc. Soc. S. Afr.*, 11, 117, 1981.
462. **Markus, M. B., Daly, T. J. M., and Biggs, H. C.,** Domestic dog as a final host of *Sarcocystis* of the mountain zebra *Equus zebra hartmannae*, *S. Afr. J. Sci.*, 79, 471, 1983.
463. **Markus, M. B., Mundy, P. J., and Daly, T. J. M.,** Vultures *Gyps* spp. as final hosts of *Sarcocystis* of the impala *Aepyceros melampus*, *S. Afr. J. Sci.*, 81, 43, 1985.
464. **Mason, F. E.,** Sarcocysts in the camel in Egypt, *J. Comp. Pathol. Ther.*, 23, 168, 1910.

465. Mathey, W. J., *Isospora buteonis* Henry 1932 in an American kestrel (*Falco sparverius*) and a golden eagle (*Aquila chrysaëtos*), *Bull. Wildl. Dis. Assoc.*, 2, 20, 1966.
466. Matuschka, F. R., Life cycle of *Sarcocystis* between poikilothermic hosts. Lizards are intermediate hosts for *S. podarcicolubris* sp. nov, snakes function as definitive hosts, *Z. Naturforsch. Teil C*, 36, 1093, 1981.
467. Matuschka, F. R., Infectivity of Sarcocystis from donkey for horse via sporocysts from dogs, *Z. Parasitenkd.*, 69, 299, 1983.
468. Matuschka, F. R. and Mehlhorn, H., Sarcocysts of *Sarcocystis podarcicolubris* from experimentally infected Tyrrhenian wall lizards (*Podarcis tiliguerta*), *S. gallotiae* from naturally infected Canarian lizards (*Gallotia galloti*) and *S. dugesii* from Madeirian lizards (*Lacerta dugesii*), *Protistologica*, 20, 133, 1984.
468a. Matuschka, F. R., Experimental investigations on the host range of *Sarcocystis podarcicolubris*, *Int. J. Parasitol.*, 15, 77, 1985.
469. Matuschka, F. R., *Sarcocystis clethrionomyelaphis* n. sp. from snakes of the genus *Elaphe* and different voles of the family Arvicolidae, *J. Parasitol.*, 72, 226, 1986.
470. Matuschka, F. R., Schnieder, T., Daugschies, A., and Rommel, M., Cyclic transmission of *Sarcocystis bertrami* Doflein, 1901 by the dog to the horse, *Protistologica*, 22, 231, 1986.
471. Matuschka, F. R., Reptiles as intermediate and/or final hosts of Sarcosporidia, *Parasitol. Res.*, 73, 22, 1987.
472. Matuschka, F. R., Heydorn, A. O., Mehlhorn, H., Abd-Al-Aal, Z., Diesing, L., and Biehler, A., Experimental transmission of *Sarcocystis muriviperae* n. sp. to laboratory mice by sporocysts from the Palestinian viper (*Vipera palaestinae*): a light and electron microscope study, *Parasitol. Res.*, 73, 33, 1987.
473. Matuschka, F. R., *Sarcocystis chalcidicolubris* n. sp.: recognition of the life cycle of skinks of the genus *Chalcides* and snakes of the genus *Coluber*, *J. Parasitol.*, 73, 1014, 1987.
474. Matuschka, F. R. and Bannert, B., Cannibalism and autotomy as predator-prey relationship for monoxenous Sarcosporidia, *Parasitol. Res.*, 74, 88, 1987.
474a. Matuschka, F. R., Mehlhorn, H., and Abd-Al-Aal, Z., Replacement of *Besnoitia* Matuschka and Häfner, 1984 by *Sarcocystis hoarensis*, *Parasitol. Res.*, 74, 94, 1987.
475. Mayhew, I. G., de Lahunta, A., Whitlock, R. H., and Pollock, R. V. H., Equine protozoal myeloencephalitis, Proc. 22nd Annu. Conv. Am. Assoc. Equine Pract., Dallas, TX, 1976, 107.
476. Mayhew, I. G., Dellers, R. W., Timoney, J. F., Kemen, M. F., Fayer, R., and Lunde, M. N., Microbiology and serology. Chapter 7, in "Spinal Cord Disease in the Horse." Mayhew, I. G., de Lahunta, A., Whitlock, R. H., Krook, L., and Tasker, J. B., Eds., *Cornell Vet.*, 68, 148, 1978.
477. Mayhew, I. G. and de Lahunta A., Neuropathology, "In Spinal Cord Disease in the Horse", Mayhew, I. G., de Lahunta, A., Whitlock, R. H., Krook, L., and Tasker, J. B., Eds., *Cornell Vet.*, 68, 106, 1978.
478. Mayhew, I. G. and Greiner, E. C., Protozoal diseases. Equine protozoal myeloencephalitis, *Vet. Clin. North Am. Equine Pract.*, 2, 439, 1986.
479. McCausland, I. P., Badman, R. T., Hides, S., and Slee, K. J., Multiple apparent *Sarcocystis* abortion in four bovine herds, *Cornell Vet.*, 74, 146, 1984.
480. McErlean, B. A., Ovine paralysis associated with spinal lesions of toxoplasmosis, *Vet. Rec.*, 94, 264, 1974.
481. McCarron, R. M., Kempski, O., Spatz, M., and McFarlin, E. E., Presentation of myelin basic protein by murine cerebral vascular endothelial cells, *J. Immunol.*, 134, 3100, 1985.
482. McKenna, P. B. and Charleston, W. A. G., Coccidia (Protozoa: Sporozoasida) of cats and dogs. I. Identity and prevalence in cats, *N.Z. Vet. J.*, 28, 86, 1980.
483. McKenna, P. B. and Charleston, W. A. G., Coccidia (Protozoa: Sporozoasida) of cats and dogs. IV. Identity and prevalence in dogs, *N.Z. Vet. J.*, 28, 128, 1980.
484. McKenna, P. B. and Charleston, W. A. G., *Sarcocystis* spp. infections in naturally infected cats and dogs: levels of sporocyst production and the influence of host, environmental and seasonal factors on the prevalence of infection, *N.Z. Vet. J.*, 31, 49, 1983.
484a. McKenna, P. B. and Charleston, W. A. G., Coccidia (*Protozoa: Sporozoasida*) of cats and dogs. II. Experimental induction of *Sarcocystis* infections in mice, *N.Z. Vet. J.*, 28, 117, 1980.
484b. McKenna, P. B. and Charleston, W. A. G., Evaluation of a concentration method for counting *Sarcocystis gigantea* sporocysts in cat faeces, *Vet. Parasitol.*, 26, 207, 1988.
484c. McKenna, P. B. and Charleston, W. A. G., Recovery of *Sarcocystis gigantea* sporocysts from cat faeces, *Vet. Parasitol.*, 26, 215, 1988.
485. Meads, E. B., Dalmeny disease — another outbreak — probably sarcocystosis, *Can. Vet. J.*, 17, 271, 1976.
485a. Mehlhorn, H., and Scholtyseck, E., Elektronenmikroskopische Untersuchungen an Cystenstadien von *Sarcocystis tenella* aus der Oesophagus-Muskulatur des Schafes., *Z. Parasitenkd.*, 41, 291, 1973.
485b. Mehlhorn, H. and Scholtyseck, E., Licht- und elektronenmikroskopische Untersuchungen an Entwicklungsstadien von *Sarcocystis tenella* aus der Darmwand der Hauskatze. I. Die Oocysten und Sporocysten, *Z. Parasitenkd.*, 43, 251, 1974.

485c. **Mehlhorn, H., Senaud, J., and Scholtyseck, E.**, Étude ultrastructurale des coccidies formant des kystes: *Toxoplasma gondii, Sarcocystis tenella, Besnoitia jellisoni* et *Frenkelia* sp.: distribution de la phosphatase acide et des polysaccharides au niveau des ultrastructures chez le parasite et chez l'hote, *Protistoligica,* 10, 21, 1974.

485d. **Mehlhorn, H. and Scholtyseck, E.**, Die Parasit-Wirtsbeziehungen bei verschiedenen Gattungen der Sporozoen (Eimeria, Toxoplasma, Sarcocystis, Frenkelia, Hepatozoon, Plasmodium and Babesia) unter Anwendung spezieller Verfahren, *Microsop. Acta,* 75, 429, 1974.

486. **Mehlhorn, H., Scholtyseck, E., and Senaud, J.**, Transmission de *Sarcocystis tenella,* chez le chat, à partir des formes kystiques parasites intramusculaires du mouton: les oocystes et les sporocystes en microscopie photonique et électronique, *C.R. Acad. Sci. Ser. D,* 278, 1111, 1974.

486a. **Mehlhorn, H. and Senaud, J.**, Action lytique des batéries présentes dans les kystes de *Sarcocystis tenella* (Sporozoa, Protozoa), *Arch. Microbiol.,* 104, 241, 1975.

486b. **Mehlhorn, H.**, Elekronenmikroskopischer Nachweis von alakalischer Poshphatase und ATP-ase in Cystenstadien von *Sarcocystis tenella* (Sporozoa, Coccidia) aus der Schlundmuskulatur von Schafen, *Z. Parasitenkd.,* 46, 95, 1975.

486c. **Mehlhorn, H., Sénaud, J., Chobotar, B., Scholtyseck, E.**, Electron microscope studies of cyst stages of *Sarcocystis tenella*: the origin of micronemes and rhoptries, *Z. Parasitenkd.,* 45, 227, 1975.

487. **Mehlhorn, H., Heydorn, A. O., and Gestrich, R.**, Licht- und elektronenmikroskopische Untersuchungen an Cysten von *Sarcocystis ovicanis* Heydorn et al. (1975) in der Muskulatur von Schafen, *Z. Parasitenkd.,* 48, 83, 1975.

488. **Mehlhorn, H., Heydorn, O., and Gestrich, R.**, Licht- und elektronenmikroskopische Untersuchungen an Cysten von *Sarcocystis fusiformis* in der Muskulatur von Kälbern nach experimenteller Infektion mit Oocysten und Sporocysten der großen Form von *Isospora bigemina* des Hundes. I. Zur Entstehung der Cyste und der "Cystenwand", *Zentralbl. Bakteriol. Parasitenkd. Infektionskr. Hyg. I Abt. Orig. Reihe A,* 232, 392, 1975.

488a. **Mehlhorn, H., Heydorn, A. O., and Gestrich, R.**, Licht- und elektronenmikroskopische Untersuchungen an Cysten von *Sarcocystis fusiformis* in der Muskulatur von Kälbern nach experimenteller Infektion mit Oocysten und Sporocysten von *Isospora hominis* Railliet et Lucet, 1891. I. Zur Entstehung der Cyste und der Cystenwand, *Zentralbl. Bakteriol. Parasitenkd. Infektionskr. Hyg. I Abt. Orig. Reihe A,* 231, 301, 1975.

489. **Mehlhorn, H., Senaud, J., Heydorn, A. O., and Gestrich, R.**, Comparaison des ultrastructures des kystes de *Sarcocystis fusiformis* Railliet, 1897 dans la musculature du boeuf, après infection naturelle et après infection expérimentale par des sporocystes d'*Isospora hominis* et par des sporocystes des grandes formes d'*Isospora bigemina* du chien et du chat, *Protistologica,* 11, 445, 1975.

490. **Mehlhorn, H., Hartley, W. J., and Heydorn, A. O.**, A comparative ultrastructural study of the cyst wall of 13 Sarcocystis species, *Protistologica,* 12, 451, 1976.

491. **Mehlhorn, H., Heydorn, A. O., and Janitschke, K.**, Light and electron microscopical study on sarcocysts from muscles of the rhesus monkey (*Macaca mulatta*), baboon (*Papio cynocephalus*) and tamarin (*Saguinus* (= *Oedipomidas*) *oedipus*), *Z. Parasitenkd.,* 51, 165, 1977.

492. **Mehlhorn, H. and Heydorn, A. O.**, Light and electron microscopic studies of *Sarcocystis suihominis*. I. The development of cysts in experimentally infected pigs, *Zentralbl. Bakteriol. Parasitenkd. Infektionskr. Hyg. I Abt. Orig. Reihe A,* 239, 124, 1977.

493. **Mehlhorn, H. and Heydorn, A.O.**, The Sarcosporidia (Protozoa, Sporozoa): life cycle and fine structure, *Adv. Parasitol.,* 16, 43, 1978.

494. **Mehlhorn, H. and Heydorn, A.O.**, Electron microscopical study on gamogony of *Sarcocystis suihominis* in human tissue cultures, *Z. Parasitenkd.,* 58, 97, 1979.

495. **Mehlhorn, H., Heydorn, A. O., Frenkel, J. K., and Göbel, E.**, Announcement of the establishment of neohepantotypes for some important *Sarcocystis* species, *Z. Parasitenkd.,* 71, 689, 1985.

496. **Mehlhorn, H. and Matuschka, F. R.**, Ultrastructural studies of the development of *Sarcocystis clethrionomyelaphis* within its final and intermediate hosts, *Protistologica,* 22, 97, 1986.

496a. **Mehlhorn, H., Düwel, D., and Raether, W.**, *Parasitologie. Diagnose und Therapie der Parasiten von Haus-, Nutz- und Heimtieren,* Gustav Fisher, Stuttgart, 1986.

497. **Meingassner, J. G. and Burtscher, H.**, Doppelinfektion des Gehirns mit *Frenkelia* species und *Toxoplasma gondii* bei *Chinchilla laniger, Vet. Pathol.,* 14, 146, 1977.

497a. **Melville, R. V.**, *Nomina dubia* and available names, *Z. Parasitenkd.,* 62, 105, 1980.

498. **Melville, R. B.**, Reply to Frenkel, Mehlhorn, and Heydorn on protozoan *Nomen dubia, J. Parasitol.,* 70, 815, 1984.

499. **Meshkov, S.**, The jackal (*Canis aureus*) as a new host of *Sarcocystis* infecting swine (In Russian), *Veterinarna Sbirka,* 78, 20, 1980.

500. Miescher, F., Über eigenthümliche Schläuche in den Muskeln einer Hausmaus, *Ber. Verh. Naturforsch. Ges.*, 5, 198, 1843.
501. Migaki, G. and Albert, T. F., Sarcosporidiosis in the ringed seal, *J. Am. Vet. Med. Assoc.*, 177, 917, 1980.
501a. Montag, T., Thietz, H. J., Brose, E., Liebenthal, C., Mann, W., Hiepe, T., Hiepe, F., and Coupek, J., The mitogenicity of extracts from *Sarcocystis gigantea* macrocysts is due to lectin(s), *Parasitol. Res.*, 74, 112, 1987.
502. Moore, S., Two types of ovine *Sarcocystis* macrocysts distinguished by periodic acid-Schiff staining of the cyst walls, *N.Z. Vet. J.*, 28, 101, 1980.
503. Morgan, G., Terlecki, S., and Bradley, R., A suspected case of Sarcocystis encephalitis in sheep, *Br. Vet. J.*, 140, 64, 1984.
504. Mori, Y., Studies on the *Sarcocystis* in Japanese black cattle (in Japanese), *Bull. Azabu. Univ. Vet. Med.*, 6, 51-65, 1985.
504a. Moulé, L. T., Des sarcosporides et de leur fréquence, principalement chez les animaux de boucherie, *Soc. Sci. Arts Vitry le-Francois*, 14, 3, 1888.
505. Muhm, R. L., Barnett, D., Bryant, D. T., Cole, J. H., and Kadel, W. L., Sarcocystosis: a case study, *Annu. Proc. Am. Assoc. Vet. Lab. Diagn.*, 22, 139, 1979.
506. Munday, B. L., Mason, R. W., and Cumming, R., Observations on diseases of the central nervous system of cattle in Tasmania, *Aust. Vet. J.*, 49, 451, 1973.
507. Munday, B. L. and Rickard, M. D., Is *Sarcocystis tenella* two species?, *Aust. Vet. J.*, 50, 558, 1974.
508. Munday, B. L., The prevalence of sarcosporidiosis in Australian meat animals, *Aust. Vet. J.*, 51, 478, 1975.
509. Munday, B. L., Barker, I. K., and Rickard, M. D., The developmental cycle of a species of *Sarcocystis* occurring in dogs and sheep, with observations on pathogenicity in the intermediate host, *Z. Parasitenkd.*, 46, 111, 1975.
510. Munday, B. L. and Black, H., Suspected *Sarcocystis* infections of the bovine placenta and foetus, *Z. Parasitenkd.*, 51, 129, 1976.
511. Munday, B. L., A species of *Sarcocystis* using owls as definitive hosts, *J. Wildl. Dis.*, 13, 205, 1977.
512. Munday, B. L., Humphrey, J. D., and Kila, V., Pathology produced by, prevalence of, and probable life-cycle of a species of Sarcocystis in the domestic fowl, *Avian Dis.*, 21, 697, 1977.
513. Munday, B. L., Cats as definitive hosts for *Sarcocystis* of sheep, *N.Z. Vet. J.*, 26, 166, 1978.
514. Munday, B. L., Mason, R. W., Hartley, W., J., Presidente, P. J. A., and Obendorf, D., *Sarcocystis* and related organisms in Australian wildlife. I. Survey findings in mammals, *J. Wildl. Dis.*, 14, 417, 1978.
515. Munday, B. L., Cats and Sarcocystis of sheep, *N.Z. Vet. J.*, 25, 19, 1979.
516. Munday, B. L., The effect of *Sarcocystis ovicanis* on growth rate and haematocrit in lambs, *Vet. Parasitol.*, 5, 129, 1979.
517. Munday, B. L., Hartley, W. J., Harrigan, K. E., Presidente, P. J. A., and Obendorf, D. L., *Sarcocystis* and related organisms in Australian wildlife. II. Survey findings in birds, reptiles, amphibians and fish, *J. Wildl. Dis.*, 15, 57, 1979.
518. Munday, B. L., Smith, D. D., and Frenkel, J. K., Sarcocystis and related organisms in Australian wildlife. IV. Studies on *Sarcocystis cuniculi* in European rabbits (*Oryctolagus cuniculus*), *J. Wildl. Dis.*, 16, 201, 1980.
519. Munday, B. L. and Mason, R. W., *Sarcocystis* and related organisms in Australian wildlife. III. *Sarcocystis murinotechis* sp. n. life cycle in rats (*Rattus, Pseudomys* and *Mastocomys* spp.) and tiger snakes (*Notechis ater*), *J. Wildl. Dis.*, 16, 83, 1980.
520. Munday, B. L., Premature parturition in ewes inoculated with *Sarcocystis ovicanis*, *Vet. Parasitol.*, 9, 17, 1981.
521. Munday, B. L. and Mason, R. W., *Sarcocystis* in rats from islands, *N.Z. J. Zool.*, 8, 563, 1981.
521a. Munday, B. L., Effect of preparturient inoculation of pregnant ewes with *Sarcocystis ovicanis* upon the susceptibility of their progeny, *Vet. Parasitol.*, 9, 273, 1982.
522. Munday, B. L., An isosporan parasite of masked owls producing sarcocysts in rats, *J. Wildl. Dis.*, 19, 146, 1983.
523. Munday, B. L. and Obendorf, D. L., Morphology of *Sarcocystis gigantea* in experimentally-infected sheep, *Vet. Parasitol,.* 16, 193, 1984.
523a. Munday, B. L., The effect of *Sarcocystis tenella* on wool growth in sheep, *Vet. Parasitol.*, 15, 91, 1984.
523b. Munday, B. L. and Obendorf, D. L., Development and growth of *Sarcocystis gigantea* in experimentally-infected sheep, *Vet. Parasitol.*, 15, 203, 1984.
523c. Munday, B. L., Demonstration of viable *Sarcocystis* sporocysts in the faeces of a lamb dosed orally, *Vet. Parasitol.*, 17, 355, 1984—1985.
524. Munday, B. L., Effects of different doses of dog-derived *Sarcocystis* sporocysts on growth rate and haematocrit in lambs, *Vet. Parasitol.*, 21, 21, 1986.

525. **Murata, K.,** A case report of *Sarcocystis* infection in a lesser flamingo, *Jpn. J. Parasitol.*, 35, 555, 1986.
526. **Nakamura, K., Shoya, S., Nakajima, Y., Shimura, K., and Ito, S.,** Pathology of experimental acute sarcocystosis in a cow, *Jpn. J. Vet. Sci.,* 44, 675, 1982.
527. **Neméseri, L., Entzeroth, R., and Scholtyseck, E.,** A magyarországi gimszarvasok *Sarcocystis* -fajának gyakorisága és ultrastrukturája, *Magy. Allatorv. Lapja,* 38, 758, 1983.
527a. **Neveu-Lemaire, M.,** Sarcosporidies des mammifères domestiques, in *Parasitologie des Animaux Domestiques et Maladies Parasitaires non Bactériennes,* 1912, 300.
527b. **Niederhäusern, D.,** Psorospermien bei der Ziege, *Z. Vet. Wiss. Bern,* 1, 79, 1873.
528. **O'Donoghue, P. J. and Weyreter, H.,** Detection of *Sarcocystis* antigens in the sera of experimentally-infected pigs and mice by an immunoenzymatic assay, *Vet. Parasitol.,* 12, 13, 1983.
529. **O'Donoghue, P. J. and Ford, G. E.,** The asexual pre-cyst development of *Sarcocystis tenella* in experimentally infected specific-pathogen-free lambs, *Int. J. Parasitol,* 14, 345, 1984.
530. **O'Donoghue, P. J. and Weyreter, H.,** Examinations on the serodiagnosis of Sarcocystis infections. II. Class-specific immunoglobulin responses in mice, pigs and sheep, *Zentralbl. Bakteriol. Parasitenkd. Infektionskr. Hyg. I Abt. Orig. Reihe A,* 257, 168, 1984.
531. **O'Donoghue, P. J., Rommel, M., Weber, M., and Weyreter, H.,** Attempted immunization of swine against acute sarcocystosis using cystozoite-derived vaccines, *Vet. Immunol. Immunopathol.,* 8, 83, 1985.
532. **O'Donoghue, P. J., Adams, M., Dixon, B. R., Ford, G. E., and Baverstock, P. R.,** Morphological and biochemical correlates in the characterization of *Sarcocystis* spp., *J. Protozool.,* 33, 114, 1986.
532a. **O'Donoghue, P. J. and Ford, G. E.,** The prevalence and intensity of *Sarcocystis* spp infections in sheep, *Aust. Vet. J.,* 63, 273, 1986.
533. **O'Donoghue, P. J., Watts, C. H. S., and Dixon, B. R.,** Ultrastructure of *Sarcocystis* spp. (Protozoa: Apicomplexa) in rodents from North Sulawesi and West Java, Indonesia, *J. Wildl. Dis.,* 23, 225, 1987.
533a. **O'Donoghue, P. J., Obendorf, D. L., O'Callaghan, M. G., Moore, E., and Dixon, B. R.,** *Sarcocystis mucosa* (Blanchard 1885) Labbé 1889 in unadorned rock wallabies (*Petrogale assimilis*) and Bennett's wallabies (*Macropus rufogriseus*), *Parasitol. Res.,* 73, 113, 1987.
533b. **O'Donoghue, P. J. and Wilkinson, R. G.,** Antibody development and cellular immune responses in sheep immunized and challenged with *Sarcocystis tenella* sporocysts, *Vet. Parasitol.,* 27, 251, 1988.
534. **Obendorf, D. L. and Munday, B. L.,** Demonstration of schizogonous stages of *Sarcocystis gigantea* in experimentally infected sheep, *Vet. Parasitol.,* 19, 35, 1986.
535. **Obendorf, D. L. and Munday, B. L.,** Experimental infection with *Sarcocystis medusiformis* in sheep, *Vet. Parasitol.,* 24, 59, 1987.
536. **Odening, K.,** Sarkozysten in einer antarktischen Robbe, *Angew. Parasitol.,* 24, 197, 1983.
537. **Ogassawara, S., Larsson, C. E., Larsson, M. H. M. A., Hagiwara, M. K., and Gouveia, G.,** Ocorrência de esporocistos de *Sarcocystis* sp. em cães na cidade de São Paulo, *Rev. Microbiol. (Sao Paulo),* 8, 21, 1977.
538. **Ogassawara, S., Benassi, S., Larson, C. E., and Hagiwara, M. K.,** *Sarcocystis* sp.: ocorrência de esporocistos em gatos na cidade de São Paulo, *Arq. Inst. Biol. Sao Paulo,* 47, 23, 1980.
539. **Olafson, P. and Monlux, W. S.,** Toxoplasma infection in animals, *Cornell Vet.,* 32, 176, 1942.
540. **Orr, M. B., Collins, G. H., and Charleston, W. A. G.,** *Sarcocystis capracanis*: experimental infection of goats. II. Pathology, *Int. Goat Sheep Res.,* 3, 202, 1984.
541. **O'Toole, D., Duffell, S. J., Upcott, D. H., and Frewin, D.,** Experimental microcyst sarcocystis infection in lambs: pathology, *Vet. Rec.,* 119, 525, 1986.
542. **O'Toole, E.,** Experimental ovine sarcocystosis: sequential untrastructural pathology in skeletal muscle, *J. Comp. Pathol.,* 97, 51, 1987.
543. **Özer, E.,** *Sarcocystis capracanis* (Fiescher, 1979) 'in biyolojisi ve patojenitesi üzerinde deneysel bir arastirma, *Ankara Univ. Vet. Fak. Derg,* 31, 431, 1984.
544. **Pacheco, N. D. and Fayer, R.,** Fine structure of *Sarcocystis cruzi* schizonts, *J. Protozool.,* 24, 382, 1977.
545. **Pacheco, N. D., Sheffield, H. G., and Fayer, R.,** Fine structure of immature cysts of *Sarcocystis cruzi, J. Parasitol.,* 64, 320, 1978.
546. **Pak, S. M., Perminova, V. V., and Yeshtokina, N. V.,** *Sarcocystis citellivulpes* sp. n. from the yellow suslik *Citellus fulvus* Lichtenstain, 1823, in "Toksoplazmidy", Beyer, T. V., Bezukladnikova, N. A., Galuzo, I. G., Konoralova, S. I., and Pak, S. M., Eds. (in Russian), *Protozool. Akad. Nauk. S.S.S.R.,* 111, 1979.
546a. **Pak, S. P.,** Occurrence of *Sarcocystis* in *Vulpes corsac* (in Russian), in x vsesoyuznaya Konferentsiya po prirodenoi ochagovosti bolezei 9—11 oktyabrya 1979, Dushanbe Tezisy dokladov 1, Alma-Ata, U.S.S.R., *Nauka Kaz. S.S.R.,* 133, 1979.
546b. **Parenzan, P.,** Sarcosporidiosi (psorospermosi) da nuova specie (Prot.: *Sarcocystis atractaspidis* n. sp.) in rettile (*Atractaspis*), *Boll. Soc. Nat. Napoli,* 55, 117, 1947.
546c. **Parish, S. M., Maag-Miller, L., Besser, T. E., Weidner, J. P., McElwain, T. F., Knowles, D. P., and Leathers, C. W.,** Myelitis associated with protozoal infection in newborn calves, *J. Am. Vet. Med. Assoc.,* 191, 1599, 1987.
546d. **Parker, R. C.,** *Methods of Tissue Culture,* 3rd ed., Hoeber, New York, 1961, 145.
547. **Pathmanathan, P. and Kan, S. P.,** Human *Sarcocystis* infection in Malaysia, *Southeast Asian J. Trop. Med. Public Health,* 12, 247, 1981.

547a. **Patton, W. S. and Hindle, E.,** Notes on three new parasites of the striped hamster (*Cricetulus griseus*), *Proc. R. Soc. London Ser. B,* 100, 387, 1926.

548. **Pereira Lorenzo, A.,** Incidencia y diagnostico de *Sarcocystis miescheriana* (Kühn, 1865) Lankester, 1882, *Rev. Iber. Parasitol.,* 79, 401, 1979.

549. **Pereira, M. J. S. and Lopes, C. W. G.,** The crab-eating fox (*Cerdocyon thous*) as a final host for *Sarcocystis capracanis* (Apicomplexa: Sarcocystidae), *Arq. Univ. Fed. Rural Rio de Janeiro,* 5, 233, 1982.

550. **Perez, S. and Cruz, M.,** Ciclo biologico de *Sarcocystis gigantea* (Railliet 1886) Ashford 1977, *Rev. Iber. Parasitol.,* 39, 601, 1979.

550a. **Perrotin, C., Graber, M., Thal, J., and Petit, J. P.,** La sarcosporidiose chez le buffle Africain (*Syncerus caffer*), *Rev. Elev. Med. Vet. Pays Trop.,* 31, 423, 1978.

551. **Pessôa, S. B.,** *Sarcocystis oliverioi* n. sp., parasita do "tium" (*Forpus passerinus* L.), *Folia Clin. Biol.,* 7, 162, 1935.

552. **Pethkar, D. K. and Shah, H. L.,** Prevalence of *Sarcocystis* in goats in Madhya Pradesh, *Indian Vet. J.,* 59, 110, 1982.

553. **Pethkar, D. K. and Shah, H. L.,** Attempted cross transmission of *Sarcocystis capracanis* of the goat to the sheep, *Indian Vet. J.,* 59, 766, 1982.

554. **Pfeiffer, L.,** Ueber einige neue Formen von Miescher'schen Schläuchen mit Mikro-, Myxo- und Sarcosporidieninhalt, *Virchows Arch. Pathol. Anat. Physiol.,* 52, 552, 1890.

555. **Phillips, P. H. and Ford, G. E.,** Clinical, haematological and plasma biochemical changes in specified-pathogen-free (Sporozoa) lambs experimentally infected with low numbers of *Sarcocystis tenella* sporocysts, *Vet. Parasitol.,* 24, 15, 1987.

556. **Piekarski, G., Heydorn, A. O., Aryeetey, M. E., Hartlapp, J. H., and Kimmig, P.,** Klinische, parasitologische und serologische Untersuchungen zur Sarkosporidiose (*Sarcocystis suihominis*) des Menschen, *Immun. Infekt.,* 6, 153, 1978.

557. **Plotkowiak, J.,** Wyniki dalszych badán nad wystepowaniem i epidemiologia inwazji *Isospora hominis* (Railliet i Lucet, 1891), *Wiad. Parazytol.,* 22, 137, 1976.

557a. **Pober, J. S., Collins, T., Gimbrone, M. A., Jr., Libby, P., and Reiss, C. S.,** Inducible expression of class II major histocompatibility complex antigens and the immunogenicity of vascular endothelium, *Transplantation,* 41, 141, 1986.

558. **Poli, A., Mancianti, F., Marconcini, A., Nigro, M., and Colagreco, R.,** Prevalence, ultrastructure of the cyst wall and infectivity for the dog and cat of *Sarcocystis* sp. from fallow deer (*Cervus dama*), *J. Wild. Dis.,* 24, 97, 1988.

559. **Pomroy, W. E. and Charleston, W. A. G.,** Prevalence of dog-derived *Sarcocystis* spp. in some New Zealand lambs, *N. Z. Vet. J.,* 35, 141, 1987.

560. **Pond, D., B. and Speer, C. A.,** *Sarcocystis* in free-ranging herbivores on the National Bison Range, *J. Wildl. Dis.,* 15, 51, 1979.

561. **Ponse Alcocer, J.,** Incidencia de Sarcocystis spp. en bovinos nonatos, *Veterinaria (Mexico City),* 4, 127, 1973.

562. **Porchet-Henneré, E., and Ponchel, G.,** Queleques précisions sur l' ultrastructure de *Sarcocystis tenella* : l'architecture du kyste et l'aspect des endozoïtes en microscopie élctronique á balayage, *C. R. Acad. Sci. Ser. D,* 279, 1179, 1974.

563. **Porchet-Hennere, E.,** Quelques précisions sur l'ultrastructure de *Sarcocystis tenella*. I. L'endozoïte (après coloration négative), *J. Protozool,*. 22, 214, 1975.

563a. **Porchet, E. and Torpier, G.,** Etude du germe infectieux de *Sarcocystis tenella* et *Toxoplasma gondii* par la technique du cryodécapage, *Z. Parasitenkd.,* 54, 101, 1977.

564. **Powell, E. C. and McCarley, J. B.,** A murine *Sarcocystis* that causes an *Isospora*-like infection in cats, *J. Parasitol.,* 61, 928, 1975.

565. **Powell, E. C., Pezeshkpour, G., Dubey, J. P., and Fayer, R.,** Types of myofibers parasitized in experimentally induced infections with *Sarcocystis cruzi* and *Sarcocystis capracanis, Am. J. Vet. Res.,* 47, 514, 1986.

566. **Prasse, K. W. and Fayer, R.,** Hematology of experimental acute *Sarcocystis bovicanis* infection in calves. II. Serum biochemistry and hemostasis studies, *Vet. Pathol.,* 18, 358, 1981.

567. **Prestwood, A. K., Cahoon, R. W., and McDaniel, H. T.,** *Sarcocystis* infections in Georgia swine, *Am. J. Vet. Res.,* 41, 1879, 1980.

568. **Proctor, S. J., Barnett, D., Stalheim, O. H. V., and Fayer, R.,** Pathology of *Sarcocystis fusiformis* in cattle, *Annu. Proc. Am. Assoc. Vet. Lab. Diagn.,* 19, 329, 1976.

569. **Pucak, G. J. and Johnson, D. K.,** Sarcocystis in a Patas monkey (*Erythrocebus patus*), *Lab. Anim. Dig.,* 8, 36, 1972.

570. **Purcherea, A., Radu, A., and Neda, M.,** Cercetâbri asupra sarcosporidiilor de la bubaline, *Lucr. Stiint. Inst. Agron. Bucuresti Ser. C,* 24, 55, 1981.

570a. **Quinn, S. C., Brooks, R. J., and Cawthorn, R. J.,** Effects of the protozoan parasite *Sarcocystis rauschorum* on open-field behaviour of its intermediate vertebrate host, *Dicrostonyx richardsoni, J. Parasitol.,* 73, 265, 1987.

570b. **Quortrup, E. R. and Shillinger, J. E.**, 3,000 wild bird autopsies on Western Lake areas, *J. Am. Vet. Med. Assoc.*, 99, 382, 1941.
571. **Rahbari, S., Bazargani, T. T., and Rak, H.**, Sarcocystosis in the camel in Iran, *J. Vet. Fac. Univ. Tehran*, 37, 1, 1981.
571a. **Railliet, A.**, [*Miescheria tenella*], *Bull. Mem. Soc. Cent. Med. Vet.*, 40, 130, 1886.
572. **Rátz, I.**, Szakosztályunk ülései, *Allattani Kozl.*, 7, 177, 1908.
573. **Rátz, I.**, Az izmokban élösködö véglények és a magyar faunában elöforduló fajaik, *Allattani Kozl.*, 8, 1, 1909.
574. **Reiten, A. C., Jense, R., and Griner, L. A.**, Eosinophilic myositis (sarcosporidiosis; sarco) in beef cattle, *Am. J. Vet. Res.*, 27, 903, 1966.
575. **Reiter, I. and Mareis, A.**, Zur Differenzierung von *Sarcocystis muris-* und *S. dispersa-*Infektionen der Maus mittels isoelektrischer Fokussierung und Immunoassays, *Dtsch. Tieraeztl. Wochenschr.*, 93, 433, 1986.
576. **Reiter, I., Weiland, G., Roscher, B., Meyer, J., and Frahm, K.**, Versuche zum serologischen Nachweis der Sarkosporidoise an experimentell mit Sarkosporidien infizierten Rindern und Schafen, *Berl. Muench. Tieraerztl. Wochenschr.*, 94, 425, 1981.
577. **Retzlaff, N. and Weise, E.**, Sarkosporidien beim Wasserbüffel (*Bubalus bubalis*) in der Türkei, *Berl. Muench. Tieraerztl. Wochenschr.*, 15, 283, 1969.
578. **Rezakhani, A., Cheema, A. H., and Edjtehadi, M.**, Second degree atrioventricular block and sarcosporidiosis in sheep, *Zentralbl. Veterinaermed. Reihe A*, 24, 258, 1977.
579. **Rickard, M. D. and Munday, B. L.**, Host specificity of *Sarcocystis* spp in sheep and cattle, *Aust. Vet. J.*, 52, 48, 1976.
580. **Rifaat, M. A., Schafia, A. S., Khalil, H. M., Magda, E. A., Abdel Baki, M. H., and Abdel Ghaffar, F. M.**, A study on the life cycle of *Sarcocystis muris*, *J. Egypt. Public Health Assoc.*, 53, 341, 1978.
581. **Rimaila-Pärnänen, E. and Nikander, S.**, Generalized eosinophilic myositis with sarcosporidiosis in a Finnish cow, *Nord. Veterinaermed.*, 32, 96, 1980.
581a. **Rodriquez Osorid, M., Gomez Garcia, V., Tomas Safont, M. J., Campos Buneno, M., and Mañas Almendros, I.**, Estudio comparativo de las technicas de digestion pepsica muscular, inmunodifusion e inmunofluorescencia indirecta en el diagnostico de la sarcosporidiasis caprina, *Rev. Iber. Parasitol.*, 38, 793, 1978.
582. **Rommel, M., Heydorn, A.O., and Gruber, F.**, Beiträge zum Lebenszyklus der Sarkosporidien I. Die Sporozyste von *S. tenella* in den Fäzes der Katze, *Berl. Muench. Tieraerztl. Wochenschr.*, 85, 101, 1972.
583. **Rommel, M. and Heydorn, A. O.**, Beiträge zum Lebenszyklus der Sarkosporidien. III. *Isospora hominis* (Railliet und Lucet, 1891) Wenyon, 1923, eine Dauerform der Sarkosporidien des Rindes und des Schweins, *Berl. Muench. Tieraerztl. Wochenschr.*, 85, 143, 1972.
584. **Rommel, M., Heydorn, A. O., Fischle, B., and Gestrich, R.**, Beiträge zum Lebenszyklus der Sarkosporidien. V. Weitere Endwirte der Sarkosporidien von Rind, Schaf und Schwein und die Bedeutung des Zwischenwirtes für die Verbreitung dieser Parasitose, *Berl. Muench. Tieraerztl. Wochenschr.*, 87, 392, 1974.
585. **Rommel, M. and Geisel, O.**, Untersuchungen über die Verbreitung und den Lebenszyklus einer Sarkosporidienart des Pferdes (*Sarcocystis equicanis* n. spec.), *Berl. Muench. Tieraerztl. Wochenschr.*, 88, 468, 1975.
586. **Rommel, M. and Krampitz, H. E.**, Beiträge zum Lebenszyklus der Frenkelien. I. Die Identität von *Isospora buteonis* aus dem Mäusebussard mit einer Frenkelienart (*F. clethrionomyobuteonis* spec. n.) aus der Rötelmaus, *Berl. Muench. Tieraerztl. Wochenschr.*, 88, 338, 1975.
587. **Rommel, M., Krampitz, H. E., and Geisel, O.**, Beiträge zum Lebenszyklus der Frenkelien III. Die sexuelle Entwicklung von *F. clethrionomyobuteonis* im Mäusebussard, *Z. Parasitenk.*, 51, 139, 1977.
588. **Rommel, M. and Krampitz, H. E.**, Weitere Untersuchungen über das Zwischenwirtsspektrum und den Entwicklungszyklus von *Frenkelia microti* aus der Erdmaus, *Zentralbl. Veterinaermed. Reihe B*, 25, 273, 1978.
589. **Rommel, M.**, Das Frettchen (*Putorius putorius furo*), ein zusätzlicher Endwirt für *Sarcocystis muris*, *Z. Parasitenkd.*, 58, 187, 1979.
590. **Rommel, M., Heydorn, A. O., and Erber, M.**, Die Sarkosporidiose der Haustiere und des Menschen, *Berl. Muench. Tieraerztl. Wochenschr.*, 92, 457, 1979.
591. **Rommel, M., Schwerdtfeger, A., and Blewaska, S.**, The *Sarcocystis muris*-infection as a model for research on the chemotherapy of acute sarcocystosis of domestic animals, *Zentralbl. Bakteriol. Parasitenkd. Infektionskr. Hyg. I Abt. Orig. Reihe A*, 250, 268, 1981.
591a. **Rommel, M., Tiemann, G., Pötters, U., and Weller, W.**, Untersuchungen zur Epizootiologie von Infektionen mit zystenbildenden Kokzidien (Toxoplasmidae, Sarcocystidae) in Katzen, Schweinen, Rindern und wildebenden Nagern, *Dtsch. tieraerztl. Wochenschr.*, 89, 57, 1982.
592. **Rommel, M.**, Integrated control of protozoan diseases of livestock, in *Tropical Parasitoses and Parasitic Zoonoses*, Proc. 10th Meet. World Assoc. for the Advancement of Veterinary Parasitology, Dunsmore, J. D., Ed., Perth, 1983, 9.
593. **Rommel, M. and Schnieder, T.**, Epizootiologie und Bekämpfung der Sarkozystose der Schweine, *Angew. Parasitol.*, 26, 39, 1985.
594. **Thorton, R.**, Protozoal abortion in cattle, *Surveillance*, 14, 15, 1988.

595. **Rooney, J. R., Prickett, M. E., Delaney, F. M., and Crowe, M. W.**, Focal myelitis-encephalitis in horses, *Cornell Vet.*, 50, 494, 1970.
596. **Rüedi, D. and Hörning, B.**, Sarkosporidiennachweis als Zufallsbefund in einem Rehwildbestand im Aargau, *Schweiz. Arch. Tierheilkd.*, 125, 155, 1983.
597. **Ruiz, A. and Frenkel, J. K.**, Recognition of cyclic transmission of *Sarcocystis muris* by cats, *J. Infect. Dis.*, 133, 409, 1976.
598. **Rzepczyk, C. M.**, Evidence of a rat-snake life cycle for *Sarcocystis*, *Int. J. Parasitol.*, 4, 447, 1974.
599. **Rzepczyk, C. and Scholtyseck, E.**, Light and electron microscope studies on the *Sarcocystis* of *Rattus fuscipes*, an Australian rat, *Z. Parasitenkd.*, 50, 137, 1976.
600. **Saha, A. K., Srivastava, P. S., and Sinha, S. R. P.**, Toxic effects of the extracts of *Sarcocystis fusiformis* to laboratory mice, *Indian J. Anim. Sci.*, 55, 656, 1985.
600a. **Sahai, B. N., Singh, S. P., Sahay, M. N., Srivastava, P. S., and Juyal, P. D.**, Note on the incidence and epidemiology of *Sarcocystis* infection in cattle, buffaloes and pigs in Bihar, *Indian J. Anim. Sci.*, 52, 1005, 1982.
600b. **Sahai, B. N., Singh, S. P., Sahay, M. N., Srivastava, P. S., and Juyal, P. D.**, Role of dogs and cats in the epidemiology of bovine sarcosporidiasis, *Indian J. Anim. Sci.*, 53, 84, 1983.
601. **Sahasrabudhe, V. K. and Shah, H. L.**, The occurrence of *Sarcocystis* sp. in the dog, *J. Protozool.*, 13, 531, 1966.
602. **Saito, M., Nakajima, T., Watanabe, A., and Itagaki, H.**, *Sarcocystis miescheriana* infection and its frequency in pigs in Japan, *Jpn. J. Vet. Sci.*, 48, 1083, 1987.
603. **Samaraweera, H. P. and Kulasiri, C. de S.**, Sarcosporidiosis in goats in Ceylon, *Ceylon J. Med. Sci.*, 18, 47, 1969.
604. **Schebitz, H., and Hartwigk, H.**, Myositis sarcosporidica beim Pferd, *Tieraerztl. Umsch.*, 5, 351, 1950.
605. **Schmitz, J. A. and Wolf, W. W.**, Spontaneous fatal sarcocystosis in a calf, *Vet. Pathol.*, 14, 527, 1977.
606. **Schnieder, T. and Rommel, M.**, Ausbildung und Dauer der Immunität gegen *Sarcocystis miescheriana* im Schwein bei kontinuierlicher Verabreichung kleiner Mengen von Sporozysten, *Berl. Muench. Tieraerztl. Wochenschr.*, 96, 167, 1983.
607. **Schnieder, T., Kaup, F. J., Drommer, W., Thiel, W., and Rommel, M.**, Zur Feinstruktur und Entwicklung von *Sarcocystis aucheniae* beim Lama, *Z. Parasitenkd.*, 70, 451, 1984.
608. **Schnieder, T., Trautwein, G., and Rommel, M.**, Untersuchungen zur Persistenz der Zysten von *Sarcocystis miescheriana* in der Muskulatur des Schweines nach ein- und mehrmaliger Infektion, *Berl. Muench. Tieraerztl. Wochenschr.*, 97, 356, 1984.
609. **Schnieder, T., Zimmermann, U., Matuschka, F. R., Bürger, H. J., and Rommel, M.**, Zur Klinik, Enzymaktivität und Antikörperbildung bei experimentell mit Sarkosporidien infizierten Pferden, *Zentralbl. Veterinaermed. Reihe B*, 32, 29, 1985.
609a. **Scholtyseck, E., Mehlhorn, H., and Müller, B. E. G.**, Identifikation von Merozoiten der vier cystenbildenden Coccidien (*Sarcocystis, Toxoplasma, Besnoitia, Frenkelia*) auf Grund feinstruktureller Kriterien, *Z. Parasitenkd.*, 42, 185, 1973.
609b. **Scholtyseck, E., Mehlhorn, H., and Müller, B. E. G.**, Feinstruktur der Cyste und Cystenwand von *Sarcocystis tenella, Besnoitia jellisoni, Frenkelia* sp. and *Toxoplasma gondii*, *J. Protozool.*, 21, 284, 1974.
609c. **Scholtyseck, E. and Mehlhorn, H.**, Ultrastructural study of characteristic organelles (paired organelles, micronemes, micropores) of Sporozoa and related organisms, *Z. Parasitenkd.*, 34, 97, 1970.
610. **Scholtyseck, E. and Hilali, M.**, Ultrastructural study of the sexual stages of *Sarcocystis fusiformis* (Railliet, 1897) in domestic cats, *Z. Parasitenkd.*, 56, 205, 1978.
611. **Scholtyseck, E., Entzeroth, R., and Chobotar, B.**, Light and electron microscopy of *Sarcocystis* sp. in the skeletal muscle of an opossum (*Didelphis virginiana*), *Protistologica*, 18, 527, 1982.
612. **Schramlová, J. and Blazek, K.**, Ultrastruktur der Cystenwand der Sarkosporidien des Rehes (*Capreolus capreolus* L.), *Z. Parasitenkd.*, 55, 43, 1978.
613. **Schulze, K. and Zimmermann, T.**, Sarkosporidienzysten im Hackfleisch, *Fleischwirtschaft*, 61, 1, 1981.
614. **Schulze, K. and Zimmermann, T.**, Sarkosporidienbefall beim Rehwild mit lebensmittel-bzw. fleischhygienischer Bedeutung, *Fleischwirtschaft*, 62, 1, 1982.
614a. **Scorza, J. V., Torrealba, J. F., and Dagert, C.**, *Klossiella tejerai* nov. sp. y *Sarcocystis didelphidis* nov. sp. parasitos de un *Didelphis marsupialis* de Venezuela, *Acta. Biol. Venez.*, 2, 97, 1957.
615. **Scott, J. W.**, Economic importance of Sarcosporidia, with especial reference to *Sarcocystis tenella*, *Univ. Wyo. Agric. Exp. Stn. Bull.*, 262, 5, 1943.
616. **Scott, J. W.**, Life history of Sarcosporidia, with particular reference to *Sarcocystis tenella*, *Univ. Wyo. Agric. Exp. Stn. Bull.*, 259, 5, 1943.
617. **Sedlaczek, J. and Zipper, J.**, *Sarcocystis alceslatrans* (Apicomplexa) bei einem paläarktischen Elch (Ruminantia), *Angew. Parasitol.*, 27, 137, 1986.
617a. **Sénaud, J.**, Contribution á l'étude des sarcosporidies et des toxoplasmes (*Toxoplasmea*), *Protistologica*, 3, 167, 1967.

618. **Sénaud, J., Chobotar, B., and Scholtyseck, E.,** Role of the micropore in nutrition of the Sporozoa. Ultrastructural study of *Plasmodium cathemerium, Eimeria ferrisi, E. stiedai, Besnoitia jellisoni,* and *Frenkelia* sp., *Tropenmed. Parasitol.,* 27, 145, 1976.

619. **Senaud, J. and Cerná, Z.,** Le cycle de développement asexué de *Sarcocystis dispersa* (Cerná, Kolarova et Sulc, 1977) chez la souris: éatude au microscope électronique, *Protistologica,* 14, 155, 1978.

620. **Seneviratna, P., Atureliya, D., and Vijayakumar, R.,** The incidence of *Sarcocystis* spp. in cattle and goats in Sri Lanka, *Ceylon Vet. J.,* 23, 11, 1975.

621. **Seneviratna, P., Edward, A. G., and DeGiusti, D. L.,** Frequency of *Sarcocystis* spp in Detroit, Metropolitan Area, Michigan, *Am. J. Vet. Res.,* 36, 337, 1975.

622. **Shaw, J. J. and Lainson, R.,** *Sarcocystis* of rodents and marsupials in Brazil, *Parasitology,* 59, 233, 1969.

623. **Sheffield, H. G., Frenkel, J. K., and Ruiz, A.,** Ultrastructure of the cyst of *Sarcocystis muris, J. Parasitol.,* 63, 629, 1977.

624. **Sheffield, H. G. and Fayer, R.,** Fertilization in the coccidia: fusion of *Sarcocystis bovicanis* gametes, *Proc. Helminthol. Soc. Wash.,* 47, 118, 1980.

625. **Shimura, K., Ito, S., and Tsunoda, K.,** Sporocysts of *Sarcocystis cruzi* in mesenteric lymph nodes of dogs, *Natl. Inst. Anim. Health Q.,* 21, 186, 1981.

626. **Sibalic, S., Tomanovic, B., and Sibalic, D.,** Nalaz sporocista *Sarcocystis* spp. u pasa, *Acta Parasitol.,* 8, 49, 1977.

627. **Sibalic, D., Bordjoski, A., Conic, V., and Djurkovic-Djakovic, O.,** A study of *Toxoplasma gondii* and *Sarcocystis* sp. infections in humans suspected of acquired toxoplasmosis, *Acta Vet. (Belgrad),* 33, 39, 1983.

627a. **Simpson, C. F.,** Electron microscopy of *Sarcocystis fusiformis, J. Parasitol,.* 52, 607, 1966.

627b. **Simpson, C. F. and Forrester, D. J.,** Electron microscopy of *Sarcocystis* sp.: cyst wall, micropores, rhoptries, and an unidentified body, *Int. J. Parasitol.,* 467, 1973.

628. **Simpson, C. F. and Mayhew, I. G.,** Evidence for *Sarcocystis* as the etiologic agent of equine protozoal myeloencephalitis, *J. Protozool.,* 27, 288, 1980.

629. **Skárková, V.,** Histopathological changes in the liver tissue of house mouse and common vole during sarcocystosis, *Folia Parasitol. (Prague),* 33, 115, 1986.

630. **Skeels, M. R., Nims, L. J., and Mann, J. M.,** Intestinal parasitosis among Southeast Asian immigrants in New Mexico, *Am. J. Public Health,* 72, 57, 1982.

631. **Smith, D. D. and Frenkel, J. K.,** Cockroaches as vectors of *Sarcocystis muris* and of other coccidia in the laboratory, *J. Parasitol.,* 64, 315, 1978.

631a. **Smith, J. H., Meier, J. L., Neill, J. G., and Box, E. D.,** Pathogenesis of *Sarcocystis falcatula* in the budgerigar. I. Early pulmonary schizogony, *Lab. Invest.,* 56, 60, 1987.

631b. **Smith, J. H., Meier, J. L., Neill, J. G., and Box, E. D.,** Pathogenesis of *Sarcocystis falcatula* in the budgerigar II. Pulmonary pathology, *Lab. Invest.,* 56, 72, 1987.

632. **Smith, T. S. and Herbert, I. V.,** Experimental microcyst sarcocystis infection in lambs: serology and immunohistochemistry, *Vet. Rec.,* 119, 547, 1986.

632a. **Sobel, R. A. and Colvin, R. B.,** Responder strain-specific enhancement of endothelial and mononuclear cell Ia in delayed hypersensitivity reactions in (strain 2 x strain 13) F, guinea pigs, *J. Immunol.,* 137, 2132, 1986.

632b. **Sobel, R. A., Van der Veen, R. C., and Lees, M. B.,** The immunopathology of chronic experimental allergic encephalomyelitis induced in rabbits with bovine proteolipid protein, *J. Immunol.,* 136, 157, 1986.

632c. **Speer, C. A., Hammond, D. M., Mahrt, J. L., and Roberts, W.,** Structure of the oocyst and oocyst walls and excystation of sporozoites of *Isospora canis, J. Parasitol.,* 59, 35, 1973.

632d. **Somvanshi, R., Koul, G. L., and Biswas, J. C.,** Sarcocystis in a leopard (*Panthera pardus*), *Indian Vet. Med. J.,* 11, 174, 1987.

633. **Speer, C. A., Pond, D. B., and Ernst, J. V.,** Development of *Sarcocystis hemionilatrantis* Hudkins and Kistner, 1977 in the small intestine of coyotes, *Proc. Helminthol. Soc. Wash.,* 47, 106, 1980.

634. **Speer, C. A. and Dubey, J. P.,** An ultrastructural study of first- and second-generation merogony in the coccidian *Sarcocystis tenella, J. Protozool.,* 28, 424, 1981.

635. **Speer, C. A. and Dubey, J. P.,** Ultrastructure of in vivo lysis of *Sarcocystis cruzi* merozoites, *J. Parasitol.,* 67, 961, 1981.

636. **Speer, C. A. and Dubey, J. P.,** Scanning and transmission electron microscopy of ovine mesenteric arteries infected with first-generation meronts of *Sarcocystis tenella, Can. J. Zool.,* 60, 203, 1982.

637. **Speer, C. A. and Dubey, J. P.,** *Sarcocystis wapiti* sp. nov. from the North American wapiti (*Cervus elaphus*), *Can. J. Zool.,* 60, 881, 1982.

637a. **Speer, C. A.,** The Coccidia, in *In Vitro Cultivation of Protozoan Parasites,* Jensen, J. B., Ed., CRC Press, Boca Raton, FL, 1983, 1.

637b. **Speer, C. A., Reduker, D. W., Burgess, D. E., Whitmire, W. M., and Splitter, G. A.,** Lymphokine-induced inhibition of growth of *Eimeria bovis* and *Eimeria papillata* (Apicomplexa) in cultured bovine monocytes, *Infect. Immun.,* 50, 566, 1985.

638. **Speer, C. A., Cawthorn, R. J., and Dubey, J. P.,** In vitro cultivation of the vascular phase of *Sarcocystis capracanis* and *Sarcocystis tenella, J. Protozool.,* 33, 486, 1986.

639. **Speer, C. A., Whitmire, W. M., Reduker, D. W., and Dubey, J. P.,** In vitro cultivation of meronts of *Sarcocystis cruzi, J. Parasitol.,* 72, 677, 1986.
640. **Speer, C. A. and Dubey, J. P.,** Vascular phase of *Sarcocystis cruzi* cultured *in vitro, Can. J. Zool.,* 64, 209, 1986.
641. **Speer, C. A. and Dubey, J. P.,** An unusual structure in the primary cyst wall of *Sarcocystis hemionilatrantis, J. Protozool.,* 33, 130, 1986.
642. **Speer, C. A. and Burgess, D. E.,** *In vitro* cultivation of *Sarcocystis* merozoites, *Parasitol. Today,* 3, 2, 1987.
643. **Speer, C. A. and Burgess, D. E.,** In vitro development and antigen analysis of *Sarcocystis, Parasitol. Today,* 4, 46, 1988.
643a. **Speer, C. A., Reduker, D. W., Burgess, D. E., Kyle, J. E. and Dubey, J. P.,** Proteins and antigens of bradyzoites, merozoites and sporozoites of *Sarcocystis cruzi,* to be published.
644. **Srivastava, P. S., Saha, A. K., and Sinha, S. R. P.,** Spontaneous sarcocystosis in indigenous goats in Bihar, India, *Acta Protozool.,* 24, 339, 1985.
644a. **Srivastava, P. S., Sahai, B. N., Sinha, S. R. P., and Saha, A. K.,** Some differential features of the developmental cycle of bubaline *Sarcocystis* sp. in canine and feline definitive hosts, *Protistologica,* 21, 385, 1985.
645. **Srivastava, P. S., Saha, A. K., and Sinha, S. R. P.,** Effects of heating and freezing on the viability of sarcocysts of *Sarcocystis levinei* from cardiac tissues of buffaloes, *Vet. Parasitol.,* 19, 329, 1986.
645a. **Stackhouse, L. L., Cawthorn, R. J., and Brooks, R. J.,** Pathogenesis of infection with *Sarcocystis rauschorum* (Apicomplexa) in experimentally infected varying lemmings (*Dicrostonyx richardsoni*), *J. Wildl. Dis.,* 23, 566, 1987.
646. **Stalheim, O. H., Proctor, S. J., Fayer, R., and Lunde, M.,** Death and abortion in cows experimentally infected with *Sarcocystis* from dogs, *Annu. Proc. Am. Assoc. Vet. Lab. Diagn.,* 19, 317, 1976.
646a. **Stiles, C. W.,** Notes on parasites. 18. On the presence of Sarcosporidia in birds, Publ. No. 3, Bulletin of the Bureau of Animal Industry, U. S. Department of Agriculture, Washington, D. C., 1893, 79.
647. **Streitel, R. H. and Dubey, J. P.,** Prevalence of *Sarcocystis* infection and other intestinal parasitisms in dogs from a humane shelter in Ohio, *J. Am. Vet. Med. Assoc.,* 168, 423, 1976.
647a. **Strohlein, D. A. and Prestwood, A. K.,** In vitro excystation and structure of *Sarcocystis suicanis* Erber, 1977, sporocysts, *J. Parasitol,.* 72, 711, 1986.
648. **Stubbings, D. P. and Jeffrey, M.,** Presumptive protozoan (sarcocystis) encephalomyelitis with paresis in lambs, *Vet. Rec.,* 116, 373, 1985.
649. **Suteu, E. and Coman, S.,** Nouvelles observations sur le cycle biologique de *Sarcocystis fusiformis, Bull. Soc. Sci. Vet. Med. Comp. Lyon,* 75, 363, 1973.
650. **Tadmor, A., Nobel, T. A., and Mindel, J. B.,** Myositis in a mule possibly due to infestation with Sarcosporidia, *Refu. Vet.,* 23, 106, 1966.
651. **Tadros, W. A., Bird, R. G., and Ellis, D. S.,** The fine structure of cysts of *Frenkelia* (the M-organism), *Folia Parasitol. (Prague),* 19, 203, 1972.
652. **Tadros, W.,** Contribution to the understanding of the life-cycle of *Sarcocystis* of the short-tailed vole *Microtus agrestis, Folia Parasitol, (Prague),* 23, 193, 1976.
653. **Tadros, W. and Laarman, J. J.,** The cat *Felis catus* as the final host of *Sarcocystis cuniculi* Brumpt, 1913 of the rabbit *Oryctolagus cuniculus, Proc. K. Ned. Akad. Wet., Amsterdam,* 80C, 351, 1977.
654. **Tadros, W. and Laarman, J. J.,** Note on the specific designation of *Sarcocystis putorii* (Railliet & Lucet, 1891) comb. nov. of the common European vole, *Microtus arvalis, Proc. K. Ned. Akad. Wet.,* 81, 466, 1978.
655. **Tadros, W. and Laarman, J. J.,** A comparative study of the light and electron microscopic structure of the walls of the muscle cysts of several species of sarcocystid eimeriid coccidia, *Proc. K. Ned. Akad. Wet.,* 81, 469, 1978.
656. **Tadros, W. and Laarman, J. J.,** Apparent congenital transmission of *Frenkelia* (Coccidia: Eimeriidae): first recorded incidence, *Z. Parasitenkd.,* 58, 41, 1978.
657. **Tadros, W., Hazelhoff, W., and Laarman, J. J.,** The detection of circulating antibodies against *Sarcocystis* in human and bovine sera by the enzyme-linked immunosorbent assay (ELISA) technique, *Acta Leiden.,* 47, 53, 1979.
658. **Tadros, W. and Laarman, J. J.,** Some observations on the gametogonic development of *Sarcocystis cuniculi* of the common European rabbit in a feline fibroblast cell line, *Acta Leiden.,* 47, 45, 1979.
659. **Tadros, W.,** Studies on Sarcosporidia of rodents with birds of prey as definitive hosts, in *Parasitological Topics,* a Presentation Volume to P. C. C. Garnham, F. R. S., on the Occasion of his 80th Birthday, 1981, Canning, E. V., Ed., Allen Press, Lawrence, KS, 248, 1981.
660. **Tadros, W. and Laarman, J. J.,** Current concepts on the biology, evolution and taxonomy of tissue cyst-forming eimeriid coccidia, *Adv. Parasitol.,* 20, 293, 1982.
661. **Takla, M.,** Akute Sarkosporidoise beim Rind. Ein Fall einer klinischen natürlichen Meningoenzephalitis durch *Sarcocystis cruzi* beim Bullen, *Tieraerztl. Prax.,* 12, 167, 1984.
661a. **Takos, M. J.,** Notes on sarcosporidia of birds in Panama, *J. Parasitol.,* 43, 183, 1957.

661b. **Tanabe, M. and Okinami, M.,** On the parasitic protozoa of the ground squirrel, *Eutamis asiaticus* Uthensis, with special reference to *Sarcocystis eutamias* sp. nov., *Keijo J. Med.,* 10, 126, 1940.

662. **Tenter, A. M.,** Comparison of dot-ELISA, ELISA, and IFAT for the detection of IgG antibodies to *Sarcocystis muris* in experimentally infected and immunized mice, *Vet. Parasitol.,* in press.

663. **Tenter, A. M.,** Comparison of enzyme-linked immunosorbent assay and indirect fluorescent antibody test for the detection of IgG antibodies to *Sarcocystis muris, Zentralbl. Bakteriol. Parasitenkd. Infektionskr. Hyg. I Abt. Orig. Reihe A,* 267, 259, 1987.

664. **Terrell, T. G. and Stookey, J. L.,** Chronic eosinophilic myositis in a rhesus monkey infected with sarcosporidiosis, *Vet. Pathol.,* 9, 266, 1972.

665. **Thomas, V. and Dissanaike, A. S.,** Antibodies to *Sarcocystis* in Malaysians, *Trans. R. Soc. Trop. Med. Hyg.,* 72, 303, 1978.

666. **Tietz, H. J., Montag, T., Brose, E., Hiepe, T., Mann, W., Hiepe, F., and Halle, H.,** Extracts from *Sarcocystis gigantea* macrocysts are mitogenic for human blood lymphocytes, *Angew. Parasitol.,* 27, 201, 1986.

667. **Tinling, S. P., Cardinet, G. H., III, Blythe, L. L., Cohen, M., and Vonderfecht, S. L.,** A light and electron microscopic study of sarcocysts in a horse, *J. Parasitol.,* 66, 458, 1980.

667a. **Todd, K. S., Jr., Gallina, A. M., and Nelson, W. B.,** *Sarcocystis* species in psittaciform birds, *J. Zoo Anim. Med.,* 6, 21, 1975.

668. **Tongson, M. S. and Molina, R. M.,** Light and electron microscope studies on Sarcocystis sp. of the Philippine buffaloes (*Bubalus bubalis*), *Philipp. J. Vet. Med.,* 18, 16, 1979.

669. **Tongson, M. S. and Pablo, L. S. M.,** Preliminary screening of the possible definitive hosts of *Sarcocystis* sp. found in Philippine buffaloes (*Bubalus bubalis*), *Philipp. J. Vet. Med.,* 18, 42, 1979.

670. **Tongson, M. S. and Calingasan, N. Y.,** Demonstration of the developmental stages of *Sarcocystis* sp. of Philippine buffaloes (*Bubalus bubalis*) in the small intestine of dogs, *Philipp. J. Vet. Med.,* 19, 52, 1980.

671. **Tuggle, B. N. and Schmeling, S. K.,** Parasites of the bald eagle (*Haliaeetus leucocephalus*) of North America, *J. Wildl. Dis.,* 18, 501, 1982.

672. **Uggla, A., Hilali, M., and Lövgren, K.,** Serological responses in *Sarcocystis cruzi* infected calves challenged with *Toxoplasma gondii, Res. Vet. Sci.,* 43, 127, 1987.

672a. **Van Knapen, F., Bouwmann, D. and Greve, E.,** Onderzoek naar het voorkomen van *Sarcocystis* spp. bij Nederlandse rundern met verschillende methoden, *Tijdschr. Diergeneeskd.,* 112, 1095, 1987.

672b. **Vercruysse, J. and Van Mark, E.,** Les sarcosporidies des petits ruminants au Sénégal, *Rev. Elev. Med. Vet. Pays Trop.,* 34, 377, 1981.

673. **Vershinin, I. I.,** Studies on the cycle of development of *Sarcocystis hirsuta* (in Russian), *Veterinariya,* 51, 77, 1974.

673a. **Vetterling, J. M., Pacheco, N. D., and Fayer, R.,** Fine structure of gametogony and oocyst formation in *Sarcocystis* sp. in cell culture, *J. Protozool.,* 20, 613, 1973.

674. **Vickers, M. C. and Brooks, H. V.,** Suspected *Sarcocystis* infection in an aborted bovine foetus, *N.Z. Vet. J.,* 31, 166, 1983.

675. **Viles, J. M. and Powell, E. C.,** The ultrastructure of the cyst wall of a murine *Sarcocystis, Z. Parasitenkd.,* 49, 127, 1976.

676. **Viles, J. M. and Powell, E. C.,** Myofiber damage accompanying intramuscular parasitism by *Sarcocystis muris, Tissue Cell,* 13, 45, 1981.

677. **Voigt, W. P. and Heydorn, A. O.,** Chemotherapy of sarcosporidiosis and theileriosis in domestic animals, *Zentralbl. Bakteriol. Parasitenkd. Infektionskr. Hyg. I Abt. Orig. Reihe A,* 250, 256, 1981.

678. **Wallace, G. D.,** Sarcocystis in mice inoculated with Toxoplasma-like oocysts from cat feces, *Science,* 180, 1375, 1973.

679. **Weber, M., Weyreter, H., O'Donoghue, P. J., Rommel, M., and Trautwein, G.,** Persistence of acquired immunity to *Sarcocystis miescheriana* infection in growing pigs, *Vet. Parasitol.,* 13, 287, 1983.

679a. **Wei, T., Chang, P. Z., Doung, M. X., Wang, X. Y., and Xia, A. Q.,** Description of two new species of *Sarcocystis* from the yak (*Poephagus grunniens*) (in Chinese), *Sci. Agric. Sin.,* 4, 80, 1985.

679b. **Weiland, G., Reiter, I., and Boch, J.,** Möglichkeiten und Grenzen des serologischen Nachweises von Sarkosporidieninfektionen, *Berl. Muench. Tieraerztl. Wochenschr.,* 95, 387, 1982.

680. **Wenzel, R., Erber, M., Boch, J., and Schellner, H. P.,** Sarkosporidien-Infektion bei Haushuhn, Fasan und Bleβhuhn, *Berl. Muench. Tieraerztl. Wochenschr.,* 95, 188, 1982.

681. **Weyreter, H., O'Donoghue, P. J., Weber, M., and Rommel, M.,** Class-specific antibody responses in pigs following immunization and challenge with sporocysts of *Sarcocystis miescheriana, Vet. Parasitol.,* 16, 201, 1984.

682. **Weyreter, H. and O'Donoghue, P. J.,** Untersuchungen zur Immunserodiagnose der Sarcocystis-Infektionen. I. Antikörperbildung bei Maus und Schwein, *Zentralbl. Bakteriol. Parasitenkd. Infektionskr. Hyg. I Abt. Orig. Reihe A,* 253, 407, 1982.

683. **Wicht, R. J.,** Transmission of *Sarcocystis rileyi* to the striped skunk (*Mephitis mephitis*), *J. Wildl. Dis.,* 17, 387, 1981.

684. **Wickham, N. and Carne, H. R.**, Toxoplasmosis in domestic animals in Australia, *Aust. Vet. J.*, 26, 1, 1950.
685. **Wobeser, G., Leighton, F. A., and Cawthorn, R. J.**, Occurrence of *Sarcocystis* Lankester, 1882, in wild geese in Saskatchewan, *Can. J. Zool.*, 59, 1621, 1981.
686. **Wobeser, G. and Cawthorn, R. J.**, Granulomatous myositis in association with *Sarcocystis* sp. infection in wild ducks, *Avian Dis.*, 26, 412, 1982.
687. **Wobeser, G., Cawthorn, R. J., and Gajadhar, A. A.**, Pathology of *Sarcocystis campestris* infection in Richardson's ground squirrels (*Spermophilus richardsonii*), *Can. J. Comp. Med.*, 47, 198, 1983.
688. **Woodmansee, D. B. and Powell, E. C.**, Cross-transmission and in vitro excystation experiments with *Sarcocystis muris*, *J. Parasitol.*, 70, 182, 1984.
689. **Yakimoff, W. L. and Sokoloff, I. I.**, Die Sarkozysten des Renntieres und des Maral. (*Sarcocystis grüneri* n. sp.), *Berl. Tieraerztl. Wochenschr.*, 50, 772, 1934.
690. **Yiming, L. and Ziqiang, L.**, Study on man-pig cyclic infection of *Sarcocystis suihominis* found in Yunnan Province, China (in Chinese), *Acta Zool. Sin.*, 32, 329, 1986.
691. **Zaman, V. and Colley, F. C.**, Fine structure of *Sarcocystis fusiformis* from the Indian water buffalo (*Bubalus bubalis*) in Singapore, *Southeast Asian J. Trop. Med. Public Health*, 3, 489, 1972.
692. **Zaman, V. and Colley, F. C.**, Light and electron microscopic observations of the life cycle of *Sarcocystis orientalis* sp. n. in the rat (*Rattus norvegicus*) and the Malaysian reticulated python (*Python reticulatus*), *Z. Parasitenkd.*, 47, 169, 1975.
693. **Zaman, V. and Colley, F. C.**, Replacement of *Sarcocystis orientalis* Zaman and Colley, 1975, by *Sarcocystis singaporensis* sp. n., *Z. Parasitenkd.*, 51, 137, 1976.
694. **Zaman, V.**, Host range of *Sarcocystis orientalis*, *Southeast Asian J. Trop. Med. Public Health*, 7, 112, 1976.
695. **Zaman, V., Robertson, T. A., and Papadimitriou, J. M.**, Scanning electron microscopy of *Sarcocystis fusiformis* from the water buffalo (*Bubalus bubalis*), *Southeast Asian J. Trop. Med. Public Health*, 11, 205, 1980.
696. **Zeve, V. H., Price, D. L., and Herman, C. M.**, Electron microscope study of *Sarcocystis* sp., *Exp. Parasitol.*, 18, 338, 1966.
697. **Zielasko, B., Petrich, J., Trautwein, G., and Rommel, M.**, Untersuchungen über pathologisch-anatomische Veränderungen und die Entwicklung der Immunität bei der *Sarcocystis suicanis*-Infektion, *Berl. Muench. Tieraerztl. Wochenschr.*, 94, 223, 1981.
698. **Zimmermann, U., Schnieder, T., and Rommel, M.**, Untersuchungen über die Dynamik der Antikörperentwicklung bei Schweinen nach mehrfacher Immunisierung mit Sporozysten von *Sarcocystis miescheriana* und einmaliger Belastungsinfektion, *Berl. Muench. Tieraerztl. Wochenschr.*, 97, 408, 1984.
699. **Zlotnik, I.**, Toxoplasma in sheep, *Lancet*, 2, 295, 1959.
700. **Zuo, Y. X., Chen, F. Q., and Li, W. Y.**, Two patients with Sarcocystis infection, in *Malaria and Other Protozoal Infections*, Proc. Chinese Society of Protozoology Zhongshan University, Guangzhou, 1982, 52.

INDEX

A

AAT, see Alanine aminotransferase
Abortion, 73—77, 82, 89, 114
Acquired Factor VII deficiency, 63
ACTH, see Adrenocorticotrophic hormone
Adrenocorticotrophic hormone (ACTH), 74—75
Alanine aminotransferase (AAT), 63
Amprolium, 91, 116
Amylopectin, 27—28
Anemia, 70—71
Anteater, 165
Antelope, 155
Antibodies, 80, 84—86
Anticoccidials, 91
Antigens, 79—82, 85—88, 97
Apical complex, 14, 29—30, 63
Apical rings, 14, 17, 29
Arkhar, 155
Armadillo, 165
Arprinocid, 92
Autophagic vacuoles, 48
Avians, 58, 157—159, see also Birds

B

Baboon, 57, 109
Badger, 165
Basal bodies, 40
Bay®G7183, 92
Bear, 145
Beef, 109
Besnoitia, 173—175
Birds, 12, 68, 157—159, see also Avians
Bison, 6, 60, 105, 155
Blood urea nitrogen (BUN), 63
Bobcat, 147
Bradyzoites
  antibodies induced by, 85
  antigens of, 88, 97
  apical complex of, 29
  apical rings of, 17
  aqueous extracts of, 78
  conoid of, 23—24
  development of, 101
  in digest, 96
  endoplasmic reticulum of, 24
  epitopes in, 82
  in goblet cell, 37, 46
  golgi complex of, 24
  inclusion bodies of, 27
  incubation of, 93—94
  micropores of, 21—22
  microtubles of, 20—21
  mitochondria of, 24
  multivesicular bodies of, 25
  organelles of, 24—28
  pellicle of, 16—17
  polar rings of, 17—20
  production of, 4, 6, 12
  proteins of, 79—80, 84
  rhoptries and micronemes of, 22—23
  ribosomes of, 24
  of *S. cruzi*, 31, 49
  of *S. muris*, 39, 80
  staining of, 5
  structure of, 55—56
  typical, 30
Budgerigars, 62, 158—159
Buffalo, 60, 137—139, 146, 147
BUN, see Blood urea nitrogen
Buzzard, 171

C

Calf
  amprolium in, 91
  blastogenic response in, 84
  encephalomyelitis in, 111
  lethal sporocyst dose in, 62
  newborn, 90
  S. cruzi in, 13—17, 73—77, 87, 106—108
Camel, 141—142, 146
Canary, 60
Canids, 62, 91, 113, 152
Carcas condemnation, 119, 130
Carnivores, 91, 145—148
Caryospora, 173—175
Cat
  as definitive host, see Host, definitive
  feces of, 147
  as intermediate host, 57, 60, 113, 147
  Sarcocystis species in, 10, 131—135
  sporocysts from, 91
  transmission to, 123
Cattle
  abortion in, 71—72
  acute sarcocystosis in, 70
  antigens in, 88
  encephalitis in, 112
  eosinophilic myositis in, 77
  IgG antibodies in, 82—83
  immunization of, 91
  as intermediate host, 6, 57—58, 146—147
  placental lesions in, 73, 90
  protective immunity in, 84—86
  Sarcocystis species in, 62—64, 70, 90, 105—112, 151
  transmission to, 60—61
Cellular responses, 84
Centrocone, 29, 50
Cervids, 149
Chameleon, 168
Chemoprophylaxis and chemotherapy, 91—92
Chicken, 60—61, 146—147, 157
Chimpanzee, 109

Chipmunk, 57, 166
Coccidia, 174—176
Colostrum, 86
Conoid, 14, 17, 20, 23—24, 29, 33
Cow
    abortion in, 72—74, 89
    eosinophilic myositis in, 77
    as intermediate host, 62, 105
    necrosis in, 81
Cowbird, 60, 158
Coyote
    as definitive host, see Host, definitive
    feces of, 147
    as intermediate host, 147
    S. cruzi in, 10—11
    S. hirsuta in, 10
    transmission by, 90
CPA cells, 99—103
CPK, see Creatinine phosphokinase
Creatinine phosphokinase (CPK), 63
Cultivation in vitro, 97—104
Cystoisospora, 173—175
Cystozoites, see Bradyzoites
Cytoplasm, 21

D

Dalmeny disease, 106
Decoquinate, 92
Deer
    fallow, 37, 57, 146
    mule, 9
        as intermediate host, 57—58, 146
        Sarcocystis species in, 7, 15, 149, 153
    red, 149, 152—153
    roe, 57, 60—62, 71, 146, 149—150
    Sarcocystis species in, 149—154
    white-tailed, 9, 41, 57—60, 146—151
Diagnosis of sarcocystosis, 86—89, 94—97
Diaveridine, 92
Dog
    as definitive host, see Host, definitive
    feces of, 95, 146
    as intermediate host, 148
    Sarcocystis species in, 12, 145—148
    transmission by, 90
Donkey, 60, 133
Duck, 4, 58, 60, 157—158, 161

E

Edema, 70
Eel, 167
ELISA test, 97
Elk, 9
EM, see Eosinophilic myositis
Encephalitis, 73, 112
Encephalomyelitis, 111, 119
Endodyogeny, 4, 11, 16, 27—29, 32—33
Endoplasmic reticulum, 24

Endopolygeny, 48
Eosinophilic myositis (EM), 77—78, 80
Epidemiology, 90—91
EPM, see Equine protozoal myeloencephalitis
Equids, 131—135
Equine protozoal myeloencephalitis (EPM), 135
Excystation, 47, 53, 98—101
Exocytosis pore, 39

F

Ferret, 60, 164
Fertilization, 7, 44—46, 51
Fetal death, 74—77
Fetal health, 71—73
α-Fetoprotein, 75
Fever, 70, 76
Finches, 60
Fishes, 167
Flagella, 40
Flagellar axonemes, 20
Fox
    crab-eating, 121
    as definitive host, 60, 67, 149, 151
    grey, 151
    red, 6, 105, 113, 121, 127, 145, 147, 166
    Sarcocystis species in, 145
    transmission by, 90
Frenkelia, 171—175

G

Gametogenesis, 37—44
Gametogony, 6—7
Gamonts, 6—7, 13, 39, 101—104
Gazelle, 146—147, 155
Gecko, 168
Geese, 157—158
Generic diagnosis, 2
Goat
    abortion in, 71—72
    acute sarcocystosis in, 70
    clinical signs in, 64
    domestic, 121—124
    halofuginone in, 92
    immunization of, 91
    immunosuppression in, 84
    as intermediate host, 57—62, 146
    lethal sporocyst dose in, 62
    mountain, 58, 121—124, 155
    placental lesions in, 73
    protective immunity in, 84—86
    Sarcocystis species in, 121—124
    sarcocyst types in, 56
    schizonts in, 74
    vaccinated-challenged, 78—79
Golgi adjuncts, 26
Golgi complex, 24, 28
Grackle, 60, 158
Guinea fowl, 60

Guinea pig, 165

H

Halofuginone, 92
Hammondia, 173—175
Hamster, 61, 165
Hartebeest, 155
Hawk, 171
Heifers, 77
Hemorrhage, 64—65, 71, 74
Hepatitis, 73
Homiothermic animals, 161—166
Horse
  eosinophilic myositis in, 77, 80
  as intermediate host, 58, 146
  Sarcocystis species in, 54, 131—135
  transmission to, 60
Host
  definitive, 67
    baboon as, 109
    badger as, 165
    birds as, 159
    buzzard as, 171
    cat as
      of *S. cuniculi*, 165
      of *S. cymruensis*, 163
      of *S. fusiformis*, 137
      of *S. gigantea*, 118
      of *S. hirsuta*, 109
      of *S. leporum*, 165
      of *S. medusiformis*, 119
      of *S. moulei*, 122
      of *S. muris*, 61, 162
      of *S. odoi*, 151
      of various *Sarcocystis* species, 67, 145, 155
    chimpanzee as, 109
    coyote as
      of *S. alceslatrans*, 154
      of *S. capracanis*, 121
      of *S. cruzi*, 6, 60, 105
      of *S. ferovis*, 154
      of *S. hemionilatrantis*, 153
      of *S. odocoileocanis*, 151
      of *S. wapiti*, 152
      of various *Sarcocystis* species, 67, 145, 155
    dog as
      of *S. alceslatrans*, 154
      of *S. arieticanis*, 118
      of *S. baibacinacanis*, 165
      of *S. bertrami*, 131
      of *S. cameli*, 141
      of *S. capracanis*, 121
      of *S. caproli*, 149
      of *S. cervicanis*, 153
      of *S. cruzi*, 6, 60, 105
      of *S. equicanis*, 131
      of *S. fayeri*, 131
      of *S. hemionilatrantis*, 153
      of *S. levinei*, 137
      of *S. miescheriana*, 127
      of *S. odocoileocanis*, 151
      of *S. sibirica*, 149
      of *S. sybillensis*, 152
      of *S. tenella*, 113
      of *S. wapiti*, 152
      of various *Sarcocystis* species, 67, 145, 155
    ferret as, 162, 164
    fox as, 149, 151
      crab-eating, 121
      grey, 151
      red, 105, 113, 121, 127, 145, 151, 166
    humans as, 109, 129
    jackal as, 127
    kestrel as, 163
    mink as, 164
    monkey as, 109
    opossum as, 158, 161
    owl as, 61, 162, 166
    primates as, 128
    raccoon as, 105, 127, 145, 151
    sexual stages in, 5—6
    skunk as, 157
    snake as, 61—62, 162—164, 167, 169
    stoat as, 164
    weasel as, 164
    wolf as, 105, 127, 151
  intermediate
    anteater as, 165
    armadillo as, 165
    asexual stages in, 5—7
    avians as, 58, 158
    baboon as, 57
    bison as, 60, 105
    budgerigar as, 62, 158
    buffalo as, 147
    camel as, 146
    cat as, 57
    cattle as, 57—58, 62, 105, 146
    chameleon as, 168
    chicken as, 146—147
    chipmunk as, 57, 166
    coyote as, 147
    deer as, 57—60, 62, 146—147
    dog as, 148
    donkey as, 60
    duck as, 58
    gazelle as, 146—147
    gecko as, 168
    goat as, 57—62, 146
    gross lesions in, 64—65
    guinea pig as, 165
    hamster as, 165
    horse as, 58, 146
    lemming as, 57, 61—62, 159, 166
    lizard as, 168
    marmot as, 165
    microscopic lesions in, 65—67
    monkey as, 57
    moose as, 57, 146

mountain lion as, 148
mouse as, 57—58, 61—62, 147, 159, 162, 169
opossum as, 57, 164
ox as, 60
pheasant as, 146
pig as, 58, 62, 146—147
python as, 168
rabbit as, 58, 147, 165
rat as, 57—58, 61, 147, 162—164, 169
reindeer as, 57—58
sheep as, 57—58, 60, 62, 146—147
skink as, 168
squirrel as, 57, 62, 165—166
suslik as, 166
turtle as, 168
vole as, 57, 61, 159, 163—164, 169, 171
wallaby as, 58, 164
wapiti as, 57—58, 60, 146
water buffalo as, 146
wolf as, 148
yak as, 57—58
experimental infection of, 93
specificity of, 60
Humans, 109, 128, 143—144
Humoral responses, 82—83
Hypoxia, 74—75

I

IgG antibodies, 83
IgM antibodies, 83
Immune regulation, 69—70
Immunity to sarcocystosis, 79—86
  antigenic structure, 79—82
  cellular responses and immunosuppression, 84
  humoral responses, 82—83
  protective, 84—86, 108, 116, 121, 127
Immunosuppression, 84
Inclusion bodies, 27
Inflammation, 65—69
Isoenzymes, 60—62
Isolation, 93—94
Isospora, 1, 173—175

J

Jackal, 6, 60, 127
Jirds, 61

K

Kestrel, 163

L

Lactic dehydrogenase (LDH), 63
Lamb, 72, 90, 117
Lasalocid, 92
LDH, see Lactic dehydrogenase
Lectins, 78
Lemming, 57, 61—62, 159, 166
Leopard, 145

Lesions, 65—67, 73—74, 77
Leukoencephalomalacia, 73
Life cycle of sarcocystosis, 5—13
  asexual stages, 5
  definitive hosts, 5—6
  intermediate hosts, 5, 8
  merozoites, 11—12
  oocysts, 7
  schizonts, 9, 12—13
  sporoblasts, 8
  sporocysts, 8
  sporozoites, 8
Lion, 145
Lizard, 167—168
Llama, 3, 155

M

MAbs, see Monoclonal antibodies
Macrogamonts, 6, 37—39, 44—47, 51, 101
Mammals
  sea, 156
  small, 61—62, 162—166
Marmot, 165
Meningitis, 73
Merogony, 8
Merozoites
  antigens of, 88
  apical complexes in, 63
  in blood, 11
  budding, 66—67
  conoid of, 23
  developing, 49, 64—65, 98—103
  elongation of, 50
  formation of, 9
  free, 73—74
  in parasitophorous vacuole, 16
  proteins of, 79—80, 83—84
  in sarcocyst formation, 12, 14
  of *S. cruzi*, 82, 85, 87
  spindle apparatus in, 63
  of *S. rauschorum*, 59
  of *S. tenella*, 26
  surface of, 68, 82
  transformation of, 20, 28
Metrocytes
  in endodyogeny, 32
  formation of, 20—21
  progeny of, 4
  staining of, 5
  structure of, 15—16, 29, 54
  transformation of, 28
  typical, 30
  in young sarcocyst, 25—26
MHC II antigens, 69—70
Mice, see Mouse
Microfilaments, 15
Microgamonts, 6, 10, 37—40, 49—50, 101
Micronemes, 14, 22—23, 28, 37, 50
Micropores, 16, 21—22, 39
Micropyle, 8
Microtubules, 15, 20—21, 24, 39

Miescher's tubules, 1
Mitochondria, 24, 40
Monensin, 92
Monkey, 57, 60, 109, 144
Monoclonal antibodies (MAbs), 81—82, 86
Mononuclear cells, 65—67, 73, 75, 77—78
Moose, 57, 146, 149, 154
Mouse
    antigen in, 83, 86
    deer, 9, 62, 68, 162, 169
    house, 57—58, 61—62, 147, 159, 162, 169
    IgG antibodies in, 82—83
    MAbs in, 81
    S. muris in, 92
    transmission to, 60—62
    white-footed, 61—62
Mule, 133
Multivesicular bodies, 25, 48
Myocarditis, 73, 76, 88
Myocytes, 14, 16, 20, 45
Myofibers, 3

N

Necrosis
    focal, 64
    in lymph node artery, 72
    in organs, 65
    placental, 89
    of *Sarcocystis* in cow, 81
    severe, 75
    tissue, 68—69

O

Oocysts
    in cultured cells, 104
    development of, 101
    ingestion of, 90
    resistance to freezing, 91
    sporulated, 7—8, 11
    structure of, 59—60
    unsporulated, 46
    wall of, 39
Oogony, 39
Opossum, 57, 158, 161, 164
Organelles, 22—24, 28
Owl, 61, 159, 162, 166
Ox, 60
Oxytetracycline, 92
Oxytocin, 75

P

Parasitophorous vacuole (PV), 4—9, 12, 37, 39
Parasitophorous vacuole membrane (Pm), 14, 30, 34, 48
Pathogenesis of sarcocystosis, 68—79
    abortion, 71—77, see also Abortion
    anemia, 70—71
    chronic sarcocystosis and toxins, 78—79
    edema, 70
    eosinophilic myositis, 77
    fever, 70
    immune regulation, 69—70
    inflammation, 69
    tissue necrosis, 68—69
Pathogenicity of *Sarcocystis*, 62—68
    arieticanis, 118
    carpacanis, 121
    fayeri, 132
    gigantea, 118
    miescheriana, 127
    suihominis, 129
    tenella, 113—116
Pellicle, 16—17, 50
$PGF_{2\alpha}$, 75—76
PGFM, 75
Pheasant, 146
Pig
    abortion in, 71—72
    acute sarcocystosis in, 70
    antigen in, 83, 86
    clinical signs in, 64
    eosinophilic myositis, 77
    IgG antibodies in, 82—83
    immunization of, 91
    as intermediate host, 58, 146—147
    lethal sporocyst dose in, 62
    protective immunity in, 84—85
    sarcocyst isolation in, 96
    Sarcocystis species in, 127—130
    transmission to, 60
Pigeon, 60—61, 158
Placental infection, 73—74, 89
Plasmalemma, 16—17, 21, 39—40, 44, 66
Platelet dysfunction, 63
Pm, see Parasitophorous vacuole membrane
Poikilothermic animals, 167—169
Polar rings, 14, 17—20, 29
Pony, 134
Poultry, 91
Predators, 147—148
Preservation, 93—94
Primaquine diphosphate, 92
Primary sarcocyst wall (Pw), 14—15, 20—21, 30
Primates, 143—144
Progesterone, 75
Pronghorn, 91, 154
Prostaglandins, 75—76
Proteins, 79, 83—84
Prothrombin time, 63
Purification, 93—94
PV, see Parasitophorous vacuole
Pw, see Primary sarcocyst wall
Pyrimethamine, 92
Python, 163, 168—169

R

Rabbit, 58, 60—61, 78, 147, 165
Raccoon
    as definitive host, 6, 60, 67, 105, 127, 145
    feces of, 147

transmission by, 90
Rat
  cotton, 5, 57
  as intermediate host, 147
  Malaysian, 57, 163
  moon, 57, 162
  multimamate, 61, 162
  Norway, 57—58, 61, 163, 169
  rice, 162
  spiny, 163
  transmission to, 60—61
Reindeer, 57—58, 150—152
Reptiles, 167—169
Rhoptries, 16, 22—23, 28, 50, 67
Ribosomes, 24
Robenidin, 92
Ruminants, 149—156

S

Salinomycin, 92, 116
Sarcocystis
  alceslatrans, 57, 145—146, 154
  americana, 153
  ammodrami, 158
  aramidis, 158
  arieticanis, see Sarcocystis arieticanis
  asinus, 131
  aucheniae, 3, 155
  balaenopteralis, 156
  bertrami, 54, 60, 131, 146
  booliati, 57
  bovicanis, 1, 105
  bovifelis, 1, 109
  bovihominis, 1, 109
  bozemanensis, 6, 57
  bubalis, 155
  cameli, 141—142, 146
  campestris, 6, 9, 29, 38, 57, 62
  capracanis, see Sarcocystis capracanis
  capreoli, 57, 145, 149—150
  capreolicanis, 149
  caprifelis, 121—122
  cernae, 159
  cervi, 154
  cervicanis, 146, 152—153
  chalcidicolubris, 167—168
  chamaeleonis, 168
  clethrionomyelaphis, 61, 169
  colii, 158
  crotali, 57, 169
  cruzi, see Sarcocystis cruzi
  cuniculi, 58, 147
  cymruensis, 57, 147
  dispersa, 42, 58, 61, 85, 159
  dugesii, 168
  equicanis, 131—133, 146
  falcatula, 12, 39, 54, 58—62, 68, 158
  fayeri, 38, 58, 131—134, 146
  ferovis, 36, 57, 60, 154
  fusiformis, 1, 137—139, 147
  gallotiae, 167—168
  gazellae, 155
  gigantea, see Sarcocystis gigantea
  gongyli, 168
  gracilis, 57, 60, 146, 149—150
  grüneri, 57, 151—152
  gusevi, 155
  hardangeri, 4, 151—152
  hemioni, 7, 9, 57, 153—154
  hemionilatrantis, 11, 15, 41, 58, 145—146, 153—154
  hircicanis, 57, 62, 121—123, 146
  hirsuta, 1, 7, 10, 57, 85, 88, 105—111, 147
  hominis, 1, 6, 58, 67, 105—106, 109—111, 143—144
  horvathi, 157, 159
  idahoensis, 6, 9, 62, 68, 169
  jacarinae, 158
  kinosterni, 168
  kirmsei, 158
  kortei, 144
  lacertae, 168
  leporum, 58, 145, 147
  leuti, 156
  levinei, 137—139, 146
  lindemani, 143
  medusiformis, 43, 54—58, 62, 91, 113—116, 119, 147
  microti, 9, 57
  miescheriana, 1—2, 53, 62—64, 83—85, 93, 127—130, 145—146
  montanaensis, 57
  moulei, 121—124
  mucosa, 3, 40, 58
  murinotechis, 169
  muris, see Sarcocystis muris
  muriviperae, 169
  nelsoni, 155
  nesbiti, 144
  nontenella, 158
  odocoileocanis, 9, 42, 58, 60—61, 145—146, 151
  odoi, 38, 57, 147, 151
  oliverioi, 158
  orientalis, 122
  ovicanis, 1, 113
  ovifelis, 1, 118
  pathogenicity, 62—68
  peromysci, 9
  platydactyli, 158, 168
  podarcicolubris, 167—168
  poephagi, 57, 156
  poephagicanis, 57, 156
  porcifelis, 127, 129, 147
  putorii, 57
  pythonis, 168
  rangi, 57, 151—152
  rangiferi, 4, 58, 151—152
  rauschorum, 20, 25—26, 34—35, 57—66, 159
  richardi, 6, 156
  richardsonii, 6
  rileyi, 4, 44, 58
  riyeli, 157
  ruandae, 155

salvelini, 167
scelopori, 168
sebeki, 57
setophagae, 158
sibirica, 149—150
sigmodontis, 36, 57
singaporensis, 14—15, 28, 32—33, 42, 58, 61, 169
suicanis, 127
suihominis, 50, 58, 62, 67, 127—130, 143—144
sulawesiensis, 36, 57
sybillensis, 9, 37, 40, 58—60, 146, 152—153
taradivulpes, 58, 151—152
tarandi, 57, 151—152
tenella, see Sarcocystis tenella
turdi, 158
utae, 168
villivillosi, 43, 58, 169
wapiti, 35, 57, 60, 146, 152
woodhousei, 155
youngi, 9, 41, 58, 153
zamani, 58, 169
Sarcocystis arieticanis
 in dog feces, 146
 lethal dose of, 62
 pathogenicity of, 118
 sarcocysts of, 8—9, 93
 sacrocyst wall of, 27, 37, 56—57, 114
 structure of, 118
 transmission of, 60, 113
 wet preparations of, 96
Sarcocystis capracanis
 clinical signs of, 64
 definitive hosts for, 145
 in dog feces, 146
 in goat, 92, 121—124
 lethal dose of, 62
 sarcocysts of, 36, 45, 78—79, 93
 sarcocyst wall of, 56—58
 schizonts of, 99
 sporocysts of, 85, 98
 transmission of, 60
Sarcocystis cruzi
 antigens and proteins of, 79—80, 85—86
 bradyzoites of, 31, 49, 56, 88, 93
 in calf, 73—77
 in cattle, 6, 62—64, 91
 in cow, 89
 in coyote, 145
 development of, 7, 108
 in dog feces, 146
 in fox, 145
 goblet cells of, 46—47
 lethal dose of, 62
 life cycle of, 10, 70, 105
 MAbs against, 81
 merozoites of, 68, 82—83, 87, 101—103
 naming of, 1
 oocysts of, 11, 51
 prepatent period for, 60
 sarcocysts of, 16—17, 88, 106
 sarcocyst wall of, 57
 schizonts of, 13—14, 59, 99—103
 spindle apparatus of, 48
 sporocysts of, 12, 46, 54, 92, 98
 sporozoites of, 8, 47, 55
 steer infected with, 69
 thin wall of, 58
 transmission of, 60, 90
 zoites of, 14
 zygote of, 52
Sarcocystis gigantea
 antigen from, 83
 bradyzoites of, 28
 cauliflower-like protrusions in, 43
 history of, 1
 pathogenicity of, 62, 118
 sarcocysts of, 4, 8, 54, 91—93, 119, 122
 sarcocyst wall in, 56, 58, 114—115
 sporocysts of, 147
 structure of, 118
 transmission of, 113
Sarcocystis muris
 antigen from, 83, 85, 88
 bradyzoites of, 39, 80
 gamonts of, 7
 history of, 1—2
 oocysts of, 104
 sarcocysts of, 91, 93
 sarcocyst wall of, 35, 57
 sporocysts of, 85, 147
 transmission of, 60—61
Sarcocystis tenella
 amprolium for, 91
 antibody against, 88
 bradyzoites of, 23—26
 definitive hosts for, 145
 developmental stages of, 116
 history of, 1
 merozoites of, 63, 67—68
 pathogenicity of, 62—64
 salinomycin for, 92
 sarcocysts of, 93
 sarcocyst wall in, 56, 58, 114—115
 schizonts of, 56, 59, 66, 71, 99
 in sheep, 72
 sporocysts of, 90, 98, 117, 146
 structure of, 17, 53
 tranmission of, 60—61, 113
 villar protrusions in, 41
 wet preparations of, 96
 zoites of, 27
Sarcocystosis
 acute, 65
 in cattle, 105—112, see also Cattle
 chronic, 78—79
 clinical, 91
 experimental techniques for, 93—104
  diagnostic, 94—97
  in vitro cultivation, 97—104
   gamonts, 101—104
   schizonts, 98—101
  isolation, purification, and preservation, 93—94
 general biology of, 1—92
  chemoprophylaxis and chemotherapy, 91—92

control, 91
diagnosis, 86—89
economic losses, 90
epidemiology, 90—91
history, 1—2
immunity, see Immunity to sarcocystosis
life cycle, 5—13
pathogenesis, 68—79, see Pathogenesis of sarcocystosis
structure, 2—5
taxonomic criteria, see Taxonomy of *Sarcocystis*
transmission, 90
ultrastructure, see Ultrastructure of *Sarcocystis*
intestinal, 145
muscular, 143—145
Sarcocysts
in cotton rat, 5
degenerate, 18, 67
formation of, 12
ingestion of, 90
in llama, 3
macroscopic, 4—5, 109, 119
mature, 15, 27, 93
in muscle, 95—97
in myocyte, 45
in pigs, 129—130
in Richardson's ground squirrel, 5
of *S. arieticanis*, 8—9, 93
of *S. cameli*, 142
of *S. capracanis*, 36, 45, 78—79, 93
of *S. cruzi*, 16—17, 88, 106
of *S. equicanis*, 132
of *S. fayeri*, 132—133
of *S. fusiformis*, 138
of *S. gigantea*, 4, 8, 54, 91—93, 119, 122
of *S. miescheriana*, 128
of *S. muris*, 91, 93
of *S. suihominis*, 128
of *S. tenella*, 93
structure of, 2—5, 13—15
taxonomy of, 54—59
unizoite, 20
young, 25—26
Sarcocyst walls, 9
aged, 24
classification of *Sarcocystis* by, 57—58
primary, see Primary sarcocyst wall
of *S. arieticanis*, 27, 37, 56—57, 114
of *S. capracanis*, 56—58
of *S. capreoli*, 150
of *S. cruzi*, 107
of *S. fusiformis*, 139
of *S. gigantea*, 56, 58, 114—115
of *S. gracilis*, 150
of *S. grüneri*, 152
of *S. hirsuta*, 110—111
of *S. hominis*, 107, 110—111
of *S. levinei*, 139
of *S. medusiformis*, 114
of *S. muris*, 35, 577
of *S. rangi*, 152
of *S. rangiferi*, 152
of *S. sibirica*, 150
of *S. tarandi*, 152
of *S. tarandivulpes*, 152
of *S. tenella*, 56, 58, 114—115
secondary, 4, 8
structure of, 56, 59
types of, 30—44
types 1 to 14, 30—32, 35—41
types 15 to 24, 32—36, 41—44
villar protrusions on, 115
Sarcotoxin, 78—79
SBDH, see Sorbitol dehydrogenase
Schizogony, 8, 11—12, 48—53
Schizonts
detecting, 88—89
excystation and cultivation of, 98—103
in fetal placenta, 81, 89
first-generation, 13, 71
free, 73—74
immature, 12
intermediate, 49, 64
location of, 48
mature, 66
nearly mature, 65
necrosis from, 68
second generation, 8—9, 11, 14
structure of, 59
third generation, 15
Serological techniques, 97
Serum bilirubin, 63
Sheep
abortion in, 71—72, 114
acute sarcocystosis in, 70
antigen in, 83, 88
bighorn, 57, 60, 154
blastogenic response in, 84
clinical signs in, 64, 116—117
eosinophilic myositis in, 77
halofuginone in, 92
IgG antibodies in, 82—83
immunization of, 91
as intermediate host, 57—58, 60, 93, 146—147
lethal sporocyst dose in, 62
placental lesions in, 73, 90
protective immunity in, 84—86
salinomycin in, 92
Sarcocystis species in, 4, 8—9, 54, 72, 90, 113—120, 151
sarcocyst types in, 56
schizonts in, 74
Skink, 168
Skunk, 145, 157
Snake, 61—62, 162—164, 167—169
Sorbitol dehydrogenase (SBDH), 63
Sparrow, 158
Spindle apparatus, 48, 50, 63
Spiramycin, 92
Sporoblast, 8
Sporocysts
collapse of, 47
elongate, 8
excysting sporozoites from, 98

feces of, 94—95
feeding, 93
ingestion of, 80, 90—91
lethal doses of, 62
and oocyst wall, 53
recovery of, 94
of *S. capracanis*, 85, 98
of *S. cruzi*, 12, 46, 54, 92, 98
of *S. gigantea*, 147
of *S. muris*, 85, 147
of *S. tenella*, 90, 98, 117, 146
structure of, 59—60
wall of, 54
Sporogony, 39
Sporozoites
development of, 98—100
epitopes in, 82
excystation of, 12, 53
formation of, 8
intracellular, 55
proteins of, 84
structure of, 23, 47—48
Squirrel, 5, 9, 57, 62, 165—166
Steer, 69, 77
Stieda body, 8
Stoat, 164
Sulfadimethoxine, 92
Sulfadoxine, 92
Sulfaquinoxaline, 92
Suslik, 166
Swine, see Pig
Synchytrium miescherianum, 1

T

Taxonomy of *Sarcocystis*, 53—63
 classification, 2
 host specificity, 60
 isoenzymes, 60—62
 oocysts and sporocysts, 59—60
 sarcocysts, 54—59
 schizonts, 59
T lymphocytes, 69—70
TNF, see Tumor necrosis factor
Tortoise, 167
Toxins, 78—79
Toxoplasma, 89, 173—174
Transmission, 60—62, 90
Trimethoprim, 92
Trout, 167
Tumor necrosis factor (TNF), 76, 79
Turkey, 157
Turtle, 168

U

Ultrastructure of *Sarcocystis*, 13—53

bradyzoites, 16—27, see also Bradyzoites
endodyogeny, 27—29
fertilization, 44—46
gametogenesis, 37—44
metrocytes, 15—16
oocyst, 46
sarcocysts, 13—15
sarcocyst wall types, 30—37, see also Sarcocyst wall
schizogony, 48—53
sporocyst, 46—47
sporozoites, 47

V

Vaccination, 84—86, 91
Vasculitis, 67—68, 72, 75
Vasopressin, 75
Villar protrusions
 formation of, 15
 myocyte, 45
 in sarcocyst wall types, 33, 38, 41, 115
 structure of, 21—24
Viper, 61, 162, 169
Vole
 European, 164, 171
 field, 61
 Frenkelia in, 171—172
 long-tailed, 163
 meadow, 9, 163
 S. cernae in, 159, 163
 S. clethrionomyelaphis in, 169
 S. sebeki in, 57
 transmission to, 61

W

Wallabies, 58, 164
Wapiti, 57—58, 60, 146, 152—153
Waterbuck, 155
Water buffalo, 137—139, 146
Weasel, 164
Whale, 145
Wolf
 as definitive host, 6, 60, 105, 127, 151
 feces of, 148
 transmission by, 90

Y

Yak, 57, 156

Z

Zebra, 131, 133
Zoites, 8, 14, 22, 27, 32—33, 56
Zoalene®, 92
Zygote, 45, 52